Current Topics in Microbiology and Immunology

Volume 419

More information about this series at http://www.springer.com/series/82

Ralph A. Tripp · S. Mark Tompkins
Editors

Roles of Host Gene and Non-coding RNA Expression in Virus Infection

Responsible series editor: Richard W. Compans

Springer

Editors
Ralph A. Tripp
Department Infectious Diseases,
College of Veterinary Medicine
University of Georgia
Athens, GA, USA

S. Mark Tompkins
Center for Vaccines and Immunology
University of Georgia
Athens, GA, USA

ISSN 0070-217X ISSN 2196-9965 (electronic)
Current Topics in Microbiology and Immunology
ISBN 978-3-030-05368-0 ISBN 978-3-030-05369-7 (eBook)
https://doi.org/10.1007/978-3-030-05369-7

Library of Congress Control Number: 2018963295

This Springer imprint is published by the registered company Springer Nature Switzerland AG
The registered company address is: Gewerbestrasse 11, 6330 Cham, Switzerland

Preface

Viruses are obligate parasites relying on the exploitation of host cell processes and resources for replication. The interplay between host and viruses remains largely unknown. Life cycles for individual viruses have been defined with functions ascribed for many viral proteins. The discovery of RNA interference (RNAi) and the subsequent development of tools to specifically silence individual cellular genes enabled genome-wide studies, interrogating gene function in a spectrum of processes, advancing developmental biology, and infectious disease. The advent of functional genomics allowed for the interrogation of the virus–host cell interactions and probing the genome for a role in the virus replication.

Libraries of arrayed siRNAs against human or mouse genomes have been available for more than a decade. More recently, microRNA (miRNA) mimic and inhibitor libraries have also become available for genome-wide screening, and most recently gene editing, e.g. CRISPR/Cas, has also become available. These approaches combined with transcriptomic and proteomic analyses have enabled the identification of new players in the host–virus interactome. Importantly, advances in recombinant technology, virology, and systems biology have allowed mapping of the interaction between cellular and viral gene products, including viral and cellular non-coding RNAs allowing for a better understanding of novel gene functions and pro- and anti-viral activities. These discoveries have provided an opportunity for the development of novel therapeutics and approaches to improve viral vaccines and vaccine production.

This volume presents a current understanding of the interplay between host cells and viruses during infection and replication. The first chapters present our knowledge of coronavirus, flavivirus, and human immunodeficiency virus (HIV), virus–cell interactions, i.e. three positive-sense RNA viruses (*Coronaviridae*, *Flaviviridae*, and *Retroviridae*), respectively. The volume then moves to address to negative-sense RNA viruses, with chapters on Ebola virus (*Filoviridae*), influenza virus (*Orthomyxoviridae*), and two viruses from the *Paramyxoviridae* family. The respiratory syncytial virus (RSV) chapter discusses the role of miRNAs in infection, while the henipavirus chapter explores diverse aspects of virus–host interactions. The volume finishes with a chapter on non-coding RNAs involved in herpesvirus infection, a double-stranded DNA virus (*Herpesviridae*). These chapters capture many aspects of viral genomes and life cycles, including segmented, integrating, and latent genomes, acute, chronic, and latent infections, as well as vector-borne viruses. This volume provides a representation of virus–host interactions and a valuable resource for advancing our understanding. We are grateful to the authors for their expertise and contributions to this remarkable volume.

Athens, USA

Ralph A. Tripp
S. Mark Tompkins

Contents

Host Factors in Coronavirus Replication

Adriaan H. de Wilde, Eric J. Snijder, Marjolein Kikkert and Martijn J. van Hemert

Abstract Coronaviruses are pathogens with a serious impact on human and animal health. They mostly cause enteric or respiratory disease, which can be severe and life threatening, e.g., in the case of the zoonotic coronaviruses causing severe acute respiratory syndrome (SARS) and Middle East Respiratory Syndrome (MERS) in humans. Despite the economic and societal impact of such coronavirus infections, and the likelihood of future outbreaks of additional pathogenic coronaviruses, our options to prevent or treat coronavirus infections remain very limited. This highlights the importance of advancing our knowledge on the replication of these viruses and their interactions with the host. Compared to other +RNA viruses, coronaviruses have an exceptionally large genome and employ a complex genome expression strategy. Next to a role in basic virus replication or virus assembly, many of the coronavirus proteins expressed in the infected cell contribute to the coronavirus-host interplay. For example, by interacting with the host cell to create an optimal environment for coronavirus replication, by altering host gene expression or by counteracting the host's antiviral defenses. These coronavirus–host interactions are key to viral pathogenesis and will ultimately determine the outcome of infection. Due to the complexity of the coronavirus proteome and replication cycle, our knowledge of host factors involved in coronavirus replication is still in an early stage compared to what is known for some other +RNA viruses. This review summarizes our current understanding of coronavirus–host interactions at the level of the infected cell, with special attention for the assembly and function of the viral RNA-synthesising machinery and the evasion of cellular innate immune responses.

A.H. de Wilde · E.J. Snijder · M. Kikkert · M.J. van Hemert (✉)
Molecular Virology Laboratory, Department of Medical Microbiology,
Leiden University Medical Center, Leiden, The Netherlands
e-mail: m.j.van_hemert@lumc.nl

Current Topics in Microbiology and Immunology (2018) 419:1–42
DOI 10.1007/82_2017_25

Published Online: 23 June 2017

Contents

1 Introduction

Around the end of 2002, an outbreak of a previously unknown severe acute respiratory syndrome (SARS) started in South East China and Hong Kong. Accelerated by air travel, the disease rapidly spread to several parts of the world and displayed pandemic potential. SARS-coronavirus (SARS-CoV) was identified as the causative agent of this zoonotic infection (Drosten et al. 2003; Ksiazek et al. 2003; Kuiken et al. 2003; Peiris et al. 2003), which resulted in >8000 laboratory-confirmed cases and 774 associated deaths worldwide (WHO 2004). Although in terms of death toll not comparable to influenza, HIV or HCV, the 2003 SARS-CoV outbreak caused worldwide public concern and seriously affected the global economy [estimated losses $30–100 billion; (Keogh-Brown and Smith 2008)]. SARS-CoV initially causes lower respiratory tract disease, which can lead to a progressive and potentially lethal atypical pneumonia with clinical symptoms that include fever, malaise, lymphopenia, and in some cases also diarrhea. Two years after the outbreak, horseshoe bats were identified as the likely reservoir of the SARS virus, whereas civet cats probably have served as intermediate host during the zoonotic transfer to humans (Lau et al. 2005; Li et al. 2005b). Adaptation to the human host required a small number of mutations in the receptor-binding domain of the SARS-CoV spike (S) protein, which mediates cell binding and entry (Li et al. 2005c) (see Chap. 2). There is increasing evidence that SARS-like coronaviruses continue to circulate in bats and that these may have the potential to readily cross the species barrier and emerge as human pathogens (Ge et al. 2013; Menachery et al. 2015). Such zoonotic scenarios therefore remain a serious public health concern.

Almost a decade after the SARS-CoV outbreak, the next zoonotic coronavirus emerged: Middle East Respiratory Syndrome coronavirus (MERS-CoV) (de Groot et al. 2013). The virus was first isolated in June 2012 from a 60-year-old Saudi Arabian male who died from acute respiratory distress syndrome (ARDS) and multiple organ failure, including renal failure (Zaki et al. 2012; van Boheemen et al. 2012). Also MERS-CoV can cause a lower respiratory tract infection with symptoms that include coughing and high fever. By the end of 2016, more than 1850 laboratory-confirmed MERS-CoV cases had been recorded, with a mortality rate of about 35% (WHO 2016). MERS-CoV is assumed to be transmitted to humans from camels and serological studies in the latter animals revealed that they have harbored MERS-CoV or MERS-CoV-like viruses for decades (Muller et al. 2014).

Besides the zoonotic SARS- and MERS-CoVs, the coronavirus family includes four 'established' human coronaviruses (HCoVs), of which HCoV-OC43 and -229E have already been known since the 1960s. These two viruses cause mild respiratory disease and, after rhinoviruses, are a leading cause of common colds (10–30% of the cases) (van der Hoek 2007; McIntosh et al. 1967; Hamre and Procknow 1966). More recently, following intensified screening for coronaviruses, two additional HCoVs were discovered, HCoV-NL63 (van der Hoek et al. 2004) and HCoV-HKU1 (Woo et al. 2005). Interestingly, recent findings suggest that also HCoV-NL63, -229E, and -OC43 originate from zoonotic transfer from bats (Huynh et al. 2012; Corman et al. 2016; Vijgen et al. 2006; Corman et al. 2015). Coronaviruses also cause a range of infectious diseases in animal species, some with serious (economical) consequences for the livestock industry. This is illustrated by the recent emergence of a novel variant of porcine epidemic diarrhea virus, which is closely related to a strain that caused a large outbreak in China in 2010, killing almost one million piglets [for a recent review, see (Lin et al. 2016)].

The economic impact of coronavirus infections, the past and likely future emergence of pathogenic zoonotic coronaviruses and the lack of effective antiviral strategies have made it painfully clear that our preparedness to treat or prevent coronavirus infections are very limited. This highlights the importance of advancing our knowledge on the replication of these viruses and their interactions with the host.

Coronaviruses are positive-stranded RNA (+RNA) viruses with, for this kind of viruses, exceptionally large genomes of $\sim$30 kb. They have a polycistronic genome organization and employ a unique transcription mechanism to generate a nested set of subgenomic (sg) mRNAs. These are used to express the open reading frames (ORFs) located downstream of the replicase ORFs 1a and 1b (see Fig. 1a), which encode structural and accessory proteins. The sg mRNAs are 3′ co-terminal but they also contain a common 5′ leader sequence. The leader and 'body' segments of the sg RNAs are joined during discontinuous negative strand RNA synthesis, which produces a subgenome-length template for each of the sg mRNAs [(Sawicki and Sawicki 1995), for a recent review, see (Sola et al. 2015)].

Following entry and uncoating, the coronavirus replicative cycle (see Fig. 1a) starts with the translation of the 5′-proximal ORFs of the viral genome (ORF1a and ORF1b), which results in the synthesis of two large replicase polyproteins (pp1a and pp1ab). Synthesis of pp1ab, a C-terminally extended form of pp1a, involves a -1 ribosomal frameshift (RFS) into ORF1b occurring near the 3′ end of ORF1a. This regulatory mechanism is thought to have evolved to downregulate expression levels of ORF1b-encoded proteins compared to ORF1a-encoded nonstructural proteins (nsps) (Brierley and Dos Ramos 2006; Brierley et al. 1989). Ultimately, 15 or 16

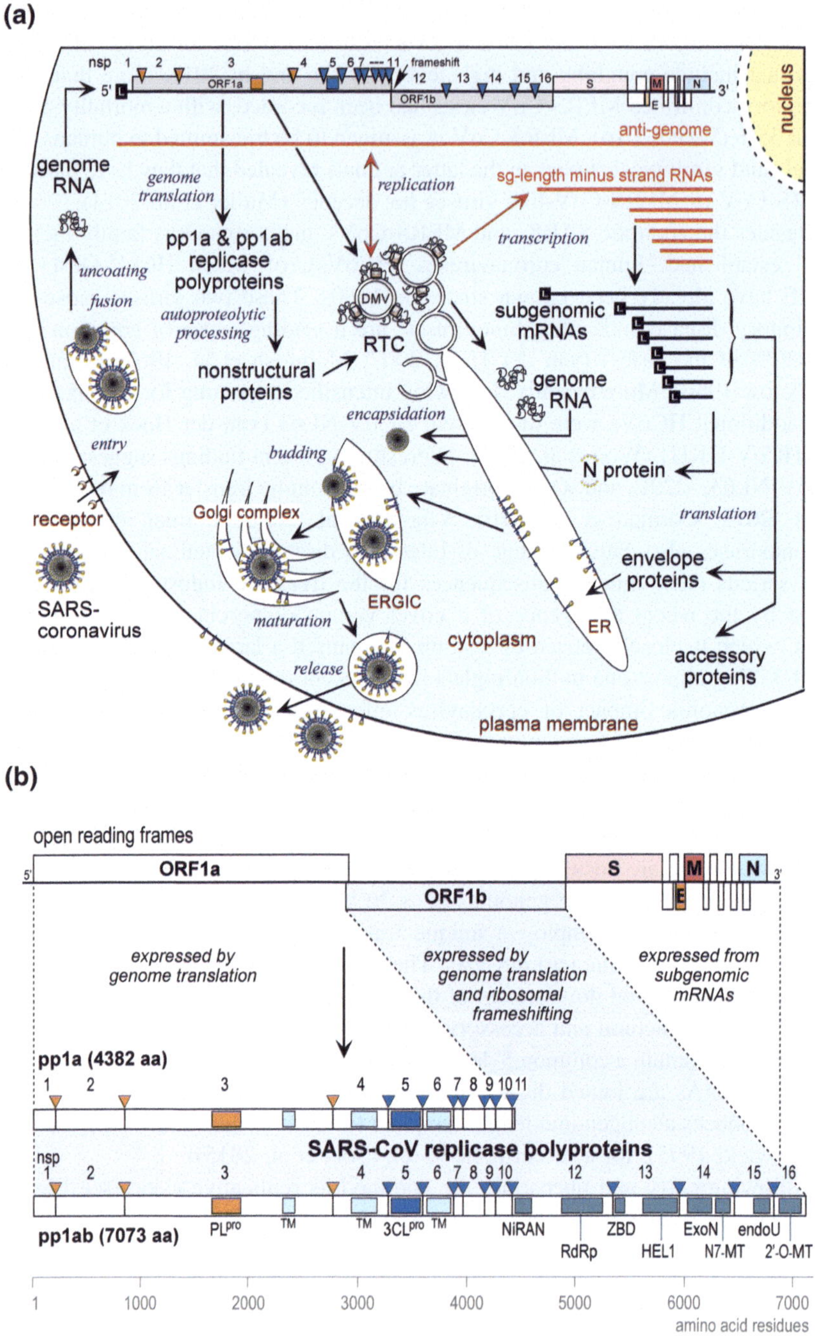
(a)
nsp
1
2
3
4
5
6
7

11
frameshift
12
13
14
15
16
5'
L
ORF1a
ORF1b
S
M
E
N
3'
nucleus
anti-genome
genome RNA
genome translation
replication
sg-length minus strand RNAs
transcription
pp1a & pp1ab replicase polyproteins
uncoating
fusion
autoproteolytic processing
DMV
RTC
subgenomic mRNAs
nonstructural proteins
genome RNA
encapsidation
entry
budding
N protein
receptor
Golgi complex
translation
envelope proteins
SARS-coronavirus
ERGIC
maturation
cytoplasm
ER
accessory proteins
release
plasma membrane
(b)
open reading frames
5'
ORF1a
ORF1b
S
M
N
E
3'
expressed by genome translation
expressed by genome translation and ribosomal frameshifting
expressed from subgenomic mRNAs
pp1a (4382 aa)
1
2
3
4
5
6
7
8
9
10
11
SARS-CoV replicase polyproteins
nsp
1
2
3
4
5
6
7
8
9
10
12
13
14
15
16
PLpro
TM
TM
3CLpro
TM
NiRAN
RdRp
ZBD
HEL1
ExoN
N7-MT
endoU
2'-O-MT
pp1ab (7073 aa)
1
1000
2000
3000
4000
5000
6000
7000
amino acid residues

◂**Fig. 1** Outline of the coronavirus replicative cycle and replicase polyprotein organization, based on SARS-CoV. **a** Schematic overview of the coronavirus replicative cycle. Following entry by receptor-mediated endocytosis and release of the genome into the cytosol, genome translation yields the pp1a and pp1ab replicase polyproteins. Following polyprotein cleavage by multiple internal proteases, the viral nsps assemble into an RTC that engages in minus-strand RNA synthesis. Both full-length and subgenome (sg)-length minus strands are produced, with the latter templating the synthesis of the sg mRNAs required to express the structural and accessory protein genes residing in the 3′-proximal quarter of the genome. Ultimately, novel genomes are packaged into nucleocapsids that become enveloped by budding from smooth intracellular membranes, after which the new virions leave the cell by following the exocytic pathway. See text for more details. **b** The 14 open reading frames in the genome are indicated, i.e., the replicase ORFs 1a and 1b, the four common CoV structural protein genes (S, E, M, and N) and the ORFs encoding so-called 'accessory proteins.' The bottom panel explains the organization and proteolytic processing of the pp1a and pp1ab replicase polyproteins, the latter being produced by −1 ribosomal frameshifting. The nsp3 (PL^{pro}) and nsp5 ($3CL^{pro}$) proteases and their cleavage sites are indicated in matching colors. The resulting 16 cleavage products [nonstructural proteins (nsps)] are indicated, as are the conserved replicase domains. Domain abbreviations and corresponding nsp numbers: PL^{pro}, papain-like proteinase (nsp3); $3CL^{pro}$, 3C-like protease (nsp5); TM, transmembrane domain (nsp3, nsp4, and nsp6); NiRAN, nidovirus RdRp-associated nucleotidyl transferase (nsp12); RdRp, RNA-dependent RNA polymerase (nsp12); ZBD, zinc-binding domain (nsp13); HEL1, superfamily 1 helicase (nsp13); ExoN, exoribonuclease (nsp14); N7-MT, N7-methyl transferase (nsp14); endoU, uridylate-specific endoribonuclease (nsp15); 2′-O-MT, 2′-O-methyl transferase (nsp16). Adopted with permission from (Snijder et al. 2016)

mature replicase proteins are released from pp1a and pp1ab due to proteolytic cleavages performed by two or three ORF1a-encoded proteases. Nsp3 contains one or two papain-like protease domains ($PL1^{pro}$ and $PL2^{pro}$, or PL^{pro} for SARS-CoV and infectious bronchitis virus) that process the nsp1-4 part of the replicase polyproteins. The remaining cleavage sites are processed by the viral main protease that resides in nsp5, a chymotrypsin-like enzyme also known as 3C-like protease (Snijder et al. 2016). A schematic overview of the proteolytic processing and domain structure of the SARS-CoV replicase is presented in Fig. 1b. The replicase proteins contain a variety of (enzymatic) activities and functions that are required for viral RNA synthesis and capping (Perlman and Netland 2009; Snijder et al. 2016), such as the RNA-dependent RNA polymerase (RdRp; nsp12), a helicase (nsp13), RNA cap-modifying methyltransferases (nsp14 and nsp16), and an exoribonuclease (nsp14). Together with recruited host cell proteins, the coronavirus nsps form membrane-associated replication and transcription complexes [RTCs; (van Hemert et al. 2008)], which localize to a network of virus-induced membrane structures in the perinuclear region of the infected cell (Knoops et al. 2008; Gosert et al. 2002; van der Meer et al. 1999; Brockway et al. 2003; Stertz et al. 2007; Ulasli et al. 2010). Many of the nsps appear to have multiple functions in the synthesis or processing of viral RNA, or in virus–host interactions aiming to create an optimal environment for coronavirus replication, for example by facilitating viral entry, gene expression, RNA synthesis or virus release. Moreover, to further enhance viral replication, host gene expression and antiviral defenses are targeted in several ways. Coronavirus–host interactions also play a decisive role in viral pathogenesis and the ultimate outcome of infection.

Due to the exceptional size of their +RNA genome and proteome, and the resulting complexity of the interactions with the host, our knowledge of host factors involved in coronavirus replication is still in an early stage compared to what is known for some other +RNA virus groups. In this review, we will summarize our current understanding of coronavirus–host interactions at the level of the infected cell, with special attention for the assembly and function of the viral RNA-synthesizing machinery and the evasion of cellular innate immune responses.

2 Host Receptors Involved in Coronavirus Entry

Entry into the target cell constitutes the first critical step in the coronavirus replication cycle. The major determinant for this step is the efficient binding of the coronavirus S glycoprotein to a protein-receptor on the cell surface. The coronavirus S protein is a type 1 glycoprotein that consists of S1 and S2 subunits and is present on the virion surface as a trimer. (Li 2016; Hulswit et al. 2016). The S1 region is involved in receptor binding and contains N- and C-terminal domains (S1-NTD and S1-CTD, respectively) (Walls et al. 2016) that may both act as receptor-binding domain (RBD), with the major determinants of cell tropism residing in S1-CTD. The elongated S2 regions form the stalk of the spike trimer and are mainly involved in triggering the fusion of the viral envelope and target cell membranes [for recent reviews on coronavirus entry and spike protein organization, see (Li 2016; Hulswit et al. 2016)].

The S1-NTD is mainly involved in facilitating virus binding and entry, by interacting with glycans on the host cell surface. Based on the crystal structure of the betacoronavirus S1-NTD and the sequence conservation among the S1-NTDs of other coronaviruses, all coronavirus S1-NTDs are thought to share a galectin fold that mediates binding to sialic acids, like N-glycolylneuraminic acid (Neu5Gc), N-acetylneuraminic acid (Neu5Ac), and/or 5-N-acetyl-9-O-acetylneuraminic acid (Neu5,9Ac2) (see (Li 2016), and references herein). An exception is the murine hepatitis virus (MHV) S1-NTD, which binds the N-terminal D1 domain of carcinoembryonic antigen-related cell adhesion molecule 1 (CEACAM1), a type-I membrane protein belonging to the immunoglobulin superfamily (Walls et al. 2016; Williams et al. 1991).

To mediate entry into host cells, the S1-CTD of most known members of the alphacoronavirus genus interacts with aminopeptidase N (APN) (for an overview and references, see Table 1). However, the alphacoronavirus HCoV-NL63 uses a different type-I membrane glycoprotein, angiotensin-converting enzyme 2 (ACE2) (Wu et al. 2009), which contains a large N-terminal ectodomain composed of two alpha-helical lobes. The same molecule, ACE2, has been identified as a receptor for the zoonotic betacoronavirus SARS-CoV (Li et al. 2003). The betacoronaviruses MERS-CoV and bat coronavirus HKU4 use yet another cellular peptidase for virus entry: dipeptidyl peptidase 4 (DPP4) (Yang et al. 2014; Raj et al. 2013). The MERS-CoV S protein has a higher affinity for human DPP4, while the HKU4 S

Table 1 Overview of known coronavirus entry receptors

Genus	Species:	S1-NTD	S1-CTD	References
Alphacoronavirus	Alphacoronavirus 1	Neu5Gc and Neu5Ac*	APN	(Tresnan et al. 1996; Delmas et al. 1992)
	PEDV	Neu5Ac	APN	(Liu et al. 2015; Li et al. 2007)
	PRCV		APN	(Schultze et al. 1996)
	HCoV-229E		APN	(Yeager et al. 1992)
	HCoV-NL63		ACE2	(Wu et al. 2009)
Betacoronavirus	Betacoronavirus 1	Neu5,9Ac2		(Schultze and Herrler 1992; Krempl et al. 1995)
	MERS-CoV		DPP4	(Raj et al. 2013)
	MHV	CEACAM1		(Williams et al. 1991)
	HKU1	Neu5,9Ac2		(Huang et al. 2015b)
	HKU4		DPP4	(Yang et al. 2014)
	SARS-CoV		ACE2	(Li et al. 2003)
Gammacoronavirus	IBV	Neu5Gc		(Schultze et al. 1993)
Deltacoronavirus	PDCV	Unknown	unknown	

(*Abbreviations PEDV* Porcine epidemic diarrhea virus; *TGEV* Transmissible gastroenteritis coronavirus; *PRCV* Porcine Respiratory coronavirus; *FCoV* Feline coronavirus; *CCoV* Canine coronavirus; *HCOV* Human coronavirus; *BCoV* Bovine coronavirus; *MHV* Murine hepatitis virus; *IBV* Infectious bronchitis virus; *PDCV* Porcine delta coronavirus). *Within the alphacoronavirus 1 species, only for TGEV the sialic acids Neu5Gc and Neu5Ac has been identified as attachment factors

protein binds more strongly to bat DPP4 (Yang et al. 2014). Chemical peptidase inhibitors do not affect virus entry, indicating that SARS-CoV and MERS-CoV receptor usage and entry are independent of the receptor's peptidase activity and merely depend on binding to these particular host receptors (Li et al. 2005c; Raj et al. 2013).

Besides the receptors discussed above, also extracellular, cell surface-associated and/or lysosomal proteases play a role in coronavirus entry by activating the fusion activity of the S protein [for a recent review, see (Li 2016)]. For SARS-CoV, fusion of the viral and cellular membrane is triggered upon cleavage of the S protein by the cell surface-associated transmembrane protease, serine 2 (TMPRSS2) (Glowacka et al. 2011). The same protease is important for cleavage and activation of the HCoV-229E and MERS-CoV S protein (Shirato et al. 2013; Bertram et al. 2013). After endocytosis, the SARS-CoV S protein is cleaved by the lysosomal proteases cathepsin L and cathepsin P in early endosomes, leading to fusion of the virus envelop with the endosome membranes and release of the viral RNA into the cytosol of the infected cell (Huang et al. 2006a, b; Simmons et al. 2005). MERS-CoV entry occurs by a similar mechanism (Shirato et al. 2013; Burkard et al. 2014), although inhibition of the cellular protease furin abolished the entry of MERS-CoV but not SARS-CoV, indicating that furin-mediated cleavage is pivotal for efficient MERS-CoV entry (Burkard et al. 2014; Follis et al. 2006). On the other hand, MHV strain A59 was shown to fuse with late endosomes and to depend on their low pH for S protein cleavage (Burkard et al. 2014). Therefore, it has been proposed that coronavirus fusion with endosomes depends on the use of a furin cleavage site just upstream of the fusion peptide (Burkard et al. 2014). However, why some coronaviruses fuse with early endosomes and others with late endosomes, and whether these events play a role in host tropism and pathogenicity, is still not completely understood (Burkard et al. 2014). The complexity of S protein cleavage is further highlighted by a recent paper by Park et al., which clearly showed that MERS-CoV entry depends on furin-mediated cleavage in virus-producing cells. Subsequently, cleaved MERS-CoV S proteins could be processed by proteases on recipient cells and virions could enter the cells via early endosomes or even by fusing with the plasma membrane. MERS-CoV virions that contain uncleaved S proteins may rather fuse with late endosomes (Park et al. 2016).

The interaction of the coronavirus S glycoprotein with its cell surface receptor is a key determinant for host tropism. In the case of SARS-CoV, only a few mutations (N479L and T487S) in the S protein's RBD sufficed to dramatically increase the affinity for human ACE2 (Li 2008). Likewise, the MERS-CoV S protein contains two mutations compared to the bat coronavirus HKU4 S protein, which can bind the human DPP4 receptor, but cannot mediate viral entry due to lack of activation by human proteases. The two mutations in the MERS-CoV S protein (S746R and N762A) enable cleavage by the human proteases and thus viral entry into human cells and may have contributed to the zoonotic transfer of MERS-CoV (Yang et al. 2015).

Several lineage A betacoronaviruses also carry a hemagglutinin-esterase (HE) protein on their surface. HE proteins contain a lectin-binding domain that mediates binding to O-acetylated sialic acids, while also possessing sialate-O-acetylesterase receptor-destroying enzyme activity targeting these same glycans on the cell surface. The receptor-destroying enzyme activity is thought to prevent attachment to non-permissive cells, while the HE protein also facilitates entry into target cells after binding to the main entry receptor (Zeng et al. 2008; Langereis et al. 2010; Bakkers et al. 2016).

3 Translation and the Unfolded Protein Response in Coronavirus-Infected Cells

All viruses depend on the host cell's translation machinery for the production of their proteins and infectious progeny. Moreover, protein synthesis is also pivotal for the host cell's response to infection by mounting an antiviral (innate) immune response. Hence, it is not surprising that many +RNA viruses modulate host protein synthesis in order to limit the translation of cellular mRNAs and favor the synthesis of viral proteins [reviewed in (Walsh and Mohr 2011; Fung et al. 2016)].

In eukaryotic cells, translation is initiated by formation of the heterotrimeric eIF2 complex, which is composed of the regulatory α-subunit, the tRNA-binding β-subunit, and a GTP-binding γ-subunit. The eIF2 complex is responsible for loading of the 40S subunit with Met-tRNAi. After mRNA binding, this 43S complex serves as a scaffold for the recruitment of several additional proteins, including eIF3, to the capped 5′ end of the mRNA. Subsequently, the cap-binding eukaryotic translation initiation factor 4F (eIF4F) joins this pre-initiation complex (48S complex), which then scans the mRNA in the 5′ to 3′ direction to localize a translation initiation codon. At this point, the 60S ribosomal subunit joins and protein synthesis starts [reviewed in (Jackson et al. 2010)]. Polyadenine-binding protein (PABP), which binds to the poly(A)-tail of mRNAs, is also involved in stimulating protein synthesis.

The eIF2 complex can be inactivated by phosphorylation of its alpha subunit (eIF2α) by one of four mammalian kinases in response to various (external) triggers. These kinases are eIF2α kinase 4 (also known as GCN2), heme-regulated inhibitor (HRI), PKR-like endoplasmic reticulum kinase (PERK), which is activated upon induction of ER stress, and double-stranded (ds) RNA-activated protein kinase (PKR).

Since several stages of the coronavirus replication cycle are closely associated with the endoplasmic reticulum (ER), ER stress is thought to occur during coronavirus infection. Indeed, expression of several coronavirus proteins, including the heavily glycosylated S protein, was shown to induce ER stress, which was also observed in coronavirus-infected cells [(Chan et al. 2006), and reviewed in (Fung et al. 2016)]. Consequently, the unfolded protein response (UPR) is induced, which alleviates the problems by inhibiting translation (by PERK-induced phosphorylation of eIF2α), stimulating protein folding, and eventually triggering apoptosis. Compared to, for example, hepatitis C virus [see review by (Chan 2014)], many details of how coronaviruses control the UPR remain unknown, but they generally seem to manipulate PERK activity to control the level of translation [reviewed by (Fung et al. 2016)].

PKR is a serine/threonine protein kinase that is activated by the presence of dsRNA, a hallmark of RNA virus infection. PKR is a key player in the innate immune response to RNA virus infection as it upregulates antiviral gene expression, including the production of interferons (IFNs). Coronaviruses have evolved various strategies to counteract PKR-mediated signaling in order to prevent the translational shut-off due to eIF2α phosphorylation. For example, infectious bronchitis virus (IBV) appears to (weakly) antagonize PKR by blocking its activation as well as inducing the expression of growth arrest and DNA-damage-inducible 34 protein (GADD34), leading to reduced eIF2α phosphorylation in IBV-infected cells (Wang et al. 2009). Upon MHV infection, sustained eIF2α phosphorylation and repression of GADD34 expression leads to translational repression of cellular mRNAs, which may be beneficial for MHV infection (Bechill et al. 2008). Recently, the MERS-CoV ORF4a protein was shown to counteract the PKR-induced formation of stress granules, probably by binding viral dsRNA to shield it from detection by PKR, thereby preventing translational inhibition (Rabouw et al. 2016). Also transmissible gastroenteritis virus (TGEV) has been reported to modulate host cell translation, in this case through its protein 7, which promotes eIF2α dephosphorylation through an interaction with protein phosphatase 1 (PP1), a key regulator of the host's antiviral response (Cruz et al. 2011). The S proteins of both SARS-CoV and IBV were found to physically interact with eIF3F, to modulate host translation, including the expression of the pro-inflammatory cytokines interleukin (IL) 6 and 8, at a later stage of infection (Xiao et al. 2008). Therefore, this interaction may play an important regulatory role in coronavirus pathogenesis.

Besides modulating eIF2α phosphorylation, coronaviruses have other ways of manipulating the translation machinery. Importantly, the nsp1 proteins of both alpha- and betacoronaviruses were identified as inhibitors of multiple steps of translation initiation (Lokugamage et al. 2012, 2015). SARS-CoV nsp1 does so by inhibiting 48S initiation complex formation and interfering with its conversion into the 80S initiation complex (Lokugamage et al. 2012). In addition, the multifunctional SARS-CoV nsp1 is able to directly bind the 40S ribosomal subunit to inhibit its function in translation (Kamitani et al. 2009). Moreover, this complex of nsp1 and the 40S subunit induces cleavage of cellular mRNAs to suppress host cell translation to an even larger extent (Kamitani et al. 2006). MERS-CoV nsp1 seems to act differently, by selectively inhibiting the translation of mRNAs produced in the nucleus, while leaving the translation of the cytosolically made viral mRNAs unaffected (Lokugamage et al. 2015). The difference with SARS-CoV nsp1 is further highlighted by the observation that MERS-CoV nsp1 does not bind to the 40S ribosomal subunit (Lokugamage et al. 2015).

Taken together, several coronavirus studies have highlighted how modulation of host protein synthesis through different—often parallel—mechanisms can have a profound effect on the cell. In this manner, viral 'translation modulators' may contribute importantly to coronavirus pathogenicity.

4 Coronavirus-Induced Modification of Host Cell Membranes

As outlined in Chap. 1, a common characteristic of +RNA viruses is that their RNA synthesis takes place in the cytoplasm and is associated with virus-induced structures derived from cellular endomembranes [reviewed in (Romero-Brey and Bartenschlager 2016; Reid et al. 2015; van der Hoeven et al. 2016)]. This is an intriguing kind of virus–host interaction and the architecture of these 'replication organelles' has now been studied in detail for quite a number of viruses. Nevertheless, their exact functions have remained largely obscure. In general, two types of +RNA virus-induced membrane structures have been recognized [recently reviewed by (van der Hoeven et al. 2016)]. The first type is characterized by single-membrane spherules, invaginations with a negative curvature formed in the membranes of organelles such as the endoplasmic reticulum (ER), peroxisomes, or endosomes, with the source of the membrane depending on the virus under study. The viral replication machinery is located within these spherules and RNA products are exported through a channel that connects the spherule's interior and the cytosol, so that they can engage in translation or particle assembly. Flaviviruses and alphaviruses are examples of virus families inducing the formation of this type of replication organelles [reviewed in (den Boon and Ahlquist 2010)].

The second type of replication structures, which includes those found in coronavirus-infected cells, is dominated by double-membrane vesicles (DMVs), often accompanied by other structures such as tubules, zippered ER and/or convoluted membranes, together forming a reticulovesicular network in the cytosol (Knoops et al. 2008; Maier et al. 2013; Ulasli et al. 2010; Hagemeijer et al. 2012; Gosert et al. 2002) (Fig. 1c). Picornaviruses, arteriviruses, and flaviviruses like hepatitis C virus (HCV) induce similar structures [reviewed in (van der Schaar et al. 2016; van der Hoeven et al. 2016; Paul et al. 2014), respectively]. It is generally thought that viral nsps that have transmembrane regions, or are otherwise anchored to membranes, drive the formation of these structures. In the case of coronaviruses, nsp3, nsp4, and nsp6 have been implicated in this process (Angelini et al. 2013; Hagemeijer et al. 2012).

In terms of host factors and pathways involved in the formation of coronavirus replication organelles, a variety of hypotheses and data sets have been put forward. Much of this data is still under debate though, making it quite difficult to formulate a consensus theory. The involvement of ER membranes seems to be generally accepted and is supported by the presence of ribosomes and ER markers such as sec61α and protein disulphide isomerase (PDI) on the surface of or inside virus-induced double-membrane structures, and the fact that the outer membrane of coronavirus-induced DMVs can be continuous with ER cisternae (Hagemeijer et al. 2014; Knoops et al. 2008, 2010; Snijder et al. 2006). In accordance with this link to the ER, studies by de Haan and co-workers (Oostra et al. 2007; Verheije et al. 2008) suggested that the secretory pathway, including coatomer protein (COP)-dependent processes and associated factors such as GBF-1, plays an important role in replication. However, no co-localization was observed between nsp4, a marker for the coronavirus-induced membrane structures, and secretory pathway markers. These observations are in agreement with the results of our own siRNA screen for host factors influencing SARS-CoV replication, which identified COPB2 (or β′-COP), a subunit of the coatomer protein complex, as a strong proviral (or 'dependency') factor (de Wilde et al. 2015). Depletion of COPB1 and GBF1, which are part of this same machinery, also severely affected SARS-CoV replication, confirming that the integrity of the secretory pathway is important for replication (de Wilde et al. 2015; Knoops et al. 2010). Several reports have described the importance of phosphatidylinositol 4-kinases (PI4Ks) in +RNA virus replication, which was first discovered through siRNA screens searching for cellular factors important for picornavirus and hepatitis C virus replication (Reiss et al. 2011; Hsu et al. 2010; Berger et al. 2009). These kinases seem to be recruited to the sites of membrane modification and stimulate the production of PI4P lipids, which together supports the formation and/or functionality of viral replication structures. The underlying mechanism is not exactly clear yet, and several hypotheses have been put forward [reviewed in (Altan-Bonnet and Balla 2012)]. Also for SARS-CoV replication one of the PI4K isoforms, PI4KIIIbeta, was shown to be important (Yang et al. 2012), although it seems to play a role in entry rather than later steps of the replication cycle. However, in our kinome-based siRNA screen, PI4Ks were not identified as cellular factors involved in SARS-CoV replication (de Wilde et al. 2015), although siRNA screens are known to yield false-negative results.

A long-standing hypothesis regarding coronavirus-induced double-membrane structures is the possible involvement of the autophagy pathway, which derives from the fact that autophagosomes also have double membranes. Some reports suggested that coronaviruses hijack the autophagy machinery for DMV biogenesis in support of their replication (de Haan and Reggiori 2008; Prentice et al. 2004; Maier and Britton 2012). However, another study showed that the essential autophagy factor Atg5 is not required at all for coronavirus replication in primary cells (Zhao et al. 2007). Molinari and co-workers then proposed that so-called EDEMosomes are being hijacked for the formation of coronavirus membrane structures (Reggiori et al. 2010). These EDEMosomes are defined as single-membrane vesicles that pinch off from the ER to remove ERAD regulators (like EDEM1 and OS-9) when this is needed to tune the ERAD machinery (Cali et al. 2008). The process seems a deviation from the autophagy pathway, with the EDEMosomes accumulating LC3-I, a form that is inactive in the canonical autophagy pathway. Reggiori and co-workers claimed that coronaviruses hijack these vesicles to form their reticulovesicular network, and this hypothesis was later extended to arteriviruses (Monastyrska et al. 2013). However, several questions have remained unanswered and other published data appear to be at odds with the EDEMosome hypothesis. For example, it has remained entirely unclear how the small single-membrane EDEMosome vesicles would be converted into the elaborate network of (much larger) DMVs, convoluted membranes (CM), and other structures that are typical of coronavirus-infected cells. Furthermore, EDEMosomes have been characterized as alternative transport vesicles that explicitly are not associated with COP-coats and are independent of the canonical secretory pathway, which—as Reggiori and co-workers argued—may explain why secretory pathway markers do not localize to replication membranes (Reggiori et al. 2010). Nonetheless, the integrity of the secretory pathway and the function of COP components clearly influences coronavirus replication (Verheije et al. 2008; Oostra et al. 2007; Knoops et al. 2010; de Wilde et al. 2015).

All in all, although a variety of studies addressed the possible host pathways and factors involved in the formation and function of the replication-associated membrane structures, our understanding of coronavirus replication organelle biogenesis is still far from complete. Possibly, the expansion of our basic knowledge regarding relevant cellular processes, such as membrane trafficking and autophagy, may provide more clues on the molecular mechanisms underlying these interesting interactions of coronaviruses with their host cells.

5 Host Proteins Interacting with the Coronavirus Genome and Its Replication or Expression

The 5′- and 3′-proximal regions of coronavirus RNAs contain key regulatory elements for their RNA synthesis [for a recent review, see (Yang and Leibowitz 2015)]. Although in general the precise role of host factors interacting with these signals is poorly understood, RNA-binding proteins have been identified as frequently used enhancers of coronaviral RNA synthesis (Table 2). Both termini of the coronavirus genome fold into higher-order RNA structures, which presumably stabilize the molecule and are also involved in inter- and intramolecular interactions that facilitate viral replication (Brian and Baric 2005). Viral and cellular proteins can bind to these structures to drive or modulate translation, replication, and subgenomic RNA synthesis.

The cellular protein polypyrimidine tract-binding protein (PTB; or heterogeneous ribonucleoprotein protein (hnRNP) I) was found to bind the 5′ leader sequence of the TGEV (Galan et al. 2009) and MHV genome (Li et al. 1999; Choi et al. 2002). In the case of MHV, PTB was found to bind a 5′-proximal pentanucleotide UCUAA repeat and to be critical for RNA synthesis. HnRNP Q, or SYNCRIP, also binds the 5′-proximal part of the MHV genome and its knockdown reduced MHV RNA synthesis and virus replication. The case for a specific role in RNA synthesis was strengthened by the observation that neither overexpression nor downregulation affected translation of MHV RNA (Choi et al. 2004). Zinc finger CCHC-type and RNA-binding motif 1 (MADP1) was shown to bind the 5′ end of the SARS-CoV and IBV genome (Tan et al. 2012). In analogy to SYNCRIP, silencing of MAPD1 reduced IBV replication by interfering with viral RNA synthesis, showing that MAPD1 plays a proviral role in the coronavirus cycle (Tan et al. 2012).

In addition to proteins that bind the 5′ UTR of the coronavirus RNA, many RNA-binding proteins have been identified that interact with the 3′ UTR and/or poly(A)-tail, although the role of these proteins is poorly understood. First, mitochondrial aconitase, stabilized by a complex of heat-shock protein (hsp) 40, hsp60, and mitochondrial hsp70, binds the 3′-terminal 42 nucleotides of the MHV 3′ UTR, just upstream the poly(A) tail (Nanda et al. 2004; Nanda and Leibowitz 2001). The protein p100 coactivator was found to bind the TGEV 3′ UTR and/or poly(A) tail (Galan et al. 2009) and the PABPs interact with the TGEV, BCoV and IBV genome to promote its efficient replication (Spagnolo and Hogue 2000; Galan et al. 2009; Emmott et al. 2013). It has been proposed that binding of PABPs to the viral RNA ensures efficient translation and mRNA stability (Enjuanes 2005). In addition, by using an RNA-affinity purification and mass spectrometry approach, Galan et al. (2009) identified several hnRNPs that bind the 3′-terminal ~500 nucleotides of the TGEV genome. Several of these are presumed RNA-binding proteins for other coronaviruses and were proposed to play a role in viral RNA synthesis (Table 2). However, the exact function and relevance of hnRNPs and other RNA-binding proteins in coronavirus replication is still not fully understood. The majority of RNA-binding proteins were identified in co-immunoprecipitation studies that made

Table 2 Overview of RNA-binding proteins that have been described to interact with the RNA of various coronaviruses

Protein	Coronavirus	Cellular function	Interactions or proposed function in viral RNA synthesis	Experimental evidence	References
Annexin A2	IBV	Cellular RNA-binding protein	Modulates IBV frameshifting efficiency	In vitro pull-down	(Kwak et al. 2011)
hnRNP A0	TGEV	Involved in RNA splicing	Binds the TGEV 3′ UTR or poly(A)-tail	In vitro pull-down	(Galan et al. 2009)
hnRNP A1	MHV	Involved in RNA transport, processing, and splicing	High affinity for the MHV (−)-strand leader RNA of the MHV genome. Also binds the intergenic region that regulates ORF7 synthesis	In vitro pull-down	(Li et al. 1997; Zhang and Lai 1995; Furuya and Lai 1993)
			Binds the MHV 3′-UTR, may mediate RNP formation to bind 5′end and 3′end of the MHV genome together with PTB.	In vitro pull-down	(Huang and Lai 1999, 2001)
			Binds the TGEV 3′ UTR or poly(A)-tail	In vitro pull-down	(Galan et al. 2009)
			hnRNP A1 regulates MHV RNA synthesis	Overexpression	(Shi et al. 2000)
hnRNP A2-B1	TGEV, MHV	Involved in RNA transport and splicing	Binds the TGEV 3′ UTR or poly(A)-tail and the (-)-strand MHV leader RNA	In vitro pull-down	(Galan et al. 2009; Shi et al. 2003)
hnRNP A-B and A3	MHV	Involved in RNA transport and splicing	Role in MHV RNA synthesis	In vitro pull-down	(Shi et al. 2003)
hnRNP Q (SYNCRIP)	TGEV, MHV	Involved in RNA processing and splicing	Binds the TGEV 3′ UTR or poly(A)-tail and the MHV 5′end.—Involved in MHV RNA synthesis	In vitro pull-down, virus infection	(Galan et al. 2009; Choi et al. 2004)
hnRNP U	TGEV	Involved in RNA processing and splicing	Binds the TGEV 3′ UTR or poly(A)-tail	In vitro pull-down	(Galan et al. 2009)

(continued)

Table 2 (continued)

Protein	Coronavirus	Cellular function	Interactions or proposed function in viral RNA synthesis	Experimental evidence	References
MADP1	SARS-CoV, IBV	Involved in RNA splicing	Binds the 5′ UTR of the viral genome. Functions in IBV RNA synthesis	In vitro pull-down, virus infection	(Tan et al. 2012)
Mitochondrial aconitase, hsp40, hsp60, mtHsp70	MHV	Mitochondrial aconitase is a component of the citric acid cycle, Hsp40: chaperone, regulate function of hsp70. Mt-hsp70: chaperone, helps to protect from cell stress. Hsp60: mitochondrial chaperone	Hsp40, hsp60, mtHsp70 stabilize complex with mitochondrial aconitase, binds the last 42 nucleotides of the 3′ UTR of MHV	In vitro pull-down, virus infection	(Yu and Leibowitz 1995; Nanda and Leibowitz 2001; Nanda et al. 2004)
NONO	IBV	RNA-binding protein which plays a role transcriptional regulation and RNA splicing	Interacts indirectly with the IBV nucleocapsid protein via viral and/or cellular RNA	In vitro pull-down, SILAC	(Emmott et al. 2013)
p100 kDa coactivator	TGEV	Involved in transcription and RNA interference	Binds the TGEV 3′ UTR or poly(A)-tail	In vitro pull-down	(Galan et al. 2009)
PABPs	BCoV, TGEV, IBV	RNA-binding protein that binds to the poly(A) tail of cellular mRNA. Involved in mRNA translation	Binds to the poly(A)-tail of the viral genome. Signal for genome replication	In vitro pull-down, SILAC	(Spagnolo and Hogue 2000; Galan et al. 2009; Emmott et al. 2013)
PTB (hnRNP I)	TGEV, MHV	Involved in RNA splicing	Binds the 5′ end leader sequence of the viral genome. (MHV: binds 5′ pentanucleotide repeat UCUAA). May form RNP complex with hnRNP A1, MHV N and viral RNA. Regulates viral transcription	In vitro pull-down, overexpression	(Galan et al. 2009; Li et al. 1999; Choi et al. 2002; Huang and Lai 1999)

(*Abbreviations PABP* poly(A)-binding protein; *PCBP* poly-r(C)-binding protein 1; *hnRNP* heterogeneous nuclear ribonucleoprotein; *PTB* polypyrimidine tract-binding protein; MADP1: zinc finger CCHC-type and RNA-binding motif 1; *DDX* DEAD-box protein; *BCoV* bovine coronavirus; *SYNCRIP* synaptotagmin-binding cytoplasmic RNA-interacting protein; *NONO* Non-POU domain-containing octamer-binding protein; *SILAC* stable isotope labeling with animo acids in cell culture) use of in vitro transcribed RNA and the importance of many of these factors has not been tested in the context of virus replication. For example, hnRNP A1 was reported to have affinity for the complement of the MHV leader sequence and the TRS that regulates mRNA7 synthesis (Li et al. 1997; Zhang and Lai 1995). Interestingly, however, Shen and Masters (Shen and Masters 2001) showed that MHV replicates equally well in hnRNP A1$^{-/-}$ CB3 cells and parental CB3 cells, which lack and express hnRNP A1, respectively. This strongly suggests that hnRNP A1 itself is not pivotal for MHV replication. On the other hand, Lai and co-workers showed that multiple hnRNPs can bind to the same viral RNA and postulated that hnRNPs may be able to functionally substitute each other in MHV infection (Shi et al. 2003). These apparent contradictions also highlight the technical complexity of dissecting the precise role of RNA-binding proteins in coronavirus replication.

On a different level, RNA-binding proteins were investigated as potential modulators of the efficiency of the coronavirus ORF1a/1b frameshift event during replicase gene translation. For example, annexin A2 binds the IBV slippery sequence, on which ribosomes are repositioned and engage in expression of ORF1b. Host proteins may also affect the activity of the viral RTC indirectly. Using an in vitro assay, the RNA-synthesizing activity of semi-purified SARS-CoV RTCs was shown to depend on an as yet unidentified cytosolic host factor, which is not directly associated with the viral RTC (van Hemert et al. 2008).

6 Host Innate Immune Responses Against Coronaviruses, and Viral Countermeasures

Cells generally respond to a virus infection by mounting an innate antiviral response to limit the spread of the infection and aid in inducing an adaptive immune response that will eventually clear the virus. Many viruses have evolved strategies to suppress and/or evade these (innate) immune responses, which can dramatically influence the course of the infection, including pathogenesis and persistence in the host. In the case of +RNA viruses, the innate immune system is often triggered by the dsRNA and 5′-triphosphate-bearing RNA molecules that arise as replication intermediates in the cytosol. These molecules are foreign to the cell and can be recognized by the intracellular sensors of the Rig-I-like receptor (RLR) family, such as retinoic acid-inducible gene 1 (RIG-I) and melanoma differentiation-associated protein 5 (MDA-5) which are expressed in almost all cells [reviewed in (Wilkins and Gale 2010; Bruns and Horvath 2014)]. For recognition of coronavirus RNAs, MDA-5 seems the most important cytosolic sensor (Zust et al. 2011; Zalinger et al. 2015; Kindler and Thiel 2014), although in some cell types RIG-I also seems to play a role (Li et al. 2010). Also the toll-like receptors which are expressed on the cell surface or reside in the endosomes of immune cells can recognize viral nucleic acids or proteins. TLR3 plays a role in the recognition of coronaviruses (Totura et al. 2015; Mazaleuskaya et al. 2012) and also TLR4 was shown to be relevant during MHV infection in mice (Khanolkar et al. 2009). Recently it was shown that the SARS-CoV M protein is recognized via a TLR-like pathway that is independent of the canonical TRAF3-mediated signaling pathway (Wang and Liu 2016). Activation of one or more of these sensors generally leads to the activation of the transcription factors IFN-regulatory factor 3 and 7 (IRF3, IRF7) and NF-κB. These stimulate the expression and excretion of Type-I IFN and pro-inflammatory cytokines, which in turn activate the JAK-STAT signaling cascade that induces the expression of a myriad of antiviral interferon-stimulated genes (ISGs). This ultimately results in an antiviral state of the infected cell, as well as neighboring cells. ISGs were shown to target virtually all steps of the viral cycle in order to restrict viral replication (Schoggins and Rice 2011). The p38 mitogen-activated protein kinases (MAPKs) play a role in the induction of inflammatory cytokines IL-6 and IL-8, and were linked to countering coronavirus infections through several studies [reviewed in (Mizutani 2007)]. IBV has evolved a strategy to counteract IL-6 and IL-8 expression by inducing the expression of dual-specificity phosphatase 1 (DUSP1), a negative regulator of p38 MAPK, although it has remained unclear which viral protein(s) is responsible (Liao et al. 2011). Innate immune and inflammatory signaling pathways are extensively regulated in order to prevent adverse effects of their over-stimulation. Apart from phosphorylation and other regulatory mechanisms, the system is controlled by ubiquitination at numerous points in the signal transduction cascade. For example, RIG-I, TANK-binding kinase 1 (TBK1), and TNF receptor-associated factor 3 (TRAF3) were shown to be activated by Lys63-linked ubiquitination (Jiang and Chen 2012).

The importance of the innate immune system in the context of coronavirus infections can be illustrated in at least three ways. First, in severe cases of SARS the pathology was associated with aberrant or hyper-activation of innate immune signaling. This resulted in the aberrant production of interferons and high levels of pro-inflammatory cytokines such as IL-1, IL-6, IL-8, CXCL-10 and TNF-alpha in the lungs [reviewed in (Totura and Baric 2012)]. Interestingly, a systems biology study that evaluated the transcriptome after infection of cultured cells with two different human MERS-CoV isolates (MERS-CoV Eng 1 vs. SA 1) suggested that viral sequence differences relate to variations in innate immune evasion, which may in turn result in different immune responses. These differences may link to differential STAT3 activation leading to activation or inhibition of, *e.g.*, IFN, NF-κB, and IRF7 (Selinger et al. 2014). Second, as described in Chap. 4, coronaviruses employ elaborate mechanisms to shield the viral replication machinery from the innate immune sensors in the cytosol. Third, besides the presumed shielding of viral PAMPs by the replication organelles, coronaviruses seem to encode numerous gene products that actively counteract or help to circumvent innate immune responses [reviewed in (Totura and Baric 2012; Kindler and Thiel 2014; Vijay and Perlman 2016)].

To illustrate the multitude of activities coronaviruses employ to actively suppress innate immunity, and the diversity of viral gene products involved, some examples are discussed below. First, besides inhibiting cellular mRNA translation (see paragraph above), SARS-CoV nsp1 was shown to block IFN signaling by reducing the amount of phosphorylated STAT1 (p-STAT1) in infected cells (Wathelet et al. 2007). Also nsp1 proteins of other alpha- and betacoronaviruses were shown to inhibit type-I IFN signaling, mostly through the host-shut-off activity of this N-terminal subunit of the coronavirus replicase polyprotein (reviewed in Narayanan et al. 2015). Further downstream in the pp1a polyprotein, the papain-like protease 2 domain ($PL2^{pro}$) of many coronaviruses and the TGEV $PL1^{pro}$ domain, which reside in nsp3, exhibit deubiquitination (DUB) activity in biochemical experiments [reviewed in (Mielech et al. 2014)]. This DUB activity may remove ubiquitin from innate immune signaling factors to suppress the induction of an antiviral state, and indeed was shown to reduce IFN signaling in biochemical experiments using $PL2^{pro}$ overexpression for several coronaviruses, including SARS-CoV (Matthews et al. 2014a; Li et al. 2016) and MERS-CoV (Chen et al. 2007; Clementz et al. 2010; Bailey-Elkin et al. 2014). Similarly, experiments suggested that MHV $PL2^{pro}$ deubiquitinates and binds TBK1, as well as IRF3 (Zheng et al. 2008; Wang et al. 2011). For the distantly related arterivirus EAV, infection with a mutant lacking a similar papain-like protease-driven DUB activity resulted in an increased innate immune response after infection, indicating that the viral DUB activity indeed has a function in suppression of innate immune response during infection (van Kasteren et al. 2013). Adjacent to the $PL2^{pro}$ domain, some coronaviruses contain a domain originally coined 'SARS-CoV unique domain' (SUD). The two domains together were recently also implicated in innate immune suppression by binding and stimulating a cellular ubiquitin E3 ligase, RCHY1, resulting in augmented degradation of p53 (Ma-Lauer et al. 2016). The main protease, nsp5, of PEDV was shown to cleave NEMO, an important innate immune regulator protein (Wang et al. 2015). When independently expressed, coronavirus nsp6, which is probably one of the most hydrophobic proteins encoded in the genome, seems to induce and/or influence autophagy. However the relevance of this observation for virus-infected cells and possible links to innate immune responses against coronaviruses remain to be investigated (Cottam et al. 2014; Cottam et al. 2011). A conserved domain in the coronavirus nsp16 directs 2′-O methylation of the viral RNA, thereby preventing its recognition by MDA5 (Menachery et al. 2014; Zust et al. 2011).

Besides conserved replicase subunits, also several of the less-conserved products of downstream open reading frames, i.e., the so-called 'accessory proteins', have innate immunity suppressing features. In fact, these proteins may have been acquired by different coronavirus lineages for this purpose, since they are generally dispensable for replication in cell culture. The phosphodiesterase (ns2) encoded by ORF2a in MHV and related viruses antagonizes RNAse L activation, and this was show to be important for replication in natural host cells (Zhao et al. 2012; Li and Weiss 2016). Observations supporting a role in blocking IFN production and/or signaling were also made for the ORF3b and ORF6 proteins of SARS-CoV (Kopecky-Bromberg et al. 2007) and the ORF3b proteins of other SARS-like coronaviruses (Zhou et al. 2012). Also the proteins encoded by MERS-CoV ORFs 4a, 4b and 5 (Siu et al. 2014; Yang et al. 2013; Niemeyer et al. 2013; Matthews et al. 2014b) all seem to block IFN signaling, although many of the initial reports were based only on overexpression experiments, which should be interpreted with caution. Recently however, more insight into the mechanisms by which some of these viral proteins suppress antiviral responses was obtained. Like MHV ns2 (see above), the MERS-CoV ORF4b protein and its homologs from other betacoronaviruses were shown to have phosphodiesterase activity. They can block activation of the RNAse L-mediated innate immune response during infection by degrading 2′,5′-oligoadenylate, which is the activator of RNAse L (Thornbrough et al. 2016). Additionally, MERS-CoV ORF4a prevents activation of the PKR-stress granule route by binding dsRNA, in the context of an infection (Rabouw et al. 2016). The SARS-CoV ORF6 protein was shown to block p-STAT1 import into the nucleus by interacting with the nuclear pore, and this block was suggested to reduce innate immune responses and the expression of genes that affect virus-induced pathogenesis (Frieman et al. 2007; Kopecky-Bromberg et al. 2007; Huang et al. 2015a). Interestingly, MERS-CoV, along with the other coronaviruses not belonging to the SARS-CoV cluster, lacks an ORF6 homolog, and since MERS-CoV is also more sensitive to treatment with type-I interferons, it was hypothesized that this could—in part—be explained by this difference (de Wilde et al. 2013b). Also the coronaviral structural proteins seem to possess innate immunity-modulating activities, although also here most studies involved overexpression experiments. The SARS-CoV and MERS-CoV M proteins bind TRAF3 to prevent its binding to TBK1 and eventually to prevent nuclear translocation of this complex, which blocks IRF3-mediated signaling (Siu et al. 2009; Lui et al. 2016). Overexpression experiments showed a negative impact of the SARS-CoV N protein on (the early stages of) innate immune signaling (Frieman et al. 2009; Kopecky-Bromberg et al. 2007; Lu et al. 2011). MHV N protein also seems to counteract type-I IFN signaling, and could do this in the context of a recombinant vaccinia virus infection (Ye et al. 2007). Purified SARS-CoV spike (S) protein stimulates inflammatory and other innate responses, possibly through TLR2 activation (Dosch et al. 2009). Finally, the SARS-CoV E protein, is a viroporin, and influences inflammatory processes by boosting the activity of the NLRP3 inflammasome, leading to IL-1beta overproduction and development of immunopathology in the host (Nieto-Torres et al. 2015).

7 Coronavirus-Induced Deregulation of the Cell Cycle

The cell cycle is a series of highly regulated events that leads to cell division. The process can be divided into four distinct phases: G1, S, G2, and M. Cell cycle regulation is critical for cell survival, as well as the prevention of uncontrolled cell division. The molecular mechanisms that control the cell cycle are ordered, directional, and controlled by cyclin-dependent protein kinases (CDKs). Cell cycle progression requires activation of different CDKs by, e.g., cyclin regulatory subunits. To reach the stage of DNA replication, CDK/cyclin complexes phosphorylate, and thereby activate or inactivate, their target proteins to coordinate progression towards the next phase of the cell cycle (Nigg 1995).

Like many other viruses [reviewed in (Bagga and Bouchard 2014)], coronaviruses have been shown to extensively manipulate and arrest cell cycle progression, to benefit from the physiological state of cells arrested in that specific phase. For example, IBV-infected cells were shown to go into cell cycle arrest in the S phase, by activating the cellular DNA damage response (Xu et al. 2011). This is beneficial to virus replication since factors that are normally needed for DNA replication and are upregulated in the S phase, can now be recruited to the cytoplasm by the virus. For example, DDX1, a cellular RNA helicase of the DExD/H family, interacts with coronavirus nsp14 (Xu et al. 2010) and was reported to be hijacked by coronaviruses to enhance their replication. DDX1 also interacts with the IBV N protein (Emmott et al. 2013) and facilitates, in complex with the phosphorylated form of the MHV-JHM N protein, the balanced synthesis of sg mRNAs and the genomic RNA (Wu et al. 2014).

Bhardwaj et al. have shown that coronavirus nsp15 interacts with and inhibits retinoblastoma protein (pRb), a tumor suppressor protein. This results in the enhanced expression of genes that are normally repressed by pRb and in an increased proportion of cells entering the S phase of the cell cycle (Bhardwaj et al. 2012). Similar effects have been observed in MHV-infected cells, which showed decreased hyperphosphorylation of pRb, an event that is necessary for the progression from G1 to S phase (Chen and Makino 2004; Chen et al. 2004). Yuan and colleagues showed that overexpression of the SARS-CoV 3a protein also leads to G1 arrest and inhibition of cell proliferation (Yuan et al. 2007).

The cyclin-dependent kinase 6 (CDK6) is downregulated upon MHV infection and seems to play a role in a virus-induced cell cycle arrest in the G0/G1 phase that promotes virus replication (Chen and Makino 2004). Similar observations have been made for SARS-CoV, with (overexpression of) the N protein limiting cell cycle progression by reducing CDK4 and CDK6 kinase activity (Surjit et al. 2006). CDK6 is a kinase involved in cell cycle progression from G1 to S phase (Jimenez-Guardeno et al. 2014) and depletion of CDK6 results in G1 phase cell cycle arrest. A host kinome-directed siRNA screen confirmed the antiviral role of CDK6 in SARS-CoV infection, as replication was enhanced in cells depleted for CDK6 (de Wilde et al. 2015).

Several coronaviruses induce activation of p53 to mediate cell cycle arrest in the S or G2/M phase. For example, TGEV N protein activates p53 which leads to accumulation of cell cycle-related kinases, like cdc-2 and cyclin B1. Synchronisation of cells in the S or G2/M phase favors TGEV RNA and virus production (Ding et al. 2013, 2014) and IBV replication (Dove et al. 2006). For MHV nsp1 a similar mechanism has been proposed: nsp1 activates p53, leading to increased p21 and decreased CDK2-cyclin E levels. This ultimately leads to hyperphosphorylation of pRb and G1 cell cycle arrest (Chen et al. 2004). In contrast, SARS-CoV infection leads to reduction of p53 expression levels. To counteract the inhibitory effect of p53, its expression is reduced by a mechanism that involves stabilization of the E3 ligase RCHY1 by SARS-CoV nsp3. RCHY1 mediates the ubiquitination and degradation of p53. However, the exact role of this mechanism in cell cycle regulation, viral replication and/or pathogenesis remains unclear (Ma-Lauer et al. 2016). Nevertheless, in cells that lack p53 expression, SARS-CoV replication was significantly enhanced (Ma-Lauer et al. 2016). Degradation of p53 by the SARS-CoV or HCoV-NL63 $PL2^{pro}$ is another mechanism to counteract the antiviral effect of p53. The deubiquitinating activity of $PL2^{pro}$ promotes the degradation of p53, thereby lowering the p53-mediated antiviral immune response (Yuan et al. 2015).

In conclusion, coronaviruses appear to favor a specific stage in the cell cycle for their replication. This stage can differ per coronavirus and may even differ per cell type. A wide variety of coronavirus proteins have been implicated in inducing cell cycle arrest, but it must be noted that many studies involved overexpression of individual viral proteins and therefore should be interpreted with caution since their expression levels are likely different than during virus infection. When expressed outside the context of virus replication, these proteins may also behave differently (*e.g.*, due to lack of viral interaction partners) and/or localize differently (*e.g.*, in the absence of virus-induced membrane structures), which might lead to less meaningful observations.

8 The Role of Cyclophilins in Coronavirus Replication

Cyclosporin A (CsA) is a well-known immunosuppressant that binds to cellular cyclophilins (Cyps), yielding a Cyp-CsA complex that inhibits calcineurin activity. This in turn prevents dephosphorylation and translocation of nuclear factor of activated T cells (NF-AT) from the cytosol into the nucleus, which prevents the transcription of immune genes, such as IL-2 [reviewed in (Tanaka et al. 2013; Davis et al. 2010; Barik 2006)]. Thus far, 17 Cyps have been identified, of which nine are targeted by CsA. Cyps are also known as peptidyl–prolyl isomerases (PPIases) and many of them have chaperone and foldase activities (Barik 2006; Davis et al. 2010) that facilitate protein folding. Cyps are involved in various signaling pathways [reviewed in (Barik 2006)], apoptosis (Schinzel et al. 2005), and RNA splicing (Teigelkamp et al. 1998; Horowitz et al. 2002).

Cyclophilins, and in particular the cytosolic CypA and the ER-associated CypB, have been implicated in the replicative cycle of many RNA viruses as essential host components [reviewed in (Baugh and Gallay 2012)]. For example, (i) Cyps are essential in the remodeling of cellular membranes into HCV replication organelles, (ii) CypA aids in HCV polyprotein processing, (iii) HIV-1 capsids are stabilized by low levels of CypA to ensure entrance into the nucleus before the virion could be destabilized in the cytosol [(Hopkins and Gallay 2015), and references herein]. The use of CsA analogs, like Alisporivir (Paeshuyse et al. 2006), that lack the immunosuppressive properties of the parental compound, has been explored in clinical trials for the treatment of chronic HCV infection, again illustrating the prominent role of Cyps in HCV replication and the druggability of Cyps (Flisiak et al. 2012).

Along the same lines, the inhibition of coronavirus replication by Cyp inhibitors like CsA and Alisporivir suggested important roles for Cyps. In cell culture infection models, low-micromolar concentrations of CsA or Alisporivir inhibit a variety of coronaviruses, including SARS-CoV, HCoV-229E, MHV, HCoV-NL63, and FCoV (Pfefferle et al. 2011; de Wilde et al. 2011; Tanaka et al. 2012; Carbajo-Lozoya et al. 2012; de Wilde et al. 2017). The mitochondrial CypD is part of the mitochondrial permeabilization transition pore and involved in caspase-independent apoptosis induced by porcine epidemic diarrea virus (PEDV) and HCoV-NL63 (Favreau et al. 2012; Kim and Lee 2014), an event that is inhibited by CsA. However, most studies on the presumed role of Cyps in coronavirus replication have focused on CypA. Initially, CypA was identified as an interaction partner of the SARS-CoV N protein (Luo et al. 2004) and by mass spectrometry it was also detected in purified SARS-CoV virions (Neuman et al. 2008). Yeast two-hybrid experiments identified multiple Cyps (CypA, CypB, CypH, and CypG) and related PPIases (FK506-binding protein 1A and 1B) as potential binding partners of SARS-CoV nsp1 (Pfefferle et al. 2011). In addition, in Caco-2 or Huh7 cells, shRNA-mediated knockdown of CypA expression to less than 3% of the normal levels was reported to near-completely block HCoV-NL63 replication and reduce HCoV-229E replication by >1 log (Carbajo-Lozoya et al. 2014; von Brunn et al. 2015). Recently, by using PPIA or PPIB knockout rather than knockdown cells, it was shown that FCoV depends on both CypA and CypB expression (>95% reduction in FCoV-infected CypA or CypB-KO cells). PPIase-deficient mutants, expressed in cells that also contained endogenous CypA and CypB, marginally reduced FCoV infection (two- to fivefold reduction in FCoV RNA compared to cells that express wt CypA or wt CypB only) (Tanaka et al. 2016). Furthermore, the work of von Brunn and colleagues suggested that HCoV-229E replication was reduced by various CypA single-nucleotide polymorphisms that affect the protein's stability and function (von Brunn et al. 2015). Although these results suggested a role for CypA in coronavirus replication, the siRNA-mediated depletion of CypA and CypB did not affect the replication of SARS-CoV (de Wilde et al. 2011), despite the fact that the same knockdown did affect the replication of the distantly related arterivirus EAV (de Wilde et al. 2013a). The differences reported using either knockout or knockdown of CypA with different efficiencies (100%, >97%, and ~75% for CypA knockout cells, and CypA shRNA- or siRNA-mediated knockdown, respectively) suggest that low CypA expression levels may suffice to support efficient coronavirus replication and that the (near-)complete depletion of CypA may be needed to inhibit virus replication.

The major role of Cyps in cellular signaling is in the NF-AT signaling pathway. Pfefferle et al. described that SARS-CoV activates the NF-AT signaling pathway (Pfefferle et al. 2011) and the replication of various coronaviruses, including SARS-CoV, HCoV-229E, and HCoV-NL63, is inhibited by the drug FK-506, which—like CsA—also blocks NF-AT signaling (Carbajo-Lozoya et al. 2012). Feline coronavirus replication seemed not to depend on a functional NF-AT signaling pathway since it was blocked by CsA concentrations that did not affect NF-AT signaling in fcwf-4 cells (Tanaka et al. 2012).

In conclusion, further studies, are needed to dissect the precise role of Cyps in coronavirus replication, with special attention for the (remaining) levels of CypA expression, and the relevance of NF-AT signaling following coronavirus infection. Such experiments should include the production and use of knockout cells for one or multiple Cyps or other members of this protein family. Although Cyp inhibitors have been shown to be potent anti-coronavirus drugs in cell culture, a first attempt to validate this effect for Alisporivir in a SARS-CoV animal model was unsuccessful (de Wilde et al. 2017). Therefore, although Cyp inhibitors may help to understand the role of host factors in coronavirus replication, they are at this moment less promising as host-directed therapeutics for the treatment of coronavirus infections.

9 Systems Biology Approaches to Identifying Host Factors in Coronavirus Replication

The application of systems biology approaches in virology has provided a wealth of information on the role of individual proteins and cellular pathways in the replication of RNA viruses. This relatively young, interdisciplinary field focuses on the complexity of the virus–host interactions that occur within the cell or even the whole organism. The aim is to provide an unbiased perspective, by applying techniques like transcriptomics, metabolomics, proteomics, and functional genomics to the infected system as a whole.

For coronaviruses, one of the first systematic studies into the role of host factors concerned an oligonucleotide microarray-based transcriptomic analysis of SARS-CoV-infected peripheral blood mononuclear cells, which revealed the upregulation of the expression of various cytokines, including IL-8 and IL-17, and the activation of macrophages and the coagulation pathway (Ng et al. 2004). A microarray analysis of lung autopsy tissue samples provided more insight into the pathogenesis of and host response to SARS-CoV infection, in particular the inflammatory and cytokine responses involved (Baas et al. 2006). MHV-JHM infected cultures of central nervous system cells showed 126 differentially expressed transcripts, the majority of which were related to intracellular regulation of innate immunity (*e.g.*, NF-κB signaling and genes involved in IFN signaling) (Rempel et al. 2005). Microarray analysis of MHV-A59-infected L cells provided insight into transcriptional changes during infection, including those related to chemokine production, RNA and protein metabolism and apoptosis (Versteeg et al. 2006). Subsequently, a genome-wide microarray analysis of MHV-infected LR7 cells revealed the downregulation of a large number of mRNAs, including many encoding proteins involved in translation, implying that the host translational shut-off that occurs in MHV-infected cells is due to a stress response and concomitant mRNA decay (Raaben et al. 2007). There is not necessarily a direct correlation between changes in mRNA levels and protein abundance in the cell.

Therefore, microarray data should be interpreted with caution and need to be validated with follow-up experiments including the direct analysis of changes in cellular protein levels.

Another approach to obtain more insight into coronavirus–host interactions is to systematically map the cellular interactome of individual coronavirus proteins. A yeast two-hybrid (Y2H) screen of this type identified subunits (BTF3 and ATF5) of the RNA polymerase complex and a subunit of cytochrome oxidase II (NADH 4L) as interaction partners of SARS-CoV nsp10 (Li et al. 2005a). The authors suggested the latter interaction to contribute to the SARS-CoV-induced cytopathic effect, but the interaction and its relevance for virus-induced cell death still awaits confirmation in the context of SARS-CoV-infected cells, as the rather artificial Y2H system is known to frequently yield false-positive hits. Indeed, another Y2H study with the SARS-CoV helicase (nsp13) demonstrated that out of the seven primary hits only one, DDX5, could be validated by independent methods as a true interactor of the helicase nsp13 (Chen et al. 2009). The functional significance of the interaction between nsp13 and the multifunctional cellular helicase DDX5, and whether the interaction is direct or mediated through RNA, remains to be determined. Using a similar Y2H screening approach, Xu and colleagues found that DDX1 bound to SARS-CoV and IBV nsp14, an interaction that enhanced IBV replication. The presence of one or more cellular helicases in the viral RTC supports their importance in virus replication. Indeed a later study reported that DDX1, regulated by the GSK-3-mediated phosphorylation of the N protein, is involved in the regulation of sg mRNA synthesis (Wu et al. 2014). A very comprehensive systematic analysis of the SARS-CoV-host interactome was performed by Pfefferle et al. using a high-throughput genome-wide Y2H screen with all 14 SARS-CoV ORFs and fragments thereof (Pfefferle et al. 2011). Network analysis revealed particularly striking interactions between nsp1 and the members of the immunophilin family, including CypA (see Chap. 8). Furthermore, the cellular E3 ubiquitin ligase RCHY1 was shown to interact with the SARS-CoV nsp3 SUD domain and might be involved in the downregulation of the antiviral factor p53 (Ma-Lauer et al. 2016).

Stable isotope labeling with amino acids in cell culture (SILAC) is a mass spectrometry-based proteomics technique to determine differences in protein abundance between samples from two different experimental conditions, *e.g.*, comparing virus-infected with uninfected cells. The first SILAC-based quantitative proteomics study of infected cells demonstrated the upregulation of NF-κB and AP-1 dependent pathways during IBV infection (Emmott et al. 2010). A combination of SILAC on 293T cells that express the IBV N protein, pull-down and mass spectrometry was used to map the cellular interactome of the IBV N protein, leading to the identification of 142 cellular proteins as potential binding partners (Emmott et al. 2013). Many of these proteins are interacting with RNA, e.g., ribosomal and nucleolar proteins, helicases, and hnRNPs (Emmott et al. 2013) and therefore likely bind the IBV N protein indirectly. Nevertheless, detailed validation and mechanistic follow-up studies confirmed the functional importance of several of the identified binding partners for IBV replication (Emmott et al. 2013).

The Baric laboratory used a systems genetics approach using the Collaborative Cross mouse panel to gain insight into the host loci that affect the outcome of SARS-CoV infection (Gralinski et al. 2015). This study—among other findings—identified the ubiquitin E3 ligase Trim55 as an important determinant of disease severity through its role in vascular cuffing and inflammation. A study by Selinger et al. (2014) documented differences in immune and inflammatory responses in MERS patients, which may codetermine the outcome of the infection and likely result from both differences in host response (genetic make-up) and MERS-CoV strain-specific properties. To better understand the molecular basis of the different immune and inflammatory responses to two different MERS-CoV isolates, a comparative transcriptome analysis was done on human airway cells infected with MERS-CoV strains SA1 and Eng1 (Selinger et al. 2014). This study suggested that differences in genome replication and/or proteins involved in innate immune evasion (PL^{pro} and ORF4a) were responsible for different transcriptional responses, resulting in the differential activation of the STAT3 pathway, which is likely involved in lung inflammation and cellular repair. These effects are mainly seen during later stages of infection, and with the MERS-CoV Eng1 strain triggering a more rapid host response than the SA1 strain.

A SILAC-based quantitative proteomics study that compared the proteome of SARS-CoV replicon-expressing BHK-21 cells with that of control cells identified 43 host proteins whose expression was upregulated and 31 that were downregulated (Zhang et al. 2010). BAG3, a multifunctional regulator of many cellular processes was identified as one of the upregulated proteins and knockdown studies revealed that BAG3 is important for efficient replication of SARS-CoV and a number of other viruses. In addition, many proteins involved in translation and the signaling proteins Cdc42 and RhoA were shown to be downregulated, as discussed in more detail in Sect. 3 (Zhang et al. 2010). A SILAC-based quantitative proteomics study of Golgi-enriched subcellular fractions revealed that upon MHV infection several proteins of the secretory pathway were depleted, while ribosomal proteins were found to be enriched (Vogels et al. 2011). SiRNA-mediated knockdown of three of the depleted proteins, C11orf59, GLG1, and sec22b, increased replication or release of infectious progeny, while overexpression of these proteins had the opposite effect. This study highlighted the importance of the secretory pathway in coronavirus replication.

A role for secretory pathway proteins was also confirmed by a proteomic analysis of purified SARS-CoV virions, which identified, besides several viral proteins including nsp2, nsp3, and nsp5, 172 host proteins in virions (Neuman et al. 2008). Several proteins from the COPI pathway were identified, which is in line with the site of virion biogenesis (ERGIC). The role, if any, of most of the identified host proteins in SARS-CoV replication or virion biogenesis remains to be elucidated.

Protein kinases are key regulators in signal transduction, control a wide variety of cellular processes, and have been shown to play important roles in the replicative cycle of many +RNA viruses. A kinome-wide siRNA screen identified a variety of host cell kinases that influence SARS-CoV replication, including 40 'proviral' proteins that promote efficient replication (de Wilde et al. 2015). Among these, proteins involved in the metabolism of complex lipids and the early secretory pathway (COPI-coated vesicles) were found to play an important role. The antiviral effect of PKR was confirmed in this study and CDK6 was identified as a novel antiviral factor. A relatively large number of antiviral hits (90 of 778 factors; ~12% of all factors tested) was identified for SARS-CoV compared to human kinome-directed screens performed with other viruses (Supekova et al. 2008; Lupberger et al. 2011; Moser et al. 2010). This might indicate that, compared to other viruses, SARS-CoV replication is more extensively influenced by cellular factors. Multiple of these factors could be linked to cellular immune responses, like interleukin (IL) signaling, which (IL-6 and -8) was also implicated in controlling coronavirus infection and coronavirus-induced inflammation in other studies (Zhang et al. 2007; Baas et al. 2006). Also several proteins from the p38 MAPK pathway were identified, which had also been implicated in coronavirus replication earlier and regulates IL-6-, IL-8-, and IL-10-mediated pro-inflammatory cytokine signaling (Chang et al. 2004; Zhang et al. 2007; Song et al. 2013). The interaction between coronaviruses and the innate immune response was already discussed in more detail in Chap. 6. A similar, genome-wide, siRNA screen identified host proteins important for the replication of IBV (Wong et al. 2015), including 83 proviral proteins, 30 of which could be mapped to networks that interact with viral proteins. Many of the identified proteins are involved in RNA binding/processing, membrane trafficking and ubiquitin conjugation. The importance of the secretory pathway that was demonstrated by de Wilde et al. (2015) was in line with an earlier study that demonstrated that MHV replication was sensitive to Brefeldin A treatment and dependent on GBF1-mediated ARF1 activation, which appear to be involved in RTC formation (Verheije et al. 2008). Similar results were also obtained for IBV (Wong et al. 2015) and this study also identified an early role in IBV infection for the valosin-containing protein (VCP) which may be involved in the maturation of virus-loaded endosomes. VCP is also important for the early stages of HCoV-229E replication (Wong et al. 2015). An RNAi screen of the druggable genome identified several endocytosis-related proteins that are required for efficient infection of HeLa cells with MHV (Burkard et al. 2014). Subsequent validation and mechanistic studies, demonstrated that—as discussed above—clathrin-mediated endocytosis and trafficking to lysosomes are crucial for MHV fusion and entry, which required the activity of lysosomal proteases. This is different for MERS-CoV, which contains a furin cleavage site upstream of the fusion peptide in the Spike protein, and therefore requires furin activity, but not lysosomal proteases (Burkard et al. 2014).

RNAi screens have revolutionized functional genetics, but major concerns are the possibility of false-positive hits due to off-target effects (including downregulation of multiple transcripts), stimulation of the immune response, or saturation of the RNAi machinery leading to a block in processing of essential cellular (mi) RNAs [reviewed in (Jackson and Linsley 2010)]. Insufficient knockdown of host factors can lead to false-negative results. Therefore, hits from siRNA screens need to be thoroughly validated, preferably using an independent technical approach. They should be considered as a mere starting point for further analysis rather than providing a definitive list of host factors involved in virus replication. Technological advancements and novel screening approaches, e.g, those based on CRISPR/Cas9-mediated genome editing or haploid genetic screens (Shalem et al. 2014; Carette et al. 2009), will likely lead to the more reliable identification of host factors important for coronavirus replication.

10 Concluding Remarks

Insight into coronavirus-host interactions, obtained, e.g., using systematic screening approaches, does not only yield valuable information on the molecular details of the replicative cycle and pathogenesis, but can also be a starting point for the development of antiviral strategies. Virus binding and entry are the first steps of the replication cycle that can be targeted with inhibitors. Several well-known inhibitors of endosomal acidification, like ammonium chloride and the FDA-approved anti-malaria drug chloroquine, have been shown to block entry of coronaviruses (Takano et al. 2013; Keyaerts et al. 2009; Krzystyniak and Dupuy 1984; Payne et al. 1990; Kono et al. 2008), including SARS-CoV and MERS-CoV (Keyaerts et al. 2004; de Wilde et al. 2014). In addition, peptides have been developed that block fusion by interfering with the interaction between the HR1 and HR2 domains of the S protein, preventing the formation of a fusogenic complex or blocking S protein oligomerisation [reviewed in (Du et al. 2009)].

Interferon (IFN) was shown to trigger the innate immune response in coronavirus-infected cells, leading to transcription of many ISGs that have a role in controlling infection (Schoggins and Rice 2011). Treatment with type-I IFNs inhibits coronavirus replication in cell culture (Garlinghouse et al. 1984; Taguchi and Siddell 1985; Haagmans et al. 2004; Paragas et al. 2005; Zheng et al. 2004; de Wilde et al. 2013b) and, for example, protected type-I pneumocytes against SARS-CoV infection in macaques (Haagmans et al. 2004). Despite the potency of IFN as an antiviral agent, its side effects like fatigue, malaise, apathy, and cognitive changes (Dusheiko 1997) emphasize the need for developing IFN-free therapeutic strategies. Besides inhibitors directed at viral enzymes (Kim et al. 2016), such therapeutic strategies could also involve host-directed approaches, based on the knowledge obtained on coronavirus–host interactions. The host-directed approach might lower the chance of development of antiviral resistance and could yield a broad-spectrum therapeutic strategy to treat infections with currently problematic coronaviruses and new variants that will undoubtedly emerge in the future.

References

Altan-Bonnet N, Balla T (2012) Phosphatidylinositol 4-kinases: hostages harnessed to build panviral replication platforms. Trends Biochem Sci 37:293–302

Angelini MM, Akhlaghpour M, Neuman BW, Buchmeier MJ (2013) Severe acute respiratory syndrome coronavirus nonstructural proteins 3, 4, and 6 induce double-membrane vesicles. MBio 4:e00524–e00513

Baas T, Taubenberger JK, Chong PY, Chui P, Katze MG (2006) SARS-CoV virus-host interactions and comparative etiologies of acute respiratory distress syndrome as determined by transcriptional and cytokine profiling of formalin-fixed paraffin-embedded tissues. J Interferon Cytokine Res: The Official Journal of The International Society for Interferon and Cytokine Research 26:309–317

Bagga S, Bouchard MJ (2014) Cell cycle regulation during viral infection. Methods Mol Biol 1170:165–227

Bailey-Elkin BA, Knaap RC, Johnson GG, Dalebout TJ, Ninaber DK, van Kasteren PB, Bredenbeek PJ, Snijder EJ, Kikkert M, Mark BL (2014) Crystal structure of the Middle East respiratory syndrome coronavirus (MERS-CoV) papain-like protease bound to ubiquitin facilitates targeted disruption of deubiquitinating activity to demonstrate its role in innate immune suppression. J Biol Chem 289:34667–34682

Bakkers MJ, Zeng Q, Feitsma LJ, Hulswit RJ, Li Z, Westerbeke A, van Kuppeveld FJ, Boons GJ, Langereis MA, Huizinga EG, de Groot RJ (2016) Coronavirus receptor switch explained from the stereochemistry of protein-carbohydrate interactions and a single mutation. Proc Natl Acad Sci U S A 113:E3111–E3119

Barik S (2006) Immunophilins: for the love of proteins. Cell Mol Life Sci 63:2889–2900

Baugh J, Gallay P (2012) Cyclophilin involvement in the replication of hepatitis C virus and other viruses. Biol Chem 393:579–587

Bechill J, Chen Z, Brewer JW, Baker SC (2008) Coronavirus infection modulates the unfolded protein response and mediates sustained translational repression. J Virol 82:4492–4501

Berger KL, Cooper JD, Heaton NS, Yoon R, Oakland TE, Jordan TX, Mateu G, Grakoui A, Randall G (2009) Roles for endocytic trafficking and phosphatidylinositol 4-kinase III alpha in hepatitis C virus replication. Proc Natl Acad Sci U S A 106:7577–7582

Bertram S, Dijkman R, Habjan M, Heurich A, Gierer S, Glowacka I, Welsch K, Winkler M, Schneider H, Hofmann-Winkler H, Thiel V, Pohlmann S (2013) TMPRSS2 activates the human coronavirus 229E for cathepsin-independent host cell entry and is expressed in viral target cells in the respiratory epithelium. J Virol 87:6150–6160

Bhardwaj K, Liu P, Leibowitz JL, Kao CC (2012) The coronavirus endoribonuclease Nsp15 interacts with retinoblastoma tumor suppressor protein. J Virol 86:4294–4304

Brian DA, Baric RS (2005) Coronavirus genome structure and replication. Curr Top Microbiol Immunol 287:1–30

Brierley I, Dos Ramos FJ (2006) Programmed ribosomal frameshifting in HIV-1 and the SARS-CoV. Virus Res 119:29–42

Brierley I, Digard P, Inglis SC (1989) Characterization of an efficient coronavirus ribosomal frameshifting signal: requirement for an RNA pseudoknot. Cell 57:537–547

Brockway SM, Clay CT, Lu XT, Denison MR (2003) Characterization of the expression, intracellular localization, and replication complex association of the putative mouse hepatitis virus RNA-dependent RNA polymerase. J Virol 77:10515–10527

Bruns AM, Horvath CM (2014) Antiviral RNA recognition and assembly by RLR family innate immune sensors. Cytokine Growth Factor Rev 25:507–512

Burkard C, Verheije MH, Wicht O, van Kasteren SI, van Kuppeveld FJ, Haagmans BL, Pelkmans L, Rottier PJ, Bosch BJ, de Haan CA (2014) Coronavirus cell entry occurs through the endo-/lysosomal pathway in a proteolysis-dependent manner. PLoS Pathog 10:e1004502

Cali T, Galli C, Olivari S, Molinari M (2008) Segregation and rapid turnover of EDEM1 by an autophagy-like mechanism modulates standard ERAD and folding activities. Biochem Biophys Res Commun 371:405–410

Carbajo-Lozoya J, Muller MA, Kallies S, Thiel V, Drosten C, von Brunn A (2012) Replication of human coronaviruses SARS-CoV, HCoV-NL63 and HCoV-229E is inhibited by the drug FK506. Virus Res 165:112–117

Carbajo-Lozoya J, Ma-Lauer Y, Malesevic M, Theuerkorn M, Kahlert V, Prell E, von Brunn B, Muth D, Baumert TF, Drosten C, Fischer G, von Brunn A (2014) Human coronavirus NL63 replication is cyclophilin A-dependent and inhibited by non-immunosuppressive cyclosporine A-derivatives including alisporivir. Virus Res 184C:44–53

Carette JE, Guimaraes CP, Varadarajan M, Park AS, Wuethrich I, Godarova A, Kotecki M, Cochran BH, Spooner E, Ploegh HL, Brummelkamp TR (2009) Haploid genetic screens in human cells identify host factors used by pathogens. Science 326:1231–1235

Chan SW (2014) Unfolded protein response in hepatitis C virus infection. Front Microbiol 5:233

Chan CP, Siu KL, Chin KT, Yuen KY, Zheng B, Jin DY (2006) Modulation of the unfolded protein response by the severe acute respiratory syndrome coronavirus spike protein. J Virol 80:9279–9287

Chang YJ, Liu CY, Chiang BL, Chao YC, Chen CC (2004) Induction of IL-8 release in lung cells via activator protein-1 by recombinant baculovirus displaying severe acute respiratory syndrome-coronavirus spike proteins: identification of two functional regions. J Immunol 173:7602–7614

Chen CJ, Makino S (2004) Murine coronavirus replication induces cell cycle arrest in G0/G1 phase. J Virol 78:5658–5669

Chen CJ, Sugiyama K, Kubo H, Huang C, Makino S (2004) Murine coronavirus nonstructural protein p28 arrests cell cycle in G0/G1 phase. J Virol 78:10410–10419

Chen Z, Wang Y, Ratia K, Mesecar AD, Wilkinson KD, Baker SC (2007) Proteolytic processing and deubiquitinating activity of papain-like proteases of human coronavirus NL63. J Virol 81:6007–6018

Chen JY, Chen WN, Poon KM, Zheng BJ, Lin X, Wang YX, Wen YM (2009) Interaction between SARS-CoV helicase and a multifunctional cellular protein (Ddx5) revealed by yeast and mammalian cell two-hybrid systems. Arch Virol 154:507–512

Choi KS, Huang P, Lai MM (2002) Polypyrimidine-tract-binding protein affects transcription but not translation of mouse hepatitis virus RNA. Virology 303:58–68

Choi KS, Mizutani A, Lai MM (2004) SYNCRIP, a member of the heterogeneous nuclear ribonucleoprotein family, is involved in mouse hepatitis virus RNA synthesis. J Virol 78:13153–13162

Clementz MA, Chen Z, Banach BS, Wang Y, Sun L, Ratia K, Baez-Santos YM, Wang J, Takayama J, Ghosh AK, Li K, Mesecar AD, Baker SC (2010) Deubiquitinating and interferon antagonism activities of coronavirus papain-like proteases. J Virol 84:4619–4629

Corman VM, Baldwin HJ, Tateno AF, Zerbinati RM, Annan A, Owusu M, Nkrumah EE, Maganga GD, Oppong S, Adu-Sarkodie Y, Vallo P, da Silva Filho LV, Leroy EM, Thiel V, van der Hoek L, Poon LL, Tschapka M, Drosten C, Drexler JF (2015) Evidence for an ancestral association of human coronavirus 229E with bats. J Virol 89:11858–11870

Corman VM, Eckerle I, Memish ZA, Liljander AM, Dijkman R, Jonsdottir H, Juma Ngeiywa KJ, Kamau E, Younan M, Al Masri M, Assiri A, Gluecks I, Musa BE, Meyer B, Muller MA, Hilali M, Bornstein S, Wernery U, Thiel V, Jores J, Drexler JF, Drosten C (2016) Link of a ubiquitous human coronavirus to dromedary camels. Proc Natl Acad Sci U S A 113:9864–9869

Cottam EM, Maier HJ, Manifava M, Vaux LC, Chandra-Schoenfelder P, Gerner W, Britton P, Ktistakis NT, Wileman T (2011) Coronavirus nsp6 proteins generate autophagosomes from the endoplasmic reticulum via an omegasome intermediate. Autophagy 7:1335–1347

Cottam EM, Whelband MC, Wileman T (2014) Coronavirus NSP6 restricts autophagosome expansion. Autophagy 10:1426–1441

Cruz JL, Sola I, Becares M, Alberca B, Plana J, Enjuanes L, Zuniga S (2011) Coronavirus gene 7 counteracts host defenses and modulates virus virulence. PLoS Pathog 7:e1002090

Davis TL, Walker JR, Campagna-Slater V, Finerty PJ, Paramanathan R, Bernstein G, MacKenzie F, Tempel W, Ouyang H, Lee WH, Eisenmesser EZ, Dhe-Paganon S (2010) Structural and biochemical characterization of the human cyclophilin family of peptidyl-prolyl isomerases. PLoS Biol 8:e1000439

de Groot RJ, Baker SC, Baric RS, Brown CS, Drosten C, Enjuanes L, Fouchier RA, Galiano M, Gorbalenya AE, Memish ZA, Perlman S, Poon LL, Snijder EJ, Stephens GM, Woo PC, Zaki AM, Zambon M, Ziebuhr J (2013) Middle East respiratory syndrome coronavirus (MERS-CoV): announcement of the coronavirus study group. J Virol 87:7790–7792

de Haan CA, Reggiori F (2008) Are nidoviruses hijacking the autophagy machinery? Autophagy 4:276–279

de Wilde AH, Zevenhoven-Dobbe JC, van der Meer Y, Thiel V, Narayanan K, Makino S, Snijder EJ, van Hemert MJ (2011) Cyclosporin A inhibits the replication of diverse coronaviruses. J Gen Virol 92:2542–2548

de Wilde AH, Li Y, van der Meer Y, Vuagniaux G, Lysek R, Fang Y, Snijder EJ, van Hemert MJ (2013a) Cyclophilin inhibitors block arterivirus replication by interfering with viral RNA synthesis. J Virol 87:1454–1464

de Wilde AH, Raj VS, Oudshoorn D, Bestebroer TM, van Nieuwkoop S, Limpens RW, Posthuma CC, van der Meer Y, Barcena M, Haagmans BL, Snijder EJ, van den Hoogen BG (2013b) MERS-coronavirus replication induces severe in vitro cytopathology and is strongly inhibited by cyclosporin A or interferon-alpha treatment. J Gen Virol 94:1749–1760

de Wilde AH, Jochmans D, Posthuma CC, Zevenhoven-Dobbe JC, van Nieuwkoop S, Bestebroer TM, van den Hoogen BG, Neyts J, Snijder EJ (2014) Screening of an FDA-approved compound library identifies four small-molecule inhibitors of Middle East respiratory syndrome coronavirus replication in cell culture. Antimicrob Agents Chemother 58:4875–4884

de Wilde AH, Wannee KF, Scholte FE, Goeman JJ, Ten Dijke P, Snijder EJ, Kikkert M, van Hemert MJ (2015) A kinome-wide siRNA screen identifies proviral and antiviral host factors in SARS-coronavirus replication, including PKR and early secretory pathway proteins. J Virol 89:8318–8333

de Wilde AH, Falzarano D, Zevenhoven-Dobbe JC, Beugeling C, Fett C, Martellaro C, Posthuma CC, Feldmann H, Perlman S, Snijder EJ (2017) Alisporivir inhibits MERS- and SARS-coronavirus replication in cell culture, but not SARS-coronavirus infection in a mouse model. Virus Res 228:7–13

Delmas B, Gelfi J, L'Haridon R, Vogel LK, Sjostrom H, Noren O, Laude H (1992) Aminopeptidase N is a major receptor for the entero-pathogenic coronavirus TGEV. Nature 357:417–420

den Boon JA, Ahlquist P (2010) Organelle-like membrane compartmentalization of positive-strand RNA virus replication factories. Annu Rev Microbiol 64:241–256

Ding L, Huang Y, Dai M, Zhao X, Du Q, Dong F, Wang L, Huo R, Zhang W, Xu X, Tong D (2013) Transmissible gastroenteritis virus infection induces cell cycle arrest at S and G2/M phases via p53-dependent pathway. Virus Res 178:241–251

Ding L, Huang Y, Du Q, Dong F, Zhao X, Zhang W, Xu X, Tong D (2014) TGEV nucleocapsid protein induces cell cycle arrest and apoptosis through activation of p53 signaling. Biochem Biophys Res Commun 445:497–503

Dosch SF, Mahajan SD, Collins AR (2009) SARS coronavirus spike protein-induced innate immune response occurs via activation of the NF-kappaB pathway in human monocyte macrophages in vitro. Virus Res 142:19–27

Dove B, Brooks G, Bicknell K, Wurm T, Hiscox JA (2006) Cell cycle perturbations induced by infection with the coronavirus infectious bronchitis virus and their effect on virus replication. J Virol 80:4147–4156

Drosten C, Gunther S, Preiser W, van der Werf S, Brodt HR, Becker S, Rabenau H, Panning M, Kolesnikova L, Fouchier RA, Berger A, Burguiere AM, Cinatl J, Eickmann M, Escriou N,

Grywna K, Kramme S, Manuguerra JC, Muller S, Rickerts V, Sturmer M, Vieth S, Klenk HD, Osterhaus AD, Schmitz H, Doerr HW (2003) Identification of a novel coronavirus in patients with severe acute respiratory syndrome. N Engl J Med 348:1967–1976

Du L, He Y, Zhou Y, Liu S, Zheng BJ, Jiang S (2009) The spike protein of SARS-CoV–a target for vaccine and therapeutic development. Nat Rev Microbiol 7:226–236

Dusheiko G (1997) Side effects of alpha interferon in chronic hepatitis C. Hepatology 26:112S–121S

Emmott E, Rodgers MA, Macdonald A, McCrory S, Ajuh P, Hiscox JA (2010) Quantitative proteomics using stable isotope labeling with amino acids in cell culture reveals changes in the cytoplasmic, nuclear, and nucleolar proteomes in Vero cells infected with the coronavirus infectious bronchitis virus. Mol Cell Proteomics 9:1920–1936

Emmott E, Munday D, Bickerton E, Britton P, Rodgers MA, Whitehouse A, Zhou EM, Hiscox JA (2013) The cellular interactome of the coronavirus infectious bronchitis virus nucleocapsid protein and functional implications for virus biology. J Virol 87:9486–9500

Enjuanes L (2005) Coronavirus replication and reverse genetics. Current Topics in Microbiology and Immunology, vol 287. Springer. doi:10.1007/b138038

Favreau DJ, Meessen-Pinard M, Desforges M, Talbot PJ (2012) Human coronavirus-induced neuronal programmed cell death is cyclophilin d dependent and potentially caspase dispensable. J Virol 86:81–93

Flisiak R, Jaroszewicz J, Flisiak I, Lapinski T (2012) Update on alisporivir in treatment of viral hepatitis C. Expert Opin Investig Drugs 21:375–382

Follis KE, York J, Nunberg JH (2006) Furin cleavage of the SARS coronavirus spike glycoprotein enhances cell-cell fusion but does not affect virion entry. Virology 350:358–369

Frieman M, Yount B, Heise M, Kopecky-Bromberg SA, Palese P, Baric RS (2007) Severe acute respiratory syndrome coronavirus ORF6 antagonizes STAT1 function by sequestering nuclear import factors on the rough endoplasmic reticulum/Golgi membrane. J Virol 81:9812–9824

Frieman M, Ratia K, Johnston RE, Mesecar AD, Baric RS (2009) Severe acute respiratory syndrome coronavirus papain-like protease ubiquitin-like domain and catalytic domain regulate antagonism of IRF3 and NF-kappaB signaling. J Virol 83:6689–6705

Fung TS, Liao Y, Liu DX (2016) Regulation of stress responses and translational control by coronavirus. Viruses 8:184

Furuya T, Lai MM (1993) Three different cellular proteins bind to complementary sites on the 5′-end-positive and 3′-end-negative strands of mouse hepatitis virus RNA. J Virol 67:7215–7222

Galan C, Sola I, Nogales A, Thomas B, Akoulitchev A, Enjuanes L, Almazan F (2009) Host cell proteins interacting with the 3′ end of TGEV coronavirus genome influence virus replication. Virology 391:304–314

Garlinghouse LE Jr, Smith AL, Holford T (1984) The biological relationship of mouse hepatitis virus (MHV) strains and interferon: in vitro induction and sensitivities. Arch Virol 82:19–29

Ge XY, Li JL, Yang XL, Chmura AA, Zhu G, Epstein JH, Mazet JK, Hu B, Zhang W, Peng C, Zhang YJ, Luo CM, Tan B, Wang N, Zhu Y, Crameri G, Zhang SY, Wang LF, Daszak P, Shi ZL (2013) Isolation and characterization of a bat SARS-like coronavirus that uses the ACE2 receptor. Nature 503:535–538

Glowacka I, Bertram S, Muller MA, Allen P, Soilleux E, Pfefferle S, Steffen I, Tsegaye TS, He Y, Gnirss K, Niemeyer D, Schneider H, Drosten C, Pohlmann S (2011) Evidence that TMPRSS2 activates the severe acute respiratory syndrome coronavirus spike protein for membrane fusion and reduces viral control by the humoral immune response. J Virol 85:4122–4134

Gosert R, Kanjanahaluethai A, Egger D, Bienz K, Baker SC (2002) RNA replication of mouse hepatitis virus takes place at double-membrane vesicles. J Virol 76:3697–3708

Gralinski LE, Ferris MT, Aylor DL, Whitmore AC, Green R, Frieman MB, Deming D, Menachery VD, Miller DR, Buus RJ, Bell TA, Churchill GA, Threadgill DW, Katze MG, McMillan L, Valdar W, Heise MT, Pardo-Manuel de Villena F, Baric RS (2015) Genome wide identification of SARS-CoV susceptibility loci using the collaborative cross. PLoS Genet 11: e1005504

Haagmans BL, Kuiken T, Martina BE, Fouchier RA, Rimmelzwaan GF, van Amerongen G, van Riel D, de Jong T, Itamura S, Chan KH, Tashiro M, Osterhaus AD (2004) Pegylated interferon-alpha protects type 1 pneumocytes against SARS coronavirus infection in macaques. Nat Med 10:290–293

Hagemeijer MC, Rottier PJ, de Haan CA (2012) Biogenesis and dynamics of the coronavirus replicative structures. Viruses 4:3245–3269

Hagemeijer MC, Monastyrska I, Griffith J, van der Sluijs P, Voortman J, van Bergen en Henegouwen PM, Vonk AM, Rottier PJ, Reggiori F, de Haan CA (2014) Membrane rearrangements mediated by coronavirus nonstructural proteins 3 and 4. Virology 458–459:125–135

Hamre D, Procknow JJ (1966) A new virus isolated from the human respiratory tract. Proc Soc Exp Biol Med 121:190–193

Hopkins S, Gallay PA (2015) The role of immunophilins in viral infection. Biochim Biophys Acta 1850:2103–2110

Horowitz DS, Lee EJ, Mabon SA, Misteli T (2002) A cyclophilin functions in pre-mRNA splicing. EMBO J 21:470–480

Hsu NY, Ilnytska O, Belov G, Santiana M, Chen YH, Takvorian PM, Pau C, van der Schaar H, Kaushik-Basu N, Balla T, Cameron CE, Ehrenfeld E, van Kuppeveld FJ, Altan-Bonnet N (2010) Viral reorganization of the secretory pathway generates distinct organelles for RNA replication. Cell 141:799–811

Huang P, Lai MM (1999) Polypyrimidine tract-binding protein binds to the complementary strand of the mouse hepatitis virus 3′ untranslated region, thereby altering RNA conformation. J Virol 73:9110–9116

Huang P, Lai MM (2001) Heterogeneous nuclear ribonucleoprotein a1 binds to the 3′-untranslated region and mediates potential 5′-3′-end cross talks of mouse hepatitis virus RNA. J Virol 75:5009–5017

Huang IC, Bosch BJ, Li F, Li W, Lee KH, Ghiran S, Vasilieva N, Dermody TS, Harrison SC, Dormitzer PR, Farzan M, Rottier PJ, Choe H (2006a) SARS coronavirus, but not human coronavirus NL63, utilizes cathepsin L to infect ACE2-expressing cells. J Biol Chem 281:3198–3203

Huang IC, Bosch BJ, Li W, Farzan M, Rottier PM, Choe H (2006b) SARS-CoV, but not HCoV-NL63, utilizes cathepsins to infect cells: viral entry. Adv Exp Med Biol 581:335–338

Huang SH, Lee TY, Lin YJ, Wan L, Lai CH, Lin CW (2015a) Phage display technique identifies the interaction of severe acute respiratory syndrome coronavirus open reading frame 6 protein with nuclear pore complex interacting protein NPIPB3 in modulating type I interferon antagonism. J Microbiol Immunol Infect. doi:10.1016/j.jmii.2015.07.002

Huang X, Dong W, Milewska A, Golda A, Qi Y, Zhu QK, Marasco WA, Baric RS, Sims AC, Pyrc K, Li W, Sui J (2015b) Human coronavirus HKU1 spike protein uses O-acetylated sialic acid as an attachment receptor determinant and employs hemagglutinin-esterase protein as a receptor-destroying enzyme. J Virol 89:7202–7213

Hulswit RJ, de Haan CA, Bosch BJ (2016) Coronavirus spike protein and tropism changes. Adv Virus Res 96:29–57

Huynh J, Li S, Yount B, Smith A, Sturges L, Olsen JC, Nagel J, Johnson JB, Agnihothram S, Gates JE, Frieman MB, Baric RS, Donaldson EF (2012) Evidence supporting a zoonotic origin of human coronavirus strain NL63. J Virol 86:12816–12825

Jackson AL, Linsley PS (2010) Recognizing and avoiding siRNA off-target effects for target identification and therapeutic application. Nat Rev Drug Discov 9:57–67

Jackson RJ, Hellen CU, Pestova TV (2010) The mechanism of eukaryotic translation initiation and principles of its regulation. Nat Rev Mol Cell Biol 11:113–127

Jiang X, Chen ZJ (2012) The role of ubiquitylation in immune defence and pathogen evasion. Nat Rev Immunol 12:35–48

Jimenez-Guardeno JM, Nieto-Torres JL, DeDiego ML, Regla-Nava JA, Fernandez-Delgado R, Castano-Rodriguez C, Enjuanes L (2014) The PDZ-binding motif of severe acute respiratory syndrome coronavirus envelope protein is a determinant of viral pathogenesis. PLoS Pathog 10:e1004320

Kamitani W, Narayanan K, Huang C, Lokugamage K, Ikegami T, Ito N, Kubo H, Makino S (2006) Severe acute respiratory syndrome coronavirus nsp1 protein suppresses host gene expression by promoting host mRNA degradation. Proc Natl Acad Sci USA 103:12885–12890

Kamitani W, Huang C, Narayanan K, Lokugamage KG, Makino S (2009) A two-pronged strategy to suppress host protein synthesis by SARS coronavirus Nsp1 protein. Nat Struct Mol Biol 16:1134–1140

Keogh-Brown MR, Smith RD (2008) The economic impact of SARS: how does the reality match the predictions? Health Policy 88:110–120

Keyaerts E, Vijgen L, Maes P, Neyts J, Van Ranst M (2004) In vitro inhibition of severe acute respiratory syndrome coronavirus by chloroquine. Biochem Biophys Res Commun 323: 264–268

Keyaerts E, Li S, Vijgen L, Rysman E, Verbeeck J, Van Ranst M, Maes P (2009) Antiviral activity of chloroquine against human coronavirus OC43 infection in newborn mice. Antimicrob Agents Chemother 53:3416–3421

Khanolkar A, Hartwig SM, Haag BA, Meyerholz DK, Harty JT, Varga SM (2009) Toll-like receptor 4 deficiency increases disease and mortality after mouse hepatitis virus type 1 infection of susceptible C3H mice. J Virol 83:8946–8956

Kim Y, Lee C (2014) Porcine epidemic diarrhea virus induces caspase-independent apoptosis through activation of mitochondrial apoptosis-inducing factor. Virology 460–461:180–193

Kim Y, Liu H, Galasiti Kankanamalage AC, Weerasekara S, Hua DH, Groutas WC, Chang KO, Pedersen NC (2016) Reversal of the progression of fatal coronavirus infection in cats by a broad-spectrum coronavirus protease inhibitor. PLoS Pathog 12:e1005531

Kindler E, Thiel V (2014) To sense or not to sense viral RNA–essentials of coronavirus innate immune evasion. Curr Opin Microbiol 20:69–75

Knoops K, Kikkert M, van den Worm SH, Zevenhoven-Dobbe JC, van der Meer Y, Koster AJ, Mommaas AM, Snijder EJ (2008) SARS-coronavirus replication is supported by a reticulovesicular network of modified endoplasmic reticulum. PLoS Biol 6:e226

Knoops K, Swett-Tapia C, van den Worm SH, te Velthuis AJ, Koster AJ, Mommaas AM, Snijder EJ, Kikkert M (2010) Integrity of the early secretory pathway promotes, but is not required for, severe acute respiratory syndrome coronavirus RNA synthesis and virus-induced remodeling of endoplasmic reticulum membranes. J Virol 84:833–846

Kono M, Tatsumi K, Imai AM, Saito K, Kuriyama T, Shirasawa H (2008) Inhibition of human coronavirus 229E infection in human epithelial lung cells (L132) by chloroquine: involvement of p38 MAPK and ERK. Antiviral Res 77:150–152

Kopecky-Bromberg SA, Martinez-Sobrido L, Frieman M, Baric RA, Palese P (2007) Severe acute respiratory syndrome coronavirus open reading frame (ORF) 3b, ORF 6, and nucleocapsid proteins function as interferon antagonists. J Virol 81:548–557

Krempl C, Schultze B, Herrler G (1995) Analysis of cellular receptors for human coronavirus OC43. Adv Exp Med Biol 380:371–374

Krzystyniak K, Dupuy JM (1984) Entry of mouse hepatitis virus 3 into cells. J Gen Virol 65: 227–231

Ksiazek TG, Erdman D, Goldsmith CS, Zaki SR, Peret T, Emery S, Tong S, Urbani C, Comer JA, Lim W, Rollin PE, Dowell SF, Ling AE, Humphrey CD, Shieh WJ, Guarner J, Paddock CD, Rota P, Fields B, DeRisi J, Yang JY, Cox N, Hughes JM, LeDuc JW, Bellini WJ, Anderson LJ, Group SW (2003) A novel coronavirus associated with severe acute respiratory syndrome. N Engl J Med 348:1953–1966

Kuiken T, Fouchier RA, Schutten M, Rimmelzwaan GF, van Amerongen G, van Riel D, Laman JD, de Jong T, van Doornum G, Lim W, Ling AE, Chan PK, Tam JS, Zambon MC, Gopal R, Drosten C, van der Werf S, Escriou N, Manuguerra JC, Stohr K, Peiris JS, Osterhaus AD (2003) Newly discovered coronavirus as the primary cause of severe acute respiratory syndrome. Lancet 362:263–270

Kwak H, Park MW, Jeong S (2011) Annexin A2 binds RNA and reduces the frameshifting efficiency of infectious bronchitis virus. PLoS ONE 6:e24067

Langereis MA, van Vliet AL, Boot W, de Groot RJ (2010) Attachment of mouse hepatitis virus to O-acetylated sialic acid is mediated by hemagglutinin-esterase and not by the spike protein. J Virol 84:8970–8974

Lau SK, Woo PC, Li KS, Huang Y, Tsoi HW, Wong BH, Wong SS, Leung SY, Chan KH, Yuen KY (2005) Severe acute respiratory syndrome coronavirus-like virus in Chinese horseshoe bats. Proc Natl Acad Sci USA 102:14040–14045

Li F (2008) Structural analysis of major species barriers between humans and palm civets for severe acute respiratory syndrome coronavirus infections. J Virol 82:6984–6991

Li F (2016) Structure, function, and evolution of coronavirus spike proteins. Annu Rev Virol 3:237–261

Li Y, Weiss SR (2016) Antagonism of RNase L Is required for murine coronavirus replication in kupffer cells and liver sinusoidal endothelial cells but not in hepatocytes. J Virol 90:9826–9832

Li HP, Zhang X, Duncan R, Comai L, Lai MM (1997) Heterogeneous nuclear ribonucleoprotein A1 binds to the transcription-regulatory region of mouse hepatitis virus RNA. Proc Natl Acad Sci USA 94:9544–9549

Li HP, Huang P, Park S, Lai MM (1999) Polypyrimidine tract-binding protein binds to the leader RNA of mouse hepatitis virus and serves as a regulator of viral transcription. J Virol 73: 772–777

Li W, Moore MJ, Vasilieva N, Sui J, Wong SK, Berne MA, Somasundaran M, Sullivan JL, Luzuriaga K, Greenough TC, Choe H, Farzan M (2003) Angiotensin-converting enzyme 2 is a functional receptor for the SARS coronavirus. Nature 426:450–454

Li Q, Wang L, Dong C, Che Y, Jiang L, Liu L, Zhao H, Liao Y, Sheng Y, Dong S, Ma S (2005a) The interaction of the SARS coronavirus non-structural protein 10 with the cellular oxido-reductase system causes an extensive cytopathic effect. J Clin Virol: The Official Publication of the Pan American Society for Clinical Virology 34:133–139

Li W, Shi Z, Yu M, Ren W, Smith C, Epstein JH, Wang H, Crameri G, Hu Z, Zhang H, Zhang J, McEachern J, Field H, Daszak P, Eaton BT, Zhang S, Wang LF (2005b) Bats are natural reservoirs of SARS-like coronaviruses. Science 310:676–679

Li W, Zhang C, Sui J, Kuhn JH, Moore MJ, Luo S, Wong SK, Huang IC, Xu K, Vasilieva N, Murakami A, He Y, Marasco WA, Guan Y, Choe H, Farzan M (2005c) Receptor and viral determinants of SARS-coronavirus adaptation to human ACE2. EMBO J 24:1634–1643

Li BX, Ge JW, Li YJ (2007) Porcine aminopeptidase N is a functional receptor for the PEDV coronavirus. Virology 365:166–172

Li J, Liu Y, Zhang X (2010) Murine coronavirus induces type I interferon in oligodendrocytes through recognition by RIG-I and MDA5. J Virol 84:6472–6482

Li SW, Wang CY, Jou YJ, Huang SH, Hsiao LH, Wan L, Lin YJ, Kung SH, Lin CW (2016) SARS coronavirus papain-like protease inhibits the TLR7 signaling pathway through removing Lys63-linked polyubiquitination of TRAF3 and TRAF6. Int J Mol Sci 17:678

Liao Y, Wang X, Huang M, Tam JP, Liu DX (2011) Regulation of the p38 mitogen-activated protein kinase and dual-specificity phosphatase 1 feedback loop modulates the induction of interleukin 6 and 8 in cells infected with coronavirus infectious bronchitis virus. Virology 420:106–116

Lin CM, Saif LJ, Marthaler D, Wang Q (2016) Evolution, antigenicity and pathogenicity of global porcine epidemic diarrhea virus strains. Virus Res 226:20–39

Liu C, Tang J, Ma Y, Liang X, Yang Y, Peng G, Qi Q, Jiang S, Li J, Du L, Li F (2015) Receptor usage and cell entry of porcine epidemic diarrhea coronavirus. J Virol 89:6121–6125

Lokugamage KG, Narayanan K, Huang C, Makino S (2012) SARS coronavirus nsp1 protein is a novel eukaryotic translation inhibitor that represses multiple steps of translation initiation. J Virol 86:13598–13608

Lokugamage KG, Narayanan K, Nakagawa K, Terasaki K, Ramirez SI, Tseng CT, Makino S (2015) Middle East respiratory syndrome coronavirus nsp1 inhibits host gene expression by selectively targeting mRNAs transcribed in the nucleus while sparing mRNAs of cytoplasmic origin. J Virol 89:10970–10981

Lu X, Pan J, Tao J, Guo D (2011) SARS-CoV nucleocapsid protein antagonizes IFN-beta response by targeting initial step of IFN-beta induction pathway, and its C-terminal region is critical for the antagonism. Virus Genes 42:37–45

Lui PY, Wong LY, Fung CL, Siu KL, Yeung ML, Yuen KS, Chan CP, Woo PC, Yuen KY, Jin DY (2016) Middle East respiratory syndrome coronavirus M protein suppresses type I interferon expression through the inhibition of TBK1-dependent phosphorylation of IRF3. Emerg Microbes Infect 5:e39

Luo C, Luo H, Zheng S, Gui C, Yue L, Yu C, Sun T, He P, Chen J, Shen J, Luo X, Li Y, Liu H, Bai D, Yang Y, Li F, Zuo J, Hilgenfeld R, Pei G, Chen K, Shen X, Jiang H (2004) Nucleocapsid protein of SARS coronavirus tightly binds to human cyclophilin A. Biochem Biophys Res Commun 321:557–565

Lupberger J, Zeisel MB, Xiao F, Thumann C, Fofana I, Zona L, Davis C, Mee CJ, Turek M, Gorke S, Royer C, Fischer B, Zahid MN, Lavillette D, Fresquet J, Cosset FL, Rothenberg SM, Pietschmann T, Patel AH, Pessaux P, Doffoel M, Raffelsberger W, Poch O, McKeating JA, Brino L, Baumert TF (2011) EGFR and EphA2 are host factors for hepatitis C virus entry and possible targets for antiviral therapy. Nat Med 17:589–595

Maier HJ, Britton P (2012) Involvement of autophagy in coronavirus replication. Viruses 4:3440–3451

Maier HJ, Hawes PC, Cottam EM, Mantell J, Verkade P, Monaghan P, Wileman T, Britton P (2013) Infectious bronchitis virus generates spherules from zippered endoplasmic reticulum membranes. MBio 4:e00801–e00813

Ma-Lauer Y, Carbajo-Lozoya J, Hein MY, Muller MA, Deng W, Lei J, Meyer B, Kusov Y, von Brunn B, Bairad DR, Hunten S, Drosten C, Hermeking H, Leonhardt H, Mann M, Hilgenfeld R, von Brunn A (2016) p53 down-regulates SARS coronavirus replication and is targeted by the SARS-unique domain and PLpro via E3 ubiquitin ligase RCHY1. Proc Natl Acad Sci USA 113:E5192–E5201

Matthews K, Schafer A, Pham A, Frieman M (2014a) The SARS coronavirus papain like protease can inhibit IRF3 at a post activation step that requires deubiquitination activity. Virol J 11:209

Matthews KL, Coleman CM, van der Meer Y, Snijder EJ, Frieman MB (2014b) The ORF4b-encoded accessory proteins of Middle East respiratory syndrome coronavirus and two related bat coronaviruses localize to the nucleus and inhibit innate immune signalling. J Gen Virol 95:874–882

Mazaleuskaya L, Veltrop R, Ikpeze N, Martin-Garcia J, Navas-Martin S (2012) Protective role of toll-like receptor 3-induced type I interferon in murine coronavirus infection of macrophages. Viruses 4:901–923

McIntosh K, Dees JH, Becker WB, Kapikian AZ, Chanock RM (1967) Recovery in tracheal organ cultures of novel viruses from patients with respiratory disease. Proc Natl Acad Sci USA 57:933–940

Menachery VD, Yount BL Jr, Josset L, Gralinski LE, Scobey T, Agnihothram S, Katze MG, Baric RS (2014) Attenuation and restoration of severe acute respiratory syndrome coronavirus mutant lacking 2′-o-methyltransferase activity. J Virol 88:4251–4264

Menachery VD, Yount BL Jr, Debbink K, Agnihothram S, Gralinski LE, Plante JA, Graham RL, Scobey T, Ge XY, Donaldson EF, Randell SH, Lanzavecchia A, Marasco WA, Shi ZL, Baric RS (2015) A SARS-like cluster of circulating bat coronaviruses shows potential for human emergence. Nat Med 21:1508–1513

Mielech AM, Chen Y, Mesecar AD, Baker SC (2014) Nidovirus papain-like proteases: multifunctional enzymes with protease, deubiquitinating and deISGylating activities. Virus Res 194:184–190

Mizutani T (2007) Signal transduction in SARS-CoV-infected cells. Ann N Y Acad Sci 1102: 86–95

Monastyrska I, Ulasli M, Rottier PJ, Guan JL, Reggiori F, de Haan CA (2013) An autophagy-independent role for LC3 in equine arteritis virus replication. Autophagy 9:164–174

Moser TS, Jones RG, Thompson CB, Coyne CB, Cherry S (2010) A kinome RNAi screen identified AMPK as promoting poxvirus entry through the control of actin dynamics. PLoS Pathog 6:e1000954
Muller MA, Corman VM, Jores J, Meyer B, Younan M, Liljander A, Bosch BJ, Lattwein E, Hilali M, Musa BE, Bornstein S, Drosten C (2014) MERS coronavirus neutralizing antibodies in camels, Eastern Africa, 1983–1997. Emerg Infect Dis 20:2093–2095
Nanda SK, Leibowitz JL (2001) Mitochondrial aconitase binds to the 3′ untranslated region of the mouse hepatitis virus genome. J Virol 75:3352–3362
Nanda SK, Johnson RF, Liu Q, Leibowitz JL (2004) Mitochondrial HSP70, HSP40, and HSP60 bind to the 3′ untranslated region of the murine hepatitis virus genome. Arch Virol 149:93–111
Narayanan K, Ramirez SI, Lokugamage KG, Makino S (2015) Coronavirus nonstructural protein 1: common and distinct functions in the regulation of host and viral gene expression. Virus Res 202:89–100
Neuman BW, Joseph JS, Saikatendu KS, Serrano P, Chatterjee A, Johnson MA, Liao L, Klaus JP, Yates JR 3rd, Wuthrich K, Stevens RC, Buchmeier MJ, Kuhn P (2008) Proteomics analysis unravels the functional repertoire of coronavirus nonstructural protein 3. J Virol 82:5279–5294
Ng LF, Hibberd ML, Ooi EE, Tang KF, Neo SY, Tan J, Murthy KR, Vega VB, Chia JM, Liu ET, Ren EC (2004) A human in vitro model system for investigating genome-wide host responses to SARS coronavirus infection. BMC Infect Dis 4:34
Niemeyer D, Zillinger T, Muth D, Zielecki F, Horvath G, Suliman T, Barchet W, Weber F, Drosten C, Muller MA (2013) Middle East respiratory syndrome coronavirus accessory protein 4a is a type I interferon antagonist. J Virol 87:12489–12495
Nieto-Torres JL, Verdia-Baguena C, Jimenez-Guardeno JM, Regla-Nava JA, Castano-Rodriguez C, Fernandez-Delgado R, Torres J, Aguilella VM, Enjuanes L (2015) Severe acute respiratory syndrome coronavirus E protein transports calcium ions and activates the NLRP3 inflammasome. Virology 485:330–339
Nigg EA (1995) Cyclin-dependent protein kinases: key regulators of the eukaryotic cell cycle. BioEssays: News and Reviews in Molecular, Cellular and Developmental Biology 17:471–480
Oostra M, te Lintelo EG, Deijs M, Verheije MH, Rottier PJ, de Haan CA (2007) Localization and membrane topology of coronavirus nonstructural protein 4: involvement of the early secretory pathway in replication. J Virol 81:12323–12336
Paeshuyse J, Kaul A, De Clercq E, Rosenwirth B, Dumont JM, Scalfaro P, Bartenschlager R, Neyts J (2006) The non-immunosuppressive cyclosporin DEBIO-025 is a potent inhibitor of hepatitis C virus replication in vitro. Hepatology 43:761–770
Paragas J, Blatt LM, Hartmann C, Huggins JW, Endy TP (2005) Interferon alfacon1 is an inhibitor of SARS-corona virus in cell-based models. Antiviral Res 66:99–102
Park JE, Li K, Barlan A, Fehr AR, Perlman S, McCray PB Jr, Gallagher T (2016) Proteolytic processing of Middle East respiratory syndrome coronavirus spikes expands virus tropism. Proc Natl Acad Sci USA 113:12262–12267
Paul D, Madan V, Bartenschlager R (2014) Hepatitis C virus RNA replication and assembly: living on the fat of the land. Cell Host Microbe 16:569–579
Payne HR, Storz J, Henk WG (1990) Initial events in bovine coronavirus infection: analysis through immunogold probes and lysosomotropic inhibitors. Arch Virol 114:175–189
Peiris JS, Lai ST, Poon LL, Guan Y, Yam LY, Lim W, Nicholls J, Yee WK, Yan WW, Cheung MT, Cheng VC, Chan KH, Tsang DN, Yung RW, Ng TK, Yuen KY, group Ss (2003) Coronavirus as a possible cause of severe acute respiratory syndrome. Lancet 361:1319–1325
Perlman S, Netland J (2009) Coronaviruses post-SARS: update on replication and pathogenesis. Nat Rev Microbiol 7:439–450
Pfefferle S, Schopf J, Kogl M, Friedel CC, Muller MA, Carbajo-Lozoya J, Stellberger T, von Dall'armi E, Herzog P, Kallies S, Niemeyer D, Ditt V, Kuri T, Zust R, Pumpor K, Hilgenfeld R, Schwarz F, Zimmer R, Steffen I, Weber F, Thiel V, Herrler G, Thiel HJ, Schwegmann-Wessels C, Pohlmann S, Haas J, Drosten C, von Brunn A (2011) The SARS-coronavirus-host interactome: identification of cyclophilins as target for pan-coronavirus inhibitors. PLoS Pathog 7:e1002331

Prentice E, Jerome WG, Yoshimori T, Mizushima N, Denison MR (2004) Coronavirus replication complex formation utilizes components of cellular autophagy. J Biol Chem 279:10136–10141

Raaben M, Groot Koerkamp MJ, Rottier PJ, de Haan CA (2007) Mouse hepatitis coronavirus replication induces host translational shutoff and mRNA decay, with concomitant formation of stress granules and processing bodies. Cell Microbiol 9:2218–2229

Rabouw HH, Langereis MA, Knaap RC, Dalebout TJ, Canton J, Sola I, Enjuanes L, Bredenbeek PJ, Kikkert M, de Groot RJ, van Kuppeveld FJ (2016) Middle East respiratory coronavirus accessory protein 4a inhibits PKR-mediated antiviral stress responses. PLoS Pathog 12:e1005982

Raj VS, Mou H, Smits SL, Dekkers DH, Muller MA, Dijkman R, Muth D, Demmers JA, Zaki A, Fouchier RA, Thiel V, Drosten C, Rottier PJ, Osterhaus AD, Bosch BJ, Haagmans BL (2013) Dipeptidyl peptidase 4 is a functional receptor for the emerging human coronavirus-EMC. Nature 495:251–254

Reggiori F, Monastyrska I, Verheije MH, Cali T, Ulasli M, Bianchi S, Bernasconi R, de Haan CA, Molinari M (2010) Coronaviruses Hijack the LC3-I-positive EDEMosomes, ER-derived vesicles exporting short-lived ERAD regulators, for replication. Cell Host Microbe 7:500–508

Reid CR, Airo AM, Hobman TC (2015) The virus-host interplay: biogenesis of +RNA replication complexes. Viruses 7:4385–4413

Reiss S, Rebhan I, Backes P, Romero-Brey I, Erfle H, Matula P, Kaderali L, Poenisch M, Blankenburg H, Hiet MS, Longerich T, Diehl S, Ramirez F, Balla T, Rohr K, Kaul A, Buhler S, Pepperkok R, Lengauer T, Albrecht M, Eils R, Schirmacher P, Lohmann V, Bartenschlager R (2011) Recruitment and activation of a lipid kinase by hepatitis C virus NS5A is essential for integrity of the membranous replication compartment. Cell Host Microbe 9:32–45

Rempel JD, Quina LA, Blakely-Gonzales PK, Buchmeier MJ, Gruol DL (2005) Viral induction of central nervous system innate immune responses. J Virol 79:4369–4381

Romero-Brey I, Bartenschlager R (2016) Endoplasmic reticulum: the favorite intracellular niche for viral replication and assembly. Viruses 8:E160

Sawicki SG, Sawicki DL (1995) Coronaviruses use discontinuous extension for synthesis of subgenome-length negative strands. Adv Exp Med Biol 380:499–506

Schinzel AC, Takeuchi O, Huang Z, Fisher JK, Zhou Z, Rubens J, Hetz C, Danial NN, Moskowitz MA, Korsmeyer SJ (2005) Cyclophilin D is a component of mitochondrial permeability transition and mediates neuronal cell death after focal cerebral ischemia. Proc Natl Acad Sci USA 102:12005–12010

Schoggins JW, Rice CM (2011) Interferon-stimulated genes and their antiviral effector functions. Curr opin virol 1:519–525

Schultze B, Herrler G (1992) Bovine coronavirus uses N-acetyl-9-O-acetylneuraminic acid as a receptor determinant to initiate the infection of cultured cells. J Gen Virol 73(Pt 4):901–906

Schultze B, Enjuanes L, Cavanagh D, Herrler G (1993) N-acetylneuraminic acid plays a critical role for the haemagglutinating activity of avian infectious bronchitis virus and porcine transmissible gastroenteritis virus. Adv Exp Med Biol 342:305–310

Schultze B, Krempl C, Ballesteros ML, Shaw L, Schauer R, Enjuanes L, Herrler G (1996) Transmissible gastroenteritis coronavirus, but not the related porcine respiratory coronavirus, has a sialic acid (N-glycolylneuraminic acid) binding activity. J Virol 70:5634–5637

Selinger C, Tisoncik-Go J, Menachery VD, Agnihothram S, Law GL, Chang J, Kelly SM, Sova P, Baric RS, Katze MG (2014) Cytokine systems approach demonstrates differences in innate and pro-inflammatory host responses between genetically distinct MERS-CoV isolates. BMC Genom 15:1161

Shalem O, Sanjana NE, Hartenian E, Shi X, Scott DA, Mikkelsen TS, Heckl D, Ebert BL, Root DE, Doench JG, Zhang F (2014) Genome-scale CRISPR-Cas9 knockout screening in human cells. Science 343:84–87

Shen X, Masters PS (2001) Evaluation of the role of heterogeneous nuclear ribonucleoprotein A1 as a host factor in murine coronavirus discontinuous transcription and genome replication. Proc Natl Acad Sci USA 98:2717–2722

Shi ST, Huang P, Li HP, Lai MM (2000) Heterogeneous nuclear ribonucleoprotein A1 regulates RNA synthesis of a cytoplasmic virus. EMBO J 19:4701–4711

Shi ST, Yu GY, Lai MM (2003) Multiple type A/B heterogeneous nuclear ribonucleoproteins (hnRNPs) can replace hnRNP A1 in mouse hepatitis virus RNA synthesis. J Virol 77:10584–10593

Shirato K, Kawase M, Matsuyama S (2013) Middle East respiratory syndrome coronavirus infection mediated by the transmembrane serine protease TMPRSS2. J Virol 87:12552–12561

Simmons G, Gosalia DN, Rennekamp AJ, Reeves JD, Diamond SL, Bates P (2005) Inhibitors of cathepsin L prevent severe acute respiratory syndrome coronavirus entry. Proc Natl Acad Sci U S A 102:11876–11881

Siu KL, Kok KH, Ng MH, Poon VK, Yuen KY, Zheng BJ, Jin DY (2009) Severe acute respiratory syndrome coronavirus M protein inhibits type I interferon production by impeding the formation of TRAF3.TANK.TBK1/IKKepsilon complex. J Biol Chem 284:16202–16209

Siu KL, Yeung ML, Kok KH, Yuen KS, Kew C, Lui PY, Chan CP, Tse H, Woo PC, Yuen KY, Jin DY (2014) Middle east respiratory syndrome coronavirus 4a protein is a double-stranded RNA-binding protein that suppresses PACT-induced activation of RIG-I and MDA5 in the innate antiviral response. J Virol 88:4866–4876

Snijder EJ, van der Meer Y, Zevenhoven-Dobbe J, Onderwater JJ, van der Meulen J, Koerten HK, Mommaas AM (2006) Ultrastructure and origin of membrane vesicles associated with the severe acute respiratory syndrome coronavirus replication complex. J Virol 80:5927–5940

Snijder EJ, Decroly E, Ziebuhr J (2016) The nonstructural proteins directing coronavirus RNA synthesis and processing. Adv Virus Res 96:59–126

Sola I, Almazan F, Zuniga S, Enjuanes L (2015) Continuous and discontinuous RNA synthesis in coronaviruses. Annu Rev Virol 2:265–288

Song S, Bi J, Wang D, Fang L, Zhang L, Li F, Chen H, Xiao S (2013) Porcine reproductive and respiratory syndrome virus infection activates IL-10 production through NF-kappaB and p38 MAPK pathways in porcine alveolar macrophages. Dev Comp Immunol 39:265–272

Spagnolo JF, Hogue BG (2000) Host protein interactions with the 3′ end of bovine coronavirus RNA and the requirement of the poly(A) tail for coronavirus defective genome replication. J Virol 74:5053–5065

Stertz S, Reichelt M, Spiegel M, Kuri T, Martinez-Sobrido L, Garcia-Sastre A, Weber F, Kochs G (2007) The intracellular sites of early replication and budding of SARS-coronavirus. Virology 361:304–315

Supekova L, Supek F, Lee J, Chen S, Gray N, Pezacki JP, Schlapbach A, Schultz PG (2008) Identification of human kinases involved in hepatitis C virus replication by small interference RNA library screening. J Biol Chem 283:29–36

Surjit M, Liu B, Chow VT, Lal SK (2006) The nucleocapsid protein of severe acute respiratory syndrome-coronavirus inhibits the activity of cyclin-cyclin-dependent kinase complex and blocks S phase progression in mammalian cells. J Biol Chem 281:10669–10681

Taguchi F, Siddell SG (1985) Difference in sensitivity to interferon among mouse hepatitis viruses with high and low virulence for mice. Virology 147:41–48

Takano T, Katoh Y, Doki T, Hohdatsu T (2013) Effect of chloroquine on feline infectious peritonitis virus infection in vitro and in vivo. Antiviral Res 99:100–107

Tan YW, Hong W, Liu DX (2012) Binding of the 5′-untranslated region of coronavirus RNA to zinc finger CCHC-type and RNA-binding motif 1 enhances viral replication and transcription. Nucleic Acids Res 40:5065–5077

Tanaka Y, Sato Y, Osawa S, Inoue M, Tanaka S, Sasaki T (2012) Suppression of feline coronavirus replication in vitro by cyclosporin A. Vet Res 43:41

Tanaka Y, Sato Y, Sasaki T (2013) Suppression of coronavirus replication by cyclophilin inhibitors. Viruses 5:1250–1260

Tanaka Y, Sato Y, Sasaki T (2016) Feline coronavirus replication is affected by both cyclophilin A and cyclophilin B. J Gen Virol 98:190–200

Teigelkamp S, Achsel T, Mundt C, Gothel SF, Cronshagen U, Lane WS, Marahiel M, Luhrmann R (1998) The 20kD protein of human [U4/U6.U5] tri-snRNPs is a novel cyclophilin that forms a complex with the U4/U6-specific 60kD and 90kD proteins. RNA 4:127–141

Thornbrough JM, Jha BK, Yount B, Goldstein SA, Li Y, Elliott R, Sims AC, Baric RS, Silverman RH, Weiss SR (2016) Middle East respiratory syndrome coronavirus NS4b protein inhibits host RNase L activation. MBio 7:e00258

Totura AL, Baric RS (2012) SARS coronavirus pathogenesis: host innate immune responses and viral antagonism of interferon. Curr Opin Virol 2:264–275

Totura AL, Whitmore A, Agnihothram S, Schafer A, Katze MG, Heise MT, Baric RS (2015) Toll-Like receptor 3 signaling via TRIF contributes to a protective innate immune response to severe acute respiratory syndrome coronavirus infection. MBio 6:e00638–e00615

Tresnan DB, Levis R, Holmes KV (1996) Feline aminopeptidase N serves as a receptor for feline, canine, porcine, and human coronaviruses in serogroup I. J Virol 70:8669–8674

Ulasli M, Verheije MH, de Haan CA, Reggiori F (2010) Qualitative and quantitative ultrastructural analysis of the membrane rearrangements induced by coronavirus. Cell Microbiol 12:844–861

van Boheemen S, de Graaf M, Lauber C, Bestebroer TM, Raj VS, Zaki AM, Osterhaus AD, Haagmans BL, Gorbalenya AE, Snijder EJ, Fouchier RA (2012) Genomic characterization of a newly discovered coronavirus associated with acute respiratory distress syndrome in humans. MBio 3:e00473–e00412

van der Hoek L (2007) Human coronaviruses: what do they cause? Antiviral therapy 12:651–658

van der Hoek L, Pyrc K, Jebbink MF, Vermeulen-Oost W, Berkhout RJ, Wolthers KC, Wertheim-van Dillen PM, Kaandorp J, Spaargaren J, Berkhout B (2004) Identification of a new human coronavirus. Nat Med 10:368–373

van der Hoeven B, Oudshoorn D, Koster AJ, Snijder EJ, Kikkert M, Barcena M (2016) Biogenesis and architecture of arterivirus replication organelles. Virus Res 220:70–90

van der Meer Y, Snijder EJ, Dobbe JC, Schleich S, Denison MR, Spaan WJ, Krijnse Locker J (1999) Localization of mouse hepatitis virus nonstructural proteins and RNA synthesis indicates a role for late endosomes in viral replication. J Virol 73:7641–7657

van der Schaar HM, Dorobantu CM, Albulescu L, Strating JR, van Kuppeveld FJ (2016) Fat(al) attraction: picornaviruses usurp lipid transfer at membrane contact sites to create replication organelles. Trends Microbiol 24:535–546

van Hemert MJ, van den Worm SH, Knoops K, Mommaas AM, Gorbalenya AE, Snijder EJ (2008) SARS-coronavirus replication/transcription complexes are membrane-protected and need a host factor for activity in vitro. PLoS Pathog 4:e1000054

van Kasteren PB, Bailey-Elkin BA, James TW, Ninaber DK, Beugeling C, Khajehpour M, Snijder EJ, Mark BL, Kikkert M (2013) Deubiquitinase function of arterivirus papain-like protease 2 suppresses the innate immune response in infected host cells. Proc Natl Acad Sci USA 110:E838–E847

Verheije MH, Raaben M, Mari M, Te Lintelo EG, Reggiori F, van Kuppeveld FJ, Rottier PJ, de Haan CA (2008) Mouse hepatitis coronavirus RNA replication depends on GBF1-mediated ARF1 activation. PLoS Pathog 4:e1000088

Versteeg GA, Slobodskaya O, Spaan WJ (2006) Transcriptional profiling of acute cytopathic murine hepatitis virus infection in fibroblast-like cells. J Gen Virol 87:1961–1975

Vijay R, Perlman S (2016) Middle East respiratory syndrome and severe acute respiratory syndrome. Curr Opin Virol 16:70–76

Vijgen L, Keyaerts E, Lemey P, Maes P, Van Reeth K, Nauwynck H, Pensaert M, Van Ranst M (2006) Evolutionary history of the closely related group 2 coronaviruses: porcine hemagglutinating encephalomyelitis virus, bovine coronavirus, and human coronavirus OC43. J Virol 80:7270–7274

Vogels MW, van Balkom BW, Kaloyanova DV, Batenburg JJ, Heck AJ, Helms JB, Rottier PJ, de Haan CA (2011) Identification of host factors involved in coronavirus replication by quantitative proteomics analysis. Proteomics 11:64–80 von Brunn A, Ciesek S, von Brunn B, Carbajo-Lozoya J (2015) Genetic deficiency and polymorphisms of cyclophilin a reveal its essential role for human coronavirus 229E replication. Curr Opin Virol 14:56–61

Walls AC, Tortorici MA, Bosch BJ, Frenz B, Rottier PJ, DiMaio F, Rey FA, Veesler D (2016) Cryo-electron microscopy structure of a coronavirus spike glycoprotein trimer. Nature 531:114–117

Walsh D, Mohr I (2011) Viral subversion of the host protein synthesis machinery. Nat Rev Microbiol 9:860–875

Wang Y, Liu L (2016) The membrane protein of severe acute respiratory syndrome coronavirus functions as a novel cytosolic pathogen-associated molecular pattern to promote beta interferon induction via a toll-like-receptor-related TRAF3-independent mechanism. MBio 7:e01872–e01815

Wang X, Liao Y, Yap PL, Png KJ, Tam JP, Liu DX (2009) Inhibition of protein kinase R activation and upregulation of GADD34 expression play a synergistic role in facilitating coronavirus replication by maintaining de novo protein synthesis in virus-infected cells. J Virol 83:12462–12472

Wang G, Chen G, Zheng D, Cheng G, Tang H (2011) PLP2 of mouse hepatitis virus A59 (MHV-A59) targets TBK1 to negatively regulate cellular type I interferon signaling pathway. PLoS ONE 6:e17192

Wang D, Fang L, Shi Y, Zhang H, Gao L, Peng G, Chen H, Li K, Xiao S (2015) Porcine epidemic diarrhea virus 3C-like protease regulates its interferon antagonism by cleaving NEMO. J Virol 90:2090–2101

Wathelet MG, Orr M, Frieman MB, Baric RS (2007) Severe acute respiratory syndrome coronavirus evades antiviral signaling: role of nsp1 and rational design of an attenuated strain. J Virol 81:11620–11633

WHO (2004) SARS website. http://www.who.int/csr/sars/country/table2004_04_21/

WHO (2016) http://www.who.int/emergencies/mers-cov/en/

Wilkins C, Gale M Jr (2010) Recognition of viruses by cytoplasmic sensors. Curr Opin Immunol 22:41–47

Williams RK, Jiang GS, Holmes KV (1991) Receptor for mouse hepatitis virus is a member of the carcinoembryonic antigen family of glycoproteins. Proc Natl Acad Sci U S A 88:5533–5536

Wong HH, Kumar P, Tay FP, Moreau D, Liu DX, Bard F (2015) Genome-wide screen reveals valosin-containing protein requirement for coronavirus exit from endosomes. J Virol 89:11116–11128

Woo PC, Lau SK, Chu CM, Chan KH, Tsoi HW, Huang Y, Wong BH, Poon RW, Cai JJ, Luk WK, Poon LL, Wong SS, Guan Y, Peiris JS, Yuen KY (2005) Characterization and complete genome sequence of a novel coronavirus, coronavirus HKU1, from patients with pneumonia. J Virol 79:884–895

Wu K, Li W, Peng G, Li F (2009) Crystal structure of NL63 respiratory coronavirus receptor-binding domain complexed with its human receptor. Proc Natl Acad Sci U S A 106:19970–19974

Wu CH, Chen PJ, Yeh SH (2014) Nucleocapsid phosphorylation and RNA helicase DDX1 recruitment enables coronavirus transition from discontinuous to continuous transcription. Cell Host Microbe 16:462–472

Xiao H, Xu LH, Yamada Y, Liu DX (2008) Coronavirus spike protein inhibits host cell translation by interaction with eIF3f. PLoS ONE 3:e1494

Xu L, Khadijah S, Fang S, Wang L, Tay FP, Liu DX (2010) The cellular RNA helicase DDX1 interacts with coronavirus nonstructural protein 14 and enhances viral replication. J Virol 84:8571–8583

Xu LH, Huang M, Fang SG, Liu DX (2011) Coronavirus infection induces DNA replication stress partly through interaction of its nonstructural protein 13 with the p125 subunit of DNA polymerase delta. J Biol Chem 286:39546–39559

Yang D, Leibowitz JL (2015) The structure and functions of coronavirus genomic 3′ and 5′ ends. Virus Res 206:120–133

Yang N, Ma P, Lang J, Zhang Y, Deng J, Ju X, Zhang G, Jiang C (2012) Phosphatidylinositol 4-kinase IIIbeta is required for severe acute respiratory syndrome coronavirus spike-mediated cell entry. J Biol Chem 287:8457–8467

Yang Y, Zhang L, Geng H, Deng Y, Huang B, Guo Y, Zhao Z, Tan W (2013) The structural and accessory proteins M, ORF 4a, ORF 4b, and ORF 5 of Middle East respiratory syndrome coronavirus (MERS-CoV) are potent interferon antagonists. Protein Cell 4:951–961

Yang Y, Du L, Liu C, Wang L, Ma C, Tang J, Baric RS, Jiang S, Li F (2014) Receptor usage and cell entry of bat coronavirus HKU4 provide insight into bat-to-human transmission of MERS coronavirus. Proc Natl Acad Sci U S A 111:12516–12521

Yang Y, Liu C, Du L, Jiang S, Shi Z, Baric RS, Li F (2015) Two mutations were critical for bat-to-human transmission of middle east respiratory syndrome coronavirus. J Virol 89: 9119–9123

Ye Y, Hauns K, Langland JO, Jacobs BL, Hogue BG (2007) Mouse hepatitis coronavirus A59 nucleocapsid protein is a type I interferon antagonist. J Virol 81:2554–2563

Yeager CL, Ashmun RA, Williams RK, Cardellichio CB, Shapiro LH, Look AT, Holmes KV (1992) Human aminopeptidase N is a receptor for human coronavirus 229E. Nature 357: 420–422

Yu W, Leibowitz JL (1995) Specific binding of host cellular proteins to multiple sites within the 3′ end of mouse hepatitis virus genomic RNA. J Virol 69:2016–2023

Yuan X, Yao Z, Wu J, Zhou Y, Shan Y, Dong B, Zhao Z, Hua P, Chen J, Cong Y (2007) G1 phase cell cycle arrest induced by SARS-CoV 3a protein via the cyclin D3/pRb pathway. Am J Respir Cell Mol Biol 37:9–19

Yuan L, Chen Z, Song S, Wang S, Tian C, Xing G, Chen X, Xiao ZX, He F, Zhang L (2015) p53 degradation by a coronavirus papain-like protease suppresses type I interferon signaling. J Biol Chem 290:3172–3182

Zaki AM, van Boheemen S, Bestebroer TM, Osterhaus AD, Fouchier RA (2012) Isolation of a novel coronavirus from a man with pneumonia in Saudi Arabia. N Engl J Med 367:1814–1820

Zalinger ZB, Elliott R, Rose KM, Weiss SR (2015) MDA5 is critical to host defense during infection with murine coronavirus. J Virol 89:12330–12340

Zeng Q, Langereis MA, van Vliet AL, Huizinga EG, de Groot RJ (2008) Structure of coronavirus hemagglutinin-esterase offers insight into corona and influenza virus evolution. Proc Natl Acad Sci U S A 105:9065–9069

Zhang X, Lai MM (1995) Interactions between the cytoplasmic proteins and the intergenic (promoter) sequence of mouse hepatitis virus RNA: correlation with the amounts of subgenomic mRNA transcribed. J Virol 69:1637–1644

Zhang X, Wu K, Wang D, Yue X, Song D, Zhu Y, Wu J (2007) Nucleocapsid protein of SARS-CoV activates interleukin-6 expression through cellular transcription factor NF-kappaB. Virology 365:324–335

Zhang L, Zhang ZP, Zhang XE, Lin FS, Ge F (2010) Quantitative proteomics analysis reveals BAG3 as a potential target to suppress severe acute respiratory syndrome coronavirus replication. J Virol 84:6050–6059

Zhao Z, Thackray LB, Miller BC, Lynn TM, Becker MM, Ward E, Mizushima NN, Denison MR, Virgin HWt (2007) Coronavirus replication does not require the autophagy gene ATG5. Autophagy 3:581–585

Zhao L, Jha BK, Wu A, Elliott R, Ziebuhr J, Gorbalenya AE, Silverman RH, Weiss SR (2012) Antagonism of the interferon-induced OAS-RNase L pathway by murine coronavirus ns2 protein is required for virus replication and liver pathology. Cell Host Microbe 11:607–616

Zheng B, He ML, Wong KL, Lum CT, Poon LL, Peng Y, Guan Y, Lin MC, Kung HF (2004) Potent inhibition of SARS-associated coronavirus (SCOV) infection and replication by type I interferons (IFN-alpha/beta) but not by type II interferon (IFN-gamma). J Interferon Cytokine Res 24:388–390

Zheng D, Chen G, Guo B, Cheng G, Tang H (2008) PLP2, a potent deubiquitinase from murine hepatitis virus, strongly inhibits cellular type I interferon production. Cell Res 18:1105–1113

Zhou P, Li H, Wang H, Wang LF, Shi Z (2012) Bat severe acute respiratory syndrome-like coronavirus ORF3b homologues display different interferon antagonist activities. J Gen Virol 93:275–281

Zust R, Cervantes-Barragan L, Habjan M, Maier R, Neuman BW, Ziebuhr J, Szretter KJ, Baker SC, Barchet W, Diamond MS, Siddell SG, Ludewig B, Thiel V (2011) Ribose 2′-O-methylation provides a molecular signature for the distinction of self and non-self mRNA dependent on the RNA sensor Mda5. Nat Immunol 12:137–143

Roles of Pro-viral Host Factors in Mosquito-Borne Flavivirus Infections

Rafael K. Campos, Mariano A. Garcia-Blanco and Shelton S. Bradrick

Abstract Identification and analysis of viral host factors is a growing area of research which aims to understand the how viruses molecularly interface with the host cell. Investigations into flavivirus–host interactions has led to new discoveries in viral and cell biology, and will potentially bolster strategies to control the important diseases caused by these pathogens. Here, we address the current knowledge of prominent host factors required for the flavivirus life-cycle and mechanisms by which they promote infection.

Contents

R.K. Campos
Department of Molecular Genetics and Microbiology,
Center for RNA Biology, Duke University, Durham, NC, USA

R.K. Campos · M.A. Garcia-Blanco (✉) · S.S. Bradrick (✉)
Department of Biochemistry and Molecular Biology,
University of Texas Medical Branch, Galveston, TX, USA
e-mail: maragarc@utmb.edu

S.S. Bradrick
e-mail: ssbradri@utmb.edu

M.A. Garcia-Blanco
Programme in Emerging Infectious Diseases,
Duke-NUS Medical School, Singapore, Singapore

Current Topics in Microbiology and Immunology (2018) 419:43–68
DOI 10.1007/82_2017_26

Published Online: 09 July 2017

1 Introduction

The flaviviruses comprise one of four genera within the *Flaviviridae*, a family of positive-strand RNA viruses that derives its name from the prototypical yellow (Latin: *flavus*) fever virus (YFV). Many flaviviruses are emerging or re-emerging global health threats (Bhatt et al. 2013; Garske et al. 2014; Weaver et al. 2016; Wilder-Smith and Byass 2016). Prominent human pathogens among the >50 flavivirus species include dengue viruses (DENV), Japanese encephalitis virus (JEV), West Nile virus (WNV), YFV and Zika virus (ZIKV). Mosquito-borne flaviviruses are distributed mainly in tropical and subtropical regions of the globe and over half of the world's population is estimated to be at risk for infection by DENV, the most widespread of the pathogenic flaviviruses (Bhatt et al. 2013). The continuing threat of DENV, the deadly YFV outbreak in Africa in 2016, and the emerging ZIKV pandemic highlight the significant burden these viruses place on humanity. Flaviviruses are transmitted mainly by mosquito or tick bites, although sexual and vertical transmission has recently been documented for ZIKV (D'Ortenzio et al. 2016; Harrower et al. 2016). In this review, we focus on flaviviruses transmitted by mosquitoes.

Flaviviruses are enveloped viruses with icosahedral-shaped particles of 40–50 nm in diameter. Although icosahedral symmetry is shared between flavivirus species, the surface topology varies significantly between viruses (Kostyuchenko et al. 2016; Zhang et al. 2013). The virion is composed of three structural proteins, two of which [membrane (M) and envelope (E)] are embedded within the lipid bilayer envelope. The remaining structural protein, capsid (C), is located within the interior of the viral particle, closely associated with the single-stranded RNA genome. The genomes of flaviviruses share a similar organization: all are ~11 kb positive-strand RNA molecules that contain a single open reading frame flanked by 5′ and 3′ untranslated regions (UTRs). Flavivirus genomes are modified with a 5′ m^7G cap structure but lack the 3′ poly (A) tail that is characteristic of most cellular mRNAs.

2 Anti- and Pro-viral Host Factors

To carry out all steps in the life-cycle, flaviviruses must rely on hundreds of host gene products and other factors. Pro-viral host factors are comprised of RNAs, proteins and lipids from humans and mosquitoes that are required for efficient flavivirus infection. In contrast, anti-viral host factors block infection and are often associated with innate immunity and interferon responses. In this review, we focus specifically on pro-viral host factors known to be required at various phases of the virus life-cycle and mechanisms underlying these requirements. Pro-viral host factors are sometimes termed dependency factors or simply host factors. Understanding what factors are required for successful infection and how they assist the virus will permit a full understanding of flavivirus biology and could yield novel targets that could be exploited to treat flaviviral diseases.

2.1 *Attachment*

The flaviviral life-cycle starts with attachment, mediated by the viral E protein, to cellular receptor(s) on the plasma membrane (Fig. 1a). A large number of candidate receptors or co-receptors that mediate flavivirus attachment have been described in different cell types (Perera-Lecoin et al. 2014), but which of these receptors and co-receptors are used by the virus during natural infections is not known. Different flaviviruses appear to have evolved to utilize distinct receptors and co-receptors, which may partly explain the divergent syndromes associated with these viruses (Jindadamrongwech et al. 2004; Thepparit and Smith 2004). It is likely that flaviviruses require multiple co-receptors to facilitate virus binding, similar to the distantly related hepatitis C virus (Zeisel et al. 2013).

2.1.1 Glycosaminoglycans

Several studies suggest that flaviviruses make initial contact with the host cell by binding to glycosaminoglycans (GAGs), which are long and unbranched sulfated polysaccharides linked to core proteins (e.g., syndecans) prominently exposed on cellular surfaces of all tissues. Implicated GAG molecules, which include heparin sulfate (Artpradit et al. 2013; Avirutnan et al. 2007; Chen et al. 1997; Dalrymple and Mackow 2011; Kroschewski et al. 2003; Lee and Lobigs 2000; Liu et al. 2004; Roehrig et al. 2013) and syndecan proteoglycans (Okamoto et al. 2012), may concentrate flavivirus particles at the target cell surface before their interaction with protein receptors. If this model is correct, GAGs in general may play an ancillary role in attachment and entry, not being absolutely required for infection but instead controlling the extent of virus interaction with the outer cell surface.

2.1.2 C-Type Lectin Receptors

Another class of molecules implicated in flavivirus attachment are C-type (calcium-dependent) lectin receptors. These are carbohydrate binding proteins and include the dendritic cell-specific intercellular adhesion molecule (ICAM)-3-grabbing non-integrin (DC-SIGN) (Navarro-Sanchez et al. 2003; Pokidysheva et al. 2006; Tassaneetrithep et al. 2003; Wang et al. 2016) expressed in macrophages, monocytes, and immature dendritic cells. Several flaviviruses, including DENV, WNV, and JEV interact with DC-SIGN, and this interaction is greatly affected by types and location of glycosylation on the envelope of the virus (Davis et al. 2006; Dejnirattisai et al. 2011). For example, DC-SIGN has higher affinity for high-mannose oligosaccharides and fucosylated glycans. These requirements for type and location of glycosylation may render viruses produced in certain cells unable to use DC-SIGN (Davis et al. 2006; Dejnirattisai et al. 2011). Internalization of DC-SIGN is not essential for DENV infection. Consequently, these lectins may

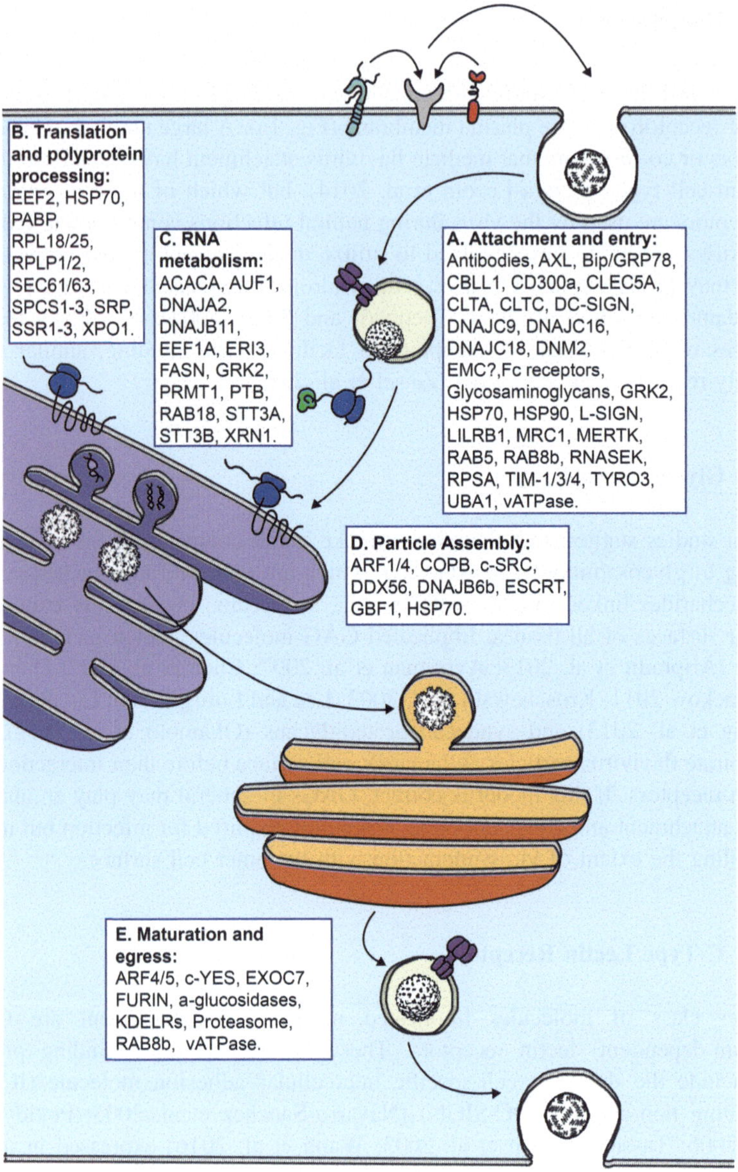
B. Translation and polyprotein processing:
EEF2, HSP70, PABP, RPL18/25, RPLP1/2, SEC61/63, SPCS1-3, SRP, SSR1-3, XPO1.
C. RNA metabolism:
ACACA, AUF1, DNAJA2, DNAJB11, EEF1A, ERI3, FASN, GRK2, PRMT1, PTB, RAB18, STT3A, STT3B, XRN1.
A. Attachment and entry:
Antibodies, AXL, Bip/GRP78, CBLL1, CD300a, CLEC5A, CLTA, CLTC, DC-SIGN, DNAJC9, DNAJC16, DNAJC18, DNM2, EMC?,Fc receptors, Glycosaminoglycans, GRK2, HSP70, HSP90, L-SIGN, LILRB1, MRC1, MERTK, RAB5, RAB8b, RNASEK, RPSA, TIM-1/3/4, TYRO3, UBA1, vATPase.
D. Particle Assembly:
ARF1/4, COPB, c-SRC, DDX56, DNAJB6b, ESCRT, GBF1, HSP70.
E. Maturation and egress:
ARF4/5, c-YES, EXOC7, FURIN, a-glucosidases, KDELRs, Proteasome, RAB8b, vATPase.

◂**Fig. 1** Host dependency factors and their implicated roles in the flaviviral life-cycle. **a.** Flaviviruses bind to co-receptors which facilitate interaction of the virus with one or more receptors. The virus enters the cell by receptor-mediated endocytosis which can occur through different endocytic pathways. Uncoating of viral RNA by fusion of the virus with the endosomal membrane requires endosomal acidification which triggers a conformational change in the virus. **b.** Translation of the viral genome may initiate in the cytosol prior to ER localization where synthesis and processing of the viral polyprotein takes place. **c.** Viral RNA metabolism involves diverse host factors. Viral replication takes place in membranous vesicles that are induced by viral nonstructural proteins. **d.** Particle assembly require host factors that associate with viral RNA, capsid or are involved in ER membrane restructuring. Some viral glycosylation events take place in the lumen of the ER. **e.** Egress of the virus requires the secretory machinery of the Golgi where additional glycosylations of viral proteins take place. After transit thought the Golgi, vesicles containing viral particles become acidified, promoting furin cleavage of prM and subsequent release from the cell by exocytosis. Pro-viral factors that do not have well defined roles at a specific phase of the life-cycle include DDX6, DNAJB7, DNAJC10, SSB and NF90

not function as receptors but rather help concentrate virus on the cell surface (Lozach et al. 2005; Miller et al. 2008).

Another protein of this family reported to bind flaviviruses is liver/lymph node-specific ICAM-3 grabbing non-integrin (L-SIGN) (Davis et al. 2006; Dejnirattisai et al. 2011). L-SIGN binds preferentially high-mannose oligosaccharides, but unlike DC-SIGN it does not seem to bind fucosylated glycans (Guo et al. 2004). Macrophages express other lectins, known as mannose receptors, which bind DENV and JEV. In contrast to DC-SIGN and L-SIGN these proteins bind specifically to terminal mannose, fucose, and N-acetyl glucosamine (Miller et al. 2008). It has been hypothesized that mannose receptors may promote step(s) subsequent to viral attachment since they have been observed to internalize during infection, unlike DC-SIGN which remains on the plasma membrane (Lozach et al. 2005; Miller et al. 2008). CLEC5A is another transmembrane protein expressed by macrophages and monocytes that is important for attachment of DENV (Chen et al. 2008; Tung et al. 2014). The binding of CLEC5A to DENV was inhibited by free mannose and fucose, suggesting these molecules are avid binders of CLEC5A. In mosquitoes, the galactose-specific binding C-type lectin (mosGCTL-1) was found to be important for WNV attachment (Cheng et al. 2010) and several mosGCTL proteins were found to interact with DENV envelope protein (Liu et al. 2014).

2.1.3 Phosphatidylserine Receptors

The lipid bilayer envelope of flaviviruses is known to contain phosphatidylserine (PS). The presence of PS in the viral envelope may mimic the surface of apoptotic bodies and induce engulfment of the viral particle in a phenomenon termed apoptotic mimicry (Moller-Tank and Maury 2014). Many PS receptors are present in antigen-presenting cells (Kobayashi et al. 2007; Seitz et al. 2007) and were identified as pro-viral host factors in a screen for genes that rendered poorly permissive 293T cells susceptible to DENV infection (Meertens et al. 2012).

Two important families of PS receptors are T-cell immunoglobulin and mucin domain (TIM) and TYRO3-AXL-MERTK (TAM). The TIM members TIM-1, TIM-3 and TIM-4, and TAM receptors (TYRO3, AXL, and MERTK), were identified as DENV, WNV and YFV entry factors (Meertens et al. 2012). For ZIKV, AXL was proposed as an entry receptor by functional genomic analysis (Savidis et al. 2016), and is important for infection of human skin cells (Hamel et al. 2015) and placental trophoblasts (Tabata et al. 2016b); however, AXL was not required for ZIKV infection of cells within the murine eye (Miner et al. 2016) or human neural progenitor cells (Wells et al. 2016), indicating that it plays a cell type-specific role. Another PS receptor that is expressed on myeloid cells, named CD300a, was also described as a DENV host factor (Carnec et al. 2016) that binds directly to DENV particles and enhances infection through recognition of phosphatidylethanolamine and, to a lesser extent, PS present in the viral envelope (Carnec et al. 2016).

2.1.4 Fc Receptors and Antibody-Dependent Enhancement of Attachment and Entry

Antibodies seem to be a double-edged sword in flavivirus infections, acting to both disrupt and enhance infection. Antibodies may contribute to flavivirus attachment and entry in a phenomenon known as antibody-dependent enhancement (ADE) (Acosta and Bartenschlager 2016). ADE occurs when cross-reactive, non-neutralizing antibodies from a prior infection with a different virus facilitate viral entry into the cell via Fc receptors, cell-surface proteins that bind to the antibody constant region. This is known to happen among serotypes of DENV, and may explain why ADE is associated with more severe disease in the context of secondary DENV infections. However, ADE can also occur between different flavivirus species. For example, anti-DENV antibodies may drive ADE of ZIKV infection (Dejnirattisai et al. 2016). Ligation of Fc gamma receptor IIB by DENV–antibody complexes may induce a signal transduction pathway leading to an anti-viral type I interferon response. To abrogate this induction, DENV–antibody complexes co-ligate the leukocyte immunoglobulin-like receptor-B1 (LILRB1), which in turn inhibits Fc gamma receptor signaling and induction of ISGs (Chan et al. 2014).

2.1.5 Other Receptor Types

Several chaperones and other factors have been implicated in DENV entry. Heat shock proteins 70 and 90 (HSP70/90) were proposed to be DENV receptors in human neuroblastoma and lymphoma cells (Reyes-Del Valle et al. 2005), although these proteins are not necessary for entry of DENV in the hepatoma cell line, HepG2 (Cabrera-Hernandez et al. 2007). The HSP70 cofactors DnaJC18, DnaJC9 and DnaJC16 were reported to be required for entry in HuH7 hepatoma cells

(Taguwa et al. 2015). It was reported that DENV-2 interacts with GRP78/BiP (Jindadamrongwech et al. 2004) while DENV-1 was suggested to bind the 37/67 kDa high-affinity laminin receptor, also known as ribosomal protein SA (RPSA) (Thepparit and Smith 2004). A subsequent study, however, observed that DENV-1, 2 and 3 all interacted with RPSA (Tio et al. 2005), but did not find evidence for DENV-2 binding to GRP78/BIP. Although several protein chaperones likely participate in the flaviviral life-cycle (see below), their roles in viral entry remain obscure.

2.2 *Endocytosis*

After attachment, the virus enters the cell by endocytosis (Fig. 1a) which, depending on the virus and cell type analyzed, is either clathrin-dependent or -independent (Acosta et al. 2009; Kalia et al. 2013; Smit et al. 2011; van der Schaar et al. 2008). For instance, JEV infects fibroblasts in a clathrin-dependent manner, but entry into neuronal cells is clathrin-independent (Kalia et al. 2013). Using single-particle tracking analysis of DENV in living cells, virions were noted to move along the cell surface, presumably over distinct attachment factors, until they bound to one or more entry receptors (van der Schaar et al. 2008). For DENV in mammalian cells, entry was independent of clathrin but dependent on dynamin (Acosta et al. 2009). In hepatoma cells, the G protein-coupled receptor kinase 2 (GRK2) was found to be required for efficient DENV entry independently of β-arrestins (Le Sommer et al. 2012), suggesting a non-canonical pathway that involves GRK2 (Evron et al. 2012).

Upon clathrin-mediated entry, DENV particles are transported to an early (Rab5 positive) endosomal compartment, which subsequently matures into late endosomes through acquisition of Rab7 and loss of Rab5 (Krishnan et al. 2007; van der Schaar et al. 2008). Acidification of the late endosome mediated by the vacuolar-type H^+-ATPase (vATPase) complex induces conformational changes in the E protein that lead to irreversible trimerization (Luca et al. 2013) and exposure of fusion peptides that promote merging of the viral envelope and endosomal membrane. This process is likely enhanced by the transmembrane protein ribonuclease kappa (RNASEK). RNASEK localizes to the plasma membrane and endosomes, and functionally associates with the vATPase proton pump to promote endocytosis of several diverse viruses (Hackett et al. 2015; Perreira et al. 2015). It has been proposed that virus particles fuse preferentially with small endosomal carrier vesicles and the release of the nucleocapsid into the cytosol depends on a second, cell-mediated membrane fusion event (Nour et al. 2013). In addition to low pH, phospholipid composition is also important to promote fusion with endosomal membranes. For example, anionic lipids, such as bis (monoacylglycero) phosphate, a lipid specific to late endosomes, are important for DENV fusion with endosomal membranes (Zaitseva et al. 2010). This highlights the role of endosomal membrane lipids and host proteins which mediate endosomal acidification as pro-viral host factors required for membrane fusion and nucleocapsid release.

Components of the endoplasmic reticulum membrane complex (EMC) were first identified in an RNAi-mediated screen for YFV host factors (Le Sommer et al. 2012) and recently confirmed in CRISPR/Cas9 screens for viral infection (DENV, WNV and ZIKV) or protection against virus-induced apoptosis (WNV) (Ma et al. 2015; Marceau et al. 2016; Savidis et al. 2016; Zhang et al. 2016). The EMC is required for early step(s) in the life-cycle of DENV and ZIKV, possibly at the entry stage (Savidis et al. 2016). Indeed, the EMC's putative role in the biogenesis of cellular membrane proteins (Richard et al. 2013; Satoh et al. 2015) suggests that expression of a flavivirus receptor may depend on the EMC. Nonetheless, it is also possible that flaviviruses have co-opted the EMC for efficient synthesis and/or processing of the viral polyprotein. Additional study will be required to define roles for the EMC in the flaviviral life-cycle.

2.3 *Uncoating*

The uncoating step is not well understood but a recent study has shed light on this phase of the life-cycle. Byk and colleagues determined that ubiquitination is crucial for DENV RNA release into the cytoplasm (Byk et al. 2016). Although ubiquitin was required, proteasome activity was dispensable for genome uncoating, and stabilization of incoming capsid protein by proteasome inhibition did not impair early viral translation (Byk and Gamarnik 2016). These data favor a model in which capsid does not need to be degraded, and may be displaced from the RNA by translating ribosomes (Garcia-Blanco et al. 2016). Inhibition of ubiquitin-activating enzyme UBA1 was found to block uncoating of the genome during infection, preventing access of the viral genome to the translation machinery (Byk et al. 2016). The ubiquitin ligase CBLL1 may play a similar role since it was found to be required for WNV entry (Krishnan et al. 2008).

2.4 *Viral Translation*

2.4.1 Translation Initiation

Flaviviral translation (Fig. 1b) is likely similar to cellular cap-dependent mRNA translation, but distinct requirements for flaviviruses have not been sufficiently explored. Flavivirus genomes possess a 5′ type 1 cap structure that is required for efficient initiation of translation (Chiu et al. 2005). The viral RNA, after uncoating in the cytosol, is thought to bind the eukaryotic initiation factor eIF4F complex via the cap structure and poly(A) binding protein via its 3′ UTR (Polacek et al. 2009). Based on translation of human mRNAs, the 43S pre-initiation complex is then recruited to scan the ~90–120 nt (depending on viral species) 5′ UTR to identify the start codon. To correctly initiate at the first AUG, DENV requires an RNA hairpin, named the cHP element, located 14 nucleotides downstream of the AUG codon, to stall the ribosome at the correct AUG codon (Clyde and Harris 2006). In a similar fashion to translation of human mRNAs, hydrolysis of GTP bound to eIF2 and release of initiation factors is thought to allow joining of the 60S ribosomal subunit to form an intact 80S ribosome poised for elongation.

The eIF4F complex is composed of eIF4E (cap-binding protein), eIF4A (helicase), and eIF4G (scaffolding protein). While flavivirus translation is generally believed to be cap-dependent, it has been reported that DENV protein synthesis remains active under conditions where cellular cap-dependent translation is repressed. Indeed, depletion of eIF4E by RNAi did not affect DENV protein synthesis or infectivity (Edgil et al. 2006). Thus, it is possible that DENV initiates translation independently of eIF4E or eIF4F altogether, at least under certain physiological conditions.

Although the genomes of flaviviruses all lack a 3′ poly(A) tail, the 3′ UTR of the DENV genome was shown to bind poly(A) binding protein (PABP) (Polacek et al. 2009), an important *trans*-acting factor that regulates mRNA stability and translation (Kahvejian et al. 2005). Interaction between PABP and eIF4G is thought to circularize cellular mRNAs and stimulate multiple aspects of translation. The addition of PABP-interacting protein 2 (PAIP2; a PABP inhibitor) reduced translation of a DENV reporter, suggesting that interaction of PABP with DENV 3′ UTR is important for DENV translation (Polacek et al. 2009). However, it has not been established whether eIF4G–PABP interaction is required for efficient translation of flaviviral genomes.

2.4.2 RNA Recruitment to the ER

Little is known about how flaviviral genomes become associated with the ER membrane, but it is likely that this depends, at least in part, on the signal recognition particle (SRP) pathway (Walter and Johnson 1994). In flavivirus genomes the first signal sequence is encoded at the C-terminal end of the capsid sequence (Lobigs et al. 2010). Once the first initiating ribosome synthesizes the signal sequence, SRP recognizes the nascent peptide and delivers the entire RNP complex to the SRP receptor and translocon embedded within ER membranes (Walter and Blobel 1981a, b; Walter et al. 1981). While this model has not been experimentally tested for flaviviruses, and recent studies find that many cellular mRNAs are recruited to the ER independently of SRP (Ast et al. 2013; Jan et al. 2014), evidence from high-throughput screens identified components of the SRP [SRP54 (Le Sommer et al. 2012), SRP9, SRP14 (Marceau et al. 2016)], translocon-associated protein complex [SSR1, SSR2 and SSR3 (Marceau et al. 2016)] and translocon [SEC61 subunits (Le Sommer et al. 2012; Marceau et al. 2016; Sessions et al. 2009; Zhang et al. 2016), SEC63 (Zhang et al. 2016)] as important flavivirus host factors, suggesting that genome recruitment to the ER depends on the SRP.

2.4.3 Translation Elongation

After 60S subunit joining, the ribosome is poised for elongation and synthesis of the viral polyprotein. Several proteins known to participate in elongation have been implicated as potent YFV host factors based on siRNA screen data (Le Sommer et al. 2012). The elongation factor eEF2, which stimulates ribosomal translocation, was a candidate YFV host factor. Exportin-T, another putative YFV host factor, is required for tRNA export from the nucleus (Kutay et al. 1998), and thus likely maintains sufficient cytoplasmic tRNA pools to enable viral translation. Among the most highly represented class of genes that are high-confidence YFV host factors are ribosomal proteins of the large subunit, such as the heterodimeric ribosomal proteins RPLP1 and RPLP2 (Campos et al. 2017), which are likely required for efficient function of the 80S ribosome during the elongation phase of translation, (see ribosomal protein section below).

2.4.4 Translation Termination and Recycling

Ribosomes terminate and dissociate at the stop codon, which precedes a relatively structured 3′ UTR in flaviviruses. Termination factors eRF1 and eRF3 recognize ribosomes in which a stop codon is located within the A-site and catalyze the final hydrolysis reaction by the 60S subunit which releases the nascent protein (Zhouravleva et al. 1995). For cellular mRNAs this process is enhanced by PABP (Ivanov et al. 2016). Ribosomes then disassemble and are recycled back into the pool of free subunits. It is currently unknown whether translation termination on flavivirus genomes occurs by the conventional mechanism or involves distinct factors not required for most cellular mRNAs.

2.4.5 Ribosomal Proteins

Functionality of specific ribosomal proteins (RPs) has become increasingly recognized as important for RNA virus infection. The majority of these approximately 80 proteins interact with rRNA and the functions of most RPs are not well understood. However, recent studies have revealed that, while some RPs are necessary for core ribosomal activities, others promote translation of specific subsets of mRNAs or viral RNAs. RPS25, an RP previously implicated in driving translation mediated by internal ribosome entry sites (Landry et al. 2009), was shown to be required for efficient RNA accumulation of DENV, WNV, and ZIKV, but not YFV (Marceau et al. 2016). Similarly, RPL18 was shown to be required for an early phase of the DENV life-cycle, possibly translation or RNA synthesis, but its knockdown did not affect general protein synthesis as measured by metabolic labeling (Cervantes-Salazar et al. 2015). The heterodimer formed by RPLP1 and RPLP2 has also been shown to be required for flavivirus infectivity (Campos et al. 2017; Le Sommer et al. 2012) and is likely required for flaviviral translation while exerting cell type-specific effects on host translation (Campos et al. 2017). It is important to note that RPs have been ascribed "moonlighting" roles in processes independent of the ribosome (Blumenthal and Carmichael 1979; Friedman et al. 1981; Kim et al. 1995; Zhou et al. 2015); their role in the flaviviral life-cycle could be unrelated to translation. For example, RPSA, also known as laminin receptor 1, may be important for flavivirus attachment (Tio et al. 2005).

2.4.6 Polyprotein Biogenesis and Processing

The flavivirus polyprotein is cleaved co- and post-translationally by NS3 and cellular proteases into three structural and seven nonstructural proteins. Most of the cleavage events, particularly those that produce mature NS proteins, are mediated by NS3 and its co-factor NS2B on the cytoplasmic face of the ER (Lindenbach et al. 2013). It was recently shown that the signal peptidase complex (SPC), which cleaves signal peptides from the N-termini of secretory and membrane proteins, is responsible for efficient cleavage at capsid-prM, prM-E, E-NS1 and 2K-NS4B junctions (Zhang et al. 2016). The SPC is composed of five subunits (Evans et al. 1986), three of which (SPCS1, SPCS2, and SPCS3) have been shown to be required for infection of human and mosquito cells by DENV and WNV (Zhang et al. 2016). SPCS1 and SPCS3 have additionally been shown to be important for infection by ZIKV, JEV, and YFV. In contrast, SPCS1 knockdown caused only modest decreases in replication of chikungunya virus (*Togaviridae*), Rift Valley fever virus (*Bunyaviridae*) and vesicular stomatitis virus (*Rhabdoviridae*). Thus, flaviviruses, but not RNA viruses representing three different families, depend on the SPC (Zhang et al. 2016).

Protein chaperones have also been identified to be important for flavivirus infection (Das et al. 2009; Padwad et al. 2010; Taguwa et al. 2015; Ye et al. 2013). HSP70 isoforms were shown to be important for DENV and JEV at multiple steps of the life-cycle (Das et al. 2009; Taguwa et al. 2015; Ye et al. 2013). Upon inhibition of HSP70, the levels of all DENV proteins analyzed were decreased, but NS5 and capsid proteins were disproportionally reduced (Taguwa et al. 2015). HSP70 appears to be required for proper folding of NS5, since the addition of proteasome inhibitors restored NS5 levels but did not rescue viral infection in HSP70-depleted cells. In addition, HSP70 proteins also associate with capsid protein and are required for viral assembly. HSP70 substrate selection depends on DNAJ proteins, many of which (DnaJA2, DnaJB6b, DnaJB7, DnaJB11 and DnaJC10) were also found to be important for DENV infection. The EMC, a host factor for ZIKV, DENV and YFV (Le Sommer et al. 2012; Marceau et al. 2016; Savidis et al. 2016; Zhang et al. 2016), has also been suggested to be important for protein folding and stability (Jonikas et al. 2009; Louie et al. 2012; Richard et al. 2013; Satoh et al. 2015). However, it is currently unknown whether it is necessary for viral polyprotein folding or processing.

2.5 *Viral RNA Synthesis*

Although the virus encodes its own helicase (NS3), polymerase (NS5) and other NS proteins that function in RNA synthesis, several host factors were identified to be required for RNA replication (Fig. 1c). A conundrum faced by positive-strand RNA viruses is the fact that the genome must serve as a template for both translation and synthesis of negative-strand RNA. Assuming that infection begins with entry of a single genome into a cell, the virus must "switch" from protein to RNA synthesis because of the incompatibility of elongating ribosomes and viral polymerase acting on the same RNA molecule. For poliovirus, this is controlled by the viral 3CD intermediate protein which inhibits translation and facilitates negative-strand RNA synthesis (Gamarnik and Andino 1998). For flaviviruses, we envision an oscillating system, in which concentrations of viral and host proteins determine the fate of the RNA for translation or replication (Garcia-Blanco et al. 2016).

2.5.1 Replication Complex (RC) Formation and RNA Synthesis

Viral translation and RNA synthesis are spatially separated. Flaviviral nonstructural proteins induce extensive rearrangements of ER membranes to form subcellular factories, known as replication complexes (RCs), which are sites of vRNA synthesis (Pena and Harris 2012). RCs are devoid of ribosomes and other translation machinery (Romero-Brey and Bartenschlager 2014), so nascent viral genomes destined for translation must relocate from the RC to nearby sites on the ER that favor protein synthesis. The RC is formed by rearranged ER membranes containing viral nonstructural proteins necessary for replication (Romero-Brey and Bartenschlager 2014). In addition to concentrating factors required for viral RNA synthesis, the RC has been postulated to protect dsRNA from detection by pattern recognition receptors (Uchida et al. 2014).

It is well known that diverse positive-strand RNA viruses induce significant membrane rearrangements associated with RNA synthesis (Diaz and Ahlquist 2012; Nagy et al. 2016). Structurally, flavivirus RCs are composed of membrane invaginations into the lumen of the ER and contain a single pore that allows access to the cytosol. Once the replication complex is formed, negative-strand synthesis ensues. The negative-strand subsequently serves as template for synthesis of multiple genomes, resulting in asymmetric abundances of positive- and negative-strand RNAs. Progeny genomes go on to serve as mediators of additional viral protein and/or negative-strand synthesis.

RCs are enriched with cellular components necessary for RNA synthesis. Certain host lipids and enzymes, such as lysophosphatidic acid, phosphatidic acid, and flippase, have been shown to favor curvature of membranes (Devaux et al. 2008; Kooijman et al. 2003) and may possibly facilitate membrane rearrangement by NS proteins. The oligosaccharyltransferase (OST) complex, located within the ER membrane, was found to be required for DENV RNA synthesis and associated with viral NS proteins. Interestingly, the catalytic function of the OST complex is not required for DENV replication, suggesting that the complex serves a structural role in formation of RCs (Marceau et al. 2016). STT3A and STT3B, the catalytic subunits that differentiate variant OST complexes were both required for DENV replication, however, only STT3A promoted YFV, ZIKV, and WNV infectivity, highlighting differences in the requirement for OST among flaviviruses (Marceau et al. 2016).

Fatty acid synthase (FASN) and acetyl-CoA carboxylase alpha (ACACA) were found to be important for DENV RNA synthesis and FASN relocalized to sites of viral RNA synthesis after infection, likely through interaction with NS3 and Rab18 (Heaton et al. 2010). Localization of FASN to sites of viral RNA synthesis may assist in RC formation since the process of membrane invagination would be hypothetically facilitated by local de novo fatty acid synthesis. Interestingly, the requirement for FASN is conserved in mosquito cells where lipid species associated with RCs are dramatically perturbed compared to endomembranes from uninfected cells (Perera et al. 2012).

The autophagy pathway has been widely implicated as important to virus infection and innate immunity. Autophagy can be either pro- or anti-viral, depending on the virus, and can regulate different steps of the life-cycle. For DENV it has been shown that processing of triglycerides within autophagosomes increases the levels of free fatty acids that are used to generate ATP through β-oxidation (Heaton and Randall 2010). This process is necessary for efficient viral RNA synthesis and a requirement for autophagy was obviated by supplementing cells with free fatty acids (Heaton and Randall 2010). Another study reported that autophagy was important for DENV morphogenesis in addition to promoting RNA synthesis (Mateo et al. 2013). Interestingly, autophagy has been variously reported to restrict WNV replication (Kobayashi et al. 2014) or have no effect at all (Beatman et al. 2012). Thus, requirements for autophagy processes appear not to be conserved among flaviviruses.

2.5.2 RNA-Binding Proteins

There are many examples of pro-viral host factors that are RNA-binding proteins and promote RNA replication. Polypyrimidine tract-binding protein was reported to be a DENV, but not YFV, host factor that interacts with NS4A and promotes viral RNA synthesis (Agis-Juarez et al. 2009; Anwar et al. 2009; Jiang et al. 2009). Exoribonuclease family member 3 (ERI3) is a host factor for both DENV and YFV. ERI3 is normally enriched in the Golgi but upon infection localizes to sites of DENV replication and enhances RNA synthesis by an unknown mechanism (Ward et al. 2016). In addition to these factors, the La protein (Garcia-Montalvo et al. 2004; Vashist et al. 2009) and NF90 (Gomila et al. 2011) have each been reported to associate with flavivirus RNA, although their roles in infection are undefined. In the context of WNV, the translation elongation factor, eEF1-α, was surprisingly found to be necessary for negative strand synthesis through binding to the 3′

terminal stem loop of genomic RNA (Davis et al. 2007). Another RNA-binding protein, AU-rich binding factor 1, p45 isoform (AUF1 p45), was reported to exert RNA chaperone activity that promotes WNV genome cyclization, which is important for RNA synthesis (Friedrich et al. 2014). AUF1 p45 affinity for WNV RNA is enhanced by methylation via arginine methyltransferase (PRMT1) which was shown to be important for WNV infection by stimulating RNA synthesis (Friedrich et al. 2016). DDX6 is an RNA helicase that was shown to bind the DENV 3′ UTR and act as a host factor by an unknown mechanism (Ward et al. 2011). Taken together, it is clear that flavivirus genomes have evolved to hijack cellular RNA-binding proteins to promote multiple phases of the life-cycle.

2.6 Particle Assembly

Assembly of virus particles (Fig. 1d) initiates with association between capsid protein and the viral genome to form the nucleocapsid. The structure of the nucleocapsid remains elusive (Kostyuchenko et al. 2016; Mukhopadhyay et al. 2003; Sirohi et al. 2016; Zhang et al. 2013). Association of the nucleocapsid with E and prM heterodimers inserted into the ER membrane precedes budding of immature viral particles into the ER lumen. This is thought to occur in close proximity to the RC pore (Junjhon et al. 2014; Welsch et al. 2009). Viral particles are transported via the secretory pathway to the Golgi apparatus where maturation and N-linked glycosylation of prM and E proteins take place. Transition from the ER to the trans-Golgi network is associated with a reduction in pH, which triggers a conformational change in prM/E spikes. Furin protease cleaves prM in this acidified compartment, converting the immature viral particle into a fully infectious virus that is subsequently released from the cell by vesicular fusion with the plasma membrane (Stadler et al. 1997).

How nascent viral genomes are extruded from RCs through the pore to nearby sites of viral assembly is unknown. However, once viral genomes destined for packaging clear the RC, they associate with capsid protein (Ivanyi-Nagy et al. 2008) located on ER membranes. Capsid also localizes to the surface of lipid droplets (LDs), organelles that function in neutral lipid storage and are often associated with ER membranes (Iglesias et al. 2015; Samsa et al. 2009), and this association is possibly important for viral particle formation. On the other hand, localization to LDs may reflect a mechanism for storage of capsid protein, preventing premature association with viral RNA and/or modulation of lipid metabolism in ways that benefit virus replication (Byk and Gamarnik 2016). Either way, mutations in capsid that disrupt targeting to LDs restrict infection (Samsa et al. 2009). Importantly, several host factors have been identified that are important for capsid localization to LDs. Golgi brefeldin A resistant guanine nucleotide exchange factor 1 (GBF1) and coatomer protein β (COPB) are two such factors (Iglesias et al. 2015). In addition, the concerted actions of ADP-ribosylation factor 1 (Arf1) and Arf4 were also required for capsid LD localization (Iglesias et al. 2015).

These factors likely mediate localization of capsid from the site of synthesis (ER) directly to LDs.

The nucleolar helicase DEAD-Box Helicase 56 (DDX56) was found to be important for morphogenesis of WNV particles (Xu and Hobman 2012). Knockdown of DDX56 inhibited WNV infection and this could be rescued by a siRNA-resistant DDX56 expression construct, but not by a mutant form lacking helicase function. DDX56-depleted cells produced and exported the same amount of capsid protein to the supernatant as control cells; however, the amount of viral RNA in the supernatant was lower due to DDX56 knockdown, indicating a defect in RNA packaging (Xu and Hobman 2012). Since DDX56 also binds to the WNV capsid (Xu and Hobman 2012), it may facilitate transfer of viral RNA from the RC to ER membranes enriched with local capsid protein.

Src kinases have also been implicated in late stages of the flavivirus life-cycle. Knockdown or chemical inhibition of c-Src inhibited the accumulation of DENV particles in the ER without affecting viral gene expression. In addition, the Src kinase, c-Yes, was reported to promote WNV trafficking through the secretory pathway (Hirsch et al. 2005). Although it is unknown how these kinases promote the late stages of flavivirus infection, they represent druggable targets that could be used as anti-viral therapies.

Several endosomal sorting complex required for transport (ESCRT) proteins were found to be important for JEV and DENV infectivity (Tabata et al. 2016a). The depletion of specific ESCRT factors strongly reduced the production of infectious virus but had no effect on a JEV replicon, indicating that ESCRT proteins are not required for RNA replication (Tabata et al. 2016a). Microscopic analyses revealed that ESCRT proteins localize to sites of virus assembly and may promote virion biogenesis by inducing membrane deformations that enable budding of viral particles into the ER lumen (Tabata et al. 2016a).

2.7 *Particle Egress*

Once the flavivirus particle buds from the ER, it enters the secretory pathway where final maturation steps take place (Fig. 1e). Multiple host factors are co-opted by flaviviruses at this late stage of infection. Proteasome function was shown to be important for the egress of DENV and the proteasome inhibitor, bortezomib, reduced DENV production in primary monocytes and DENV-associated pathology in mice (Choy et al. 2015). Notably, this drug was recently identified to potently antagonize ZIKV infection (Barrows et al. 2016), although the mechanism of action for ZIKV is unknown.

Ras-related in brain (Rab) proteins are known to play roles in vesicular trafficking and are important for egress in addition to viral entry. Rab8b was identified in a siRNA screen of 18 Rab genes to promote WNV egress (Kobayashi et al. 2016). The ADP-ribosylation proteins Arf4 and Arf5, which play important roles in endomembrane trafficking and metabolism, were found to interact with prM and stimulate DENV and YFV egress (Kudelko et al. 2012). DENV prM has also been shown to interact with KDEL receptors (KDELR), transmembrane proteins that cycle between ER and Golgi to prevent ER-resident factors from "leaking" into the Golgi. Disruption of KDELR–prM interaction decreased virus egress at the stage of viral particle transport from the ER to Golgi (Li et al. 2015).

2.7.1 Glycosylation

Flaviviruses encode three proteins that are glycosylated by host machinery: prM, E, and NS1. Glycosylation of prM may promote correct folding and heterodimerization with E protein, leading to enhanced virion morphogenesis and infectivity (Courageot et al. 2000). The N-linked glycosylation of E and prM is important for viral entry (Davis et al. 2006; Dejnirattisai et al. 2011; Guirakhoo et al. 1993) and pathogenicity in animal models (Beasley et al. 2005; Kim et al. 2008). In addition, glycosylation of particular asparagine residues on E and prM differentially promotes infection of animal and mosquito cells (Hanna et al. 2005), indicating species-specific functions for viral glycosylation (Mondotte et al. 2007). For NS1, glycosylation has been shown to be important for DENV and YFV replication and pathogenesis (Crabtree et al. 2005; Muylaert et al. 1996). Furthermore, glycosylation of NS1 stimulates secretion and stabilization of the secreted hexameric form (Crabtree et al. 2005; Flamand et al. 1999; Somnuke et al. 2011). In summary, flaviviruses strongly depend on glycosylation for infection of both human and mosquito hosts.

2.7.2 Furin-like Protease Cleavage

Shortly before the virus exits the cell, the viral structural proteins undergo reversible conformational changes caused by low pH in the exocytic compartment. The vATPase, a flavivirus entry host factor, is required for egress of DENV, due to its role in acidification of exosomes (Duan et al. 2008). Structural changes that occur due to reduced pH expose the furin cleavage site on prM (Li et al. 2008; Stadler et al. 1997; Yu et al. 2008). Cleavage by furin or furin-like proteases causes the conformational changes to become irreversible. After release from the cell by exosome fusion with the plasma membrane, pr peptides dissociate in the neutral pH of the extracellular milieu, converting the particle into a fully infectious virus (Li et al. 2008; Zybert et al. 2008).

Acknowledgements We thank our colleagues from the Bradrick and Garcia-Blanco laboratory, University of Texas Medical Branch for their support. This work was supported by NIH grants R01-AI089526 and R01-AI101431 (MAG-B) and startup funds from the University of Texas Medical Branch.

References

Acosta EG, Bartenschlager R (2016) Paradoxical role of antibodies in dengue virus infections: considerations for prophylactic vaccine development. Expert Rev Vaccines 15:467–482

Acosta EG, Castilla V, Damonte EB (2009) Alternative infectious entry pathways for dengue virus serotypes into mammalian cells. Cell Microbiol 11:1533–1549

Agis-Juarez RA, Galvan I, Medina F, Daikoku T, Padmanabhan R, Ludert JE, del Angel RM (2009) Polypyrimidine tract-binding protein is relocated to the cytoplasm and is required during dengue virus infection in Vero cells. J Gen Virol 90:2893–2901

Anwar A, Leong KM, Ng ML, Chu JJ, Garcia-Blanco MA (2009) The polypyrimidine tract-binding protein is required for efficient dengue virus propagation and associates with the viral replication machinery. J Biol Chem 284:17021–170219

Artpradit C, Robinson LN, Gavrilov BK, Rurak TT, Ruchirawat M, Sasisekharan R (2013) Recognition of heparan sulfate by clinical strains of dengue virus serotype 1 using recombinant subviral particles. Virus Res 176:69–77

Ast T, Cohen G, Schuldiner M (2013) A network of cytosolic factors targets SRP-independent proteins to the endoplasmic reticulum. Cell 152:1134–1145

Avirutnan P, Zhang L, Punyadee N, Manuyakorn A, Puttikhunt C, Kasinrerk W, Malasit P, Atkinson JP, Diamond MS (2007) Secreted NS1 of dengue virus attaches to the surface of cells via interactions with heparan sulfate and chondroitin sulfate E. PLoS Pathog 3:e183

Barrows NJ, Campos RK, Powell ST, Prasanth KR, Schott-Lerner G, Soto-Acosta R, Galarza-Munoz G, McGrath EL, Urrabaz-Garza R, Gao J, Wu P, Menon R, Saade G, Fernandez-Salas I, Rossi SL, Vasilakis N, Routh A, Bradrick SS, Garcia-Blanco MA (2016) A screen of FDA-approved drugs for inhibitors of Zika Virus infection. Cell Host Microbe 20:259–270

Beasley DW, Whiteman MC, Zhang S, Huang CY, Schneider BS, Smith DR, Gromowski GD, Higgs S, Kinney RM, Barrett AD (2005) Envelope protein glycosylation status influences mouse neuroinvasion phenotype of genetic lineage 1 West Nile virus strains. J Virol 79:8339–8347

Beatman E, Oyer R, Shives KD, Hedman K, Brault AC, Tyler KL, Beckham JD (2012) West Nile virus growth is independent of autophagy activation. Virology 433:262–272

Bhatt S, Gething PW, Brady OJ, Messina JP, Farlow AW, Moyes CL, Drake JM, Brownstein JS, Hoen AG, Sankoh O, Myers MF, George DB, Jaenisch T, Wint GR, Simmons CP, Scott TW, Farrar JJ, Hay SI (2013) The global distribution and burden of dengue. Nature 496:504–507

Blumenthal T, Carmichael GG (1979) RNA replication: function and structure of Qbeta-replicase. Annu Rev Biochem 48:525–548

Byk LA, Gamarnik AV (2016) Properties and functions of the Dengue Virus Capsid Protein. Annu Rev Virol 3:263–281

Byk LA, Iglesias NG, De Maio FA, Gebhard LG, Rossi M, Gamarnik AV (2016) Dengue virus genome uncoating requires ubiquitination. MBio 7(3):e00804–e00816

Cabrera-Hernandez A, Thepparit C, Suksanpaisan L, Smith DR (2007) Dengue virus entry into liver (HepG2) cells is independent of hsp90 and hsp70. J Med Virol 79:386–392

Campos RK, Wong B, Xie X, Lu YF, Shi PY, Pompon J, Garcia-Blanco MA, Bradrick SS (2017) RPLP1 and RPLP2 are essential flavivirus host factors that promote early viral protein accumulation. J Virol 91(4):e01706–e01716

Carnec X, Meertens L, Dejarnac O, Perera-Lecoin M, Hafirassou ML, Kitaura J, Ramdasi R, Schwartz O, Amara A (2016) The Phosphatidylserine and Phosphatidylethanolamine receptor CD300a Binds Dengue Virus and enhances infection. J Virol 90:92–102

Cervantes-Salazar M, Angel-Ambrocio AH, Soto-Acosta R, Bautista-Carbajal P, Hurtado-Monzon AM, Alcaraz-Estrada SL, Ludert JE, del Angel RM (2015) Dengue virus NS1 protein interacts with the ribosomal protein RPL18: this interaction is required for viral translation and replication in Huh-7 cells. Virology 484:113–126

Chan KR, Ong EZ, Tan HC, Zhang SL, Zhang Q, Tang KF, Kaliaperumal N, Lim AP, Hibberd ML, Chan SH, Connolly JE, Krishnan MN, Lok SM, Hanson BJ, Lin CN, Ooi EE (2014) Leukocyte immunoglobulin-like receptor B1 is critical for antibody-dependent dengue. Proc Natl Acad Sci U S A 111:2722–2727

Chen ST, Lin YL, Huang MT, Wu MF, Cheng SC, Lei HY, Lee CK, Chiou TW, Wong CH, Hsieh SL (2008) CLEC5A is critical for dengue-virus-induced lethal disease. Nature 453: 672–676

Chen Y, Maguire T, Hileman RE, Fromm JR, Esko JD, Linhardt RJ, Marks RM (1997) Dengue virus infectivity depends on envelope protein binding to target cell heparan sulfate. Nat Med 3:866–871

Cheng G, Cox J, Wang P, Krishnan MN, Dai J, Qian F, Anderson JF, Fikrig E (2010) A C-type lectin collaborates with a CD45 phosphatase homolog to facilitate West Nile virus infection of mosquitoes. Cell 142:714–725

Chiu WW, Kinney RM, Dreher TW (2005) Control of translation by the 5′- and 3′-terminal regions of the dengue virus genome. J Virol 79:8303–8315

Choy MM, Zhang SL, Costa VV, Tan HC, Horrevorts S, Ooi EE (2015) Proteasome inhibition suppresses Dengue Virus egress in antibody dependent infection. PLoS Negl Trop Dis 9: e0004058

Clyde K, Harris E (2006) RNA secondary structure in the coding region of dengue virus type 2 directs translation start codon selection and is required for viral replication. J Virol 80: 2170–2182

Courageot MP, Frenkiel MP, dos Santos CD, Deubel V, Despres P (2000) Alpha-glucosidase inhibitors reduce dengue virus production by affecting the initial steps of virion morphogenesis in the endoplasmic reticulum. J Virol 74:564–572

Crabtree MB, Kinney RM, Miller BR (2005) Deglycosylation of the NS1 protein of dengue 2 virus, strain 16681: construction and characterization of mutant viruses. Arch Virol 150:771–786

D'Ortenzio E, Matheron S, Yazdanpanah Y, de Lamballerie X, Hubert B, Piorkowski G, Maquart M, Descamps D, Damond F, Leparc-Goffart I (2016) Evidence of sexual transmission of Zika Virus. N Engl J Med 374:2195–2198

Dalrymple N, Mackow ER (2011) Productive dengue virus infection of human endothelial cells is directed by heparan sulfate-containing proteoglycan receptors. J Virol 85:9478–9485

Das S, Laxminarayana SV, Chandra N, Ravi V, Desai A (2009) Heat shock protein 70 on Neuro2a cells is a putative receptor for Japanese encephalitis virus. Virology 385:47–57

Davis CW, Mattei LM, Nguyen HY, Ansarah-Sobrinho C, Doms RW, Pierson TC (2006) The location of asparagine-linked glycans on West Nile virions controls their interactions with CD209 (dendritic cell-specific ICAM-3 grabbing nonintegrin). J Biol Chem 281:37183–37194

Davis WG, Blackwell JL, Shi PY, Brinton MA (2007) Interaction between the cellular protein eEF1A and the 3'-terminal stem-loop of West Nile virus genomic RNA facilitates viral minus-strand RNA synthesis. J Virol 81:10172–10187

Dejnirattisai W, Supasa P, Wongwiwat W, Rouvinski A, Barba-Spaeth G, Duangchinda T, Sakuntabhai A, Cao-Lormeau VM, Malasit P, Rey FA, Mongkolsapaya J, Screaton GR (2016) Dengue virus sero-cross-reactivity drives antibody-dependent enhancement of infection with zika virus. Nat Immunol 17:1102–1108

Dejnirattisai W, Webb AI, Chan V, Jumnainsong A, Davidson A, Mongkolsapaya J, Screaton G (2011) Lectin switching during dengue virus infection. J Infect Dis 203:1775–1783

Devaux PF, Herrmann A, Ohlwein N, Kozlov MM (2008) How lipid flippases can modulate membrane structure. Biochim Biophys Acta 1778:1591–1600

Diaz A, Ahlquist P (2012) Role of host reticulon proteins in rearranging membranes for positive-strand RNA virus replication. Curr Opin Microbiol 15:519–524

Duan X, Lu X, Li J, Liu Y (2008) Novel binding between pre-membrane protein and vacuolar ATPase is required for efficient dengue virus secretion. Biochem Biophys Res Commun 373:319–324

Edgil D, Polacek C, Harris E (2006) Dengue virus utilizes a novel strategy for translation initiation when cap-dependent translation is inhibited. J Virol 80:2976–8296

Evans EA, Gilmore R, Blobel G (1986) Purification of microsomal signal peptidase as a complex. Proc Natl Acad Sci U S A 83:581–585

Evron T, Daigle TL, Caron MG (2012) GRK2: multiple roles beyond G protein-coupled receptor desensitization. Trends Pharmacol Sci 33:154–164

Flamand M, Megret F, Mathieu M, Lepault J, Rey FA, Deubel V (1999) Dengue virus type 1 nonstructural glycoprotein NS1 is secreted from mammalian cells as a soluble hexamer in a glycosylation-dependent fashion. J Virol 73:6104–6110

Friedman DI, Schauer AT, Baumann MR, Baron LS, Adhya SL (1981) Evidence that ribosomal protein S10 participates in control of transcription termination. Proc Natl Acad Sci U S A 78:1115–1118

Friedrich S, Schmidt T, Geissler R, Lilie H, Chabierski S, Ulbert S, Liebert UG, Golbik RP, Behrens SE (2014) AUF1 p45 promotes West Nile virus replication by an RNA chaperone activity that supports cyclization of the viral genome. J Virol 88:11586–11599

Friedrich S, Schmidt T, Schierhorn A, Lilie H, Szczepankiewicz G, Bergs S, GOLBIK RP, Behrens SE (2016) Arginine methylation enhances the RNA chaperone activity of the West Nile virus host factor AUF1 p45. Rna 22(10):1574–1591

Gamarnik AV, Andino R (1998) Switch from translation to RNA replication in a positive-stranded RNA virus. Genes Dev 12:2293–2304

Garcia-Blanco MA, Vasudevan SG, Bradrick SS, Nicchitta C (2016) Flavivirus RNA transactions from viral entry to genome replication. Antiviral Res 134:244–249

Garcia-Montalvo BM, Medina F, del Angel RM (2004) La protein binds to NS5 and NS3 and to the 5′ and 3′ ends of Dengue 4 virus RNA. Virus Res 102:141–150

Garske T, van Kerkhove MD, Yactayo S, Ronveaux O, Lewis RF, Staples JE, Perea W, Ferguson NM, Yellow Fever Expert Committee (2014) Yellow Fever in Africa: estimating the burden of disease and impact of mass vaccination from outbreak and serological data. PLoS Med 11:e1001638

Gomila RC, Martin GW, Gehrke L (2011) NF90 binds the dengue virus RNA 3′ terminus and is a positive regulator of dengue virus replication. PLoS ONE 6:e16687

Guirakhoo F, Hunt AR, Lewis JG, Roehrig JT (1993) Selection and partial characterization of dengue 2 virus mutants that induce fusion at elevated pH. Virology 194:219–223

Guo Y, Feinberg H, Conroy E, Mitchell DA, Alvarez R, Blixt O, Taylor ME, Weis WI, Drickamer K (2004) Structural basis for distinct ligand-binding and targeting properties of the receptors DC-SIGN and DC-SIGNR. Nat Struct Mol Biol 11:591–598

Hackett BA, Yasunaga A, Panda D, Tartell MA, Hopkins KC, Hensley SE, Cherry S (2015) RNASEK is required for internalization of diverse acid-dependent viruses. Proc Natl Acad Sci U S A 112:7797–7802

Hamel R, Dejarnac O, Wichit S, Ekchariyawat P, Neyret A, Luplertlop N, Perera-Lecoin M, Surasombatpattana P, Talignani L, Thomas F, Cao-Lormeau VM, Choumet V, Briant L, Despres P, Amara A, Yssel H, Misse D (2015) Biology of Zika Virus infection in human skin cells. J Virol 89:8880–8896

Hanna SL, Pierson TC, Sanchez MD, Ahmed AA, Murtadha MM, Doms RW (2005) N-linked glycosylation of west nile virus envelope proteins influences particle assembly and infectivity. J Virol 79:13262–13274

Harrower J, Kiedrzynski T, Baker S, Upton A, Rahnama F, Sherwood J, Huang QS, Todd A, Pulford D (2016) Sexual Transmission of Zika Virus and persistence in Semen, New Zealand, 2016. Emerg Infect Dis 22:1855–1857

Heaton NS, Perera R, Berger KL, Khadka S, Lacount DJ, Kuhn RJ, Randall G (2010) Dengue virus nonstructural protein 3 redistributes fatty acid synthase to sites of viral replication and increases cellular fatty acid synthesis. Proc Natl Acad Sci U S A 107:17345–17350

Heaton NS, Randall G (2010) Dengue virus-induced autophagy regulates lipid metabolism. Cell Host Microbe 8:422–432

Hirsch AJ, Medigeshi GR, Meyers HL, Defilippis V, Fruh K, Briese T, Lipkin WI, Nelson JA (2005) The Src family kinase c-Yes is required for maturation of West Nile virus particles. J Virol 79:11943–11951

Iglesias NG, Mondotte JA, Byk LA, de Maio FA, Samsa MM, Alvarez C, Gamarnik AV (2015) Dengue Virus uses a non-canonical function of the host GBF1-Arf-COPI system for capsid protein accumulation on lipid droplets. Traffic 16:962–977

Ivanov A, Mikhailova T, Eliseev B, Yeramala L, Sokolova E, Susorov D, Shuvalov A, Schaffitzel C, Alkalaeva E (2016) PABP enhances release factor recruitment and stop codon recognition during translation termination. Nucleic Acids Res 44:7766–7776

Ivanyi-Nagy R, Lavergne JP, Gabus C, Ficheux D, Darlix JL (2008) RNA chaperoning and intrinsic disorder in the core proteins of Flaviviridae. Nucleic Acids Res 36:712–725

Jan CH, Williams CC, Weissman JS (2014) Principles of ER cotranslational translocation revealed by proximity-specific ribosome profiling. Science 346:1257521

Jiang L, Yao H, Duan X, Lu X, Liu Y (2009) Polypyrimidine tract-binding protein influences negative strand RNA synthesis of dengue virus. Biochem Biophys Res Commun 385:187–192

Jindadamrongwech S, Thepparit C, Smith DR (2004) Identification of GRP 78 (BiP) as a liver cell expressed receptor element for dengue virus serotype 2. Arch Virol 149:915–927

Jonikas MC, Collins SR, Denic V, Oh E, Quan EM, Schmid V, Weibezahn J, Schwappach B, Walter P, Weissman JS, Schuldiner M (2009) Comprehensive characterization of genes required for protein folding in the endoplasmic reticulum. Science 323:1693–1697

Junjhon J, Pennington JG, Edwards TJ, Perera R, Lanman J, Kuhn RJ (2014) Ultrastructural characterization and three-dimensional architecture of replication sites in dengue virus-infected mosquito cells. J Virol 88:4687–4697

Kahvejian A, Svitkin YV, Sukarieh R, M'Boutchou MN, Sonenberg N (2005) Mammalian poly (A)-binding protein is a eukaryotic translation initiation factor, which acts via multiple mechanisms. Genes Dev 19:104–113

Kalia M, Khasa R, Sharma M, Nain M, Vrati S (2013) Japanese encephalitis virus infects neuronal cells through a clathrin-independent endocytic mechanism. J Virol 87:148–162

Kim J, Chubatsu LS, Admon A, Stahl J, Fellous R, Linn S (1995) Implication of mammalian ribosomal protein S3 in the processing of DNA damage. J Biol Chem 270:13620–13629

Kim JM, Yun SI, Song BH, Hahn YS, Lee CH, Oh HW, Lee YM (2008) A single N-linked glycosylation site in the Japanese encephalitis virus prM protein is critical for cell type-specific prM protein biogenesis, virus particle release, and pathogenicity in mice. J Virol 82:7846–7862

Kobayashi N, Karisola P, Pena-Cruz V, Dorfman DM, Jinushi M, Umetsu SE, Butte MJ, Nagumo H, Chernova I, Zhu B, Sharpe AH, Ito S, Dranoff G, Kaplan GG, Casasnovas JM, Umetsu DT, Dekruyff RH, Freeman GJ (2007) TIM-1 and TIM-4 glycoproteins bind phosphatidylserine and mediate uptake of apoptotic cells. Immunity 27:927–940

Kobayashi S, Orba Y, Yamaguchi H, Takahashi K, Sasaki M, Hasebe R, Kimura T, Sawa H (2014) Autophagy inhibits viral genome replication and gene expression stages in West Nile virus infection. Virus Res 191:83–91

Kobayashi S, Suzuki T, Kawaguchi A, Phongphaew W, Yoshii K, Iwano T, Harada A, Kariwa H, Orba Y, Sawa H (2016) Rab8b Regulates Transport of West Nile Virus Particles from Recycling Endosomes. J Biol Chem 291:6559–6568

Kooijman EE, Chupin V, de Kruijff B, Burger KN (2003) Modulation of membrane curvature by phosphatidic acid and lysophosphatidic acid. Traffic 4:162–174

Kostyuchenko VA, Lim EX, Zhang S, Fibriansah G, Ng TS, Ooi JS, Shi J, Lok SM (2016) Structure of the thermally stable Zika virus. Nature 533:425–428

Krishnan MN, Ng A, Sukumaran B, Gilfoy FD, Uchil PD, Sultana H, Brass AL, Adametz R, Tsui M, Qian F, Montgomery RR, Lev S, Mason PW, Koski RA, Elledge SJ, Xavier RJ, Agaisse H, Fikrig E (2008) RNA interference screen for human genes associated with West Nile virus infection. Nature 455:242–245

Krishnan MN, Sukumaran B, Pal U, Agaisse H, Murray JL, Hodge TW, Fikrig E (2007) Rab 5 is required for the cellular entry of dengue and West Nile viruses. J Virol 81:4881–4885

Kroschewski H, Allison SL, Heinz FX, Mandl CW (2003) Role of heparan sulfate for attachment and entry of tick-borne encephalitis virus. Virology 308:92–100
Kudelko M, Brault JB, Kwok K, Li MY, Pardigon N, Peiris JS, Bruzzone R, Despres P, Nal B, Wang PG (2012) Class II ADP-ribosylation factors are required for efficient secretion of dengue viruses. J Biol Chem 287:767–777
Kutay U, Lipowsky G, Izaurralde E, Bischoff FR, Schwarzmaier P, Hartmann E, Gorlich D (1998) Identification of a tRNA-specific nuclear export receptor. Mol Cell 1:359–369
Landry DM, Hertz MI, Thompson SR (2009) RPS25 is essential for translation initiation by the Dicistroviridae and hepatitis C viral IRESs. Genes Dev 23:2753–2764
le Sommer C, Barrows NJ, Bradrick SS, Pearson JL, Garcia-Blanco MA (2012) G protein-coupled receptor kinase 2 promotes flaviviridae entry and replication. PLoS Negl Trop Dis 6:e1820
Lee E, Lobigs M (2000) Substitutions at the putative receptor-binding site of an encephalitic flavivirus alter virulence and host cell tropism and reveal a role for glycosaminoglycans in entry. J Virol 74:8867–8875
Li L, Lok SM, Yu IM, Zhang Y, Kuhn RJ, Chen J, Rossmann MG (2008) The flavivirus precursor membrane-envelope protein complex: structure and maturation. Science 319:1830–1834
Li MY, Grandadam M, Kwok K, Lagache T, Siu YL, Zhang JS, Bruzzone R, Wang PG (2015) KDEL receptors assist dengue virus exit from the endoplasmic reticulum. Cell reports 10 (9):1496–1507
LINDENBACH B, THIEL H-J, RICE C (2013) Flaviviridae: the viruses and their replication. In: Fields BN, Knipe DM, Howley PM. (eds) Fields virology. 6th edn. Wolters Kluwer Health/Lippincott, Williams & Wilkins, Philadelphia
Liu H, Chiou SS, Chen WJ (2004) Differential binding efficiency between the envelope protein of Japanese encephalitis virus variants and heparan sulfate on the cell surface. J Med Virol 72:618–624
Liu Y, Zhang F, Liu J, Xiao X, Zhang S, Qin C, Xiang Y, Wang P, Cheng G (2014) Transmission-blocking antibodies against mosquito C-type lectins for dengue prevention. PLoS Pathog 10:e1003931
Lobigs M, Lee E, Ng ML, Pavy M, Lobigs P (2010) A flavivirus signal peptide balances the catalytic activity of two proteases and thereby facilitates virus morphogenesis. Virology 401:80–89
Louie RJ, Guo J, Rodgers JW, White R, Shah N, Pagant S, Kim P, Livstone M, Dolinski K, McKinney BA, Hong J, Sorscher EJ, Bryan J, Miller EA, Hartman JLT (2012) A yeast phenomic model for the gene interaction network modulating CFTR-DeltaF508 protein biogenesis. Genome Med 4:103
Lozach PY, Burleigh L, Staropoli I, Navarro-Sanchez E, Harriague J, Virelizier JL, Rey FA, Despres P, Arenzana-Seisdedos F, Amara A (2005) Dendritic cell-specific intercellular adhesion molecule 3-grabbing non-integrin (DC-SIGN)-mediated enhancement of dengue virus infection is independent of DC-SIGN internalization signals. J Biol Chem 280:23698–23708
Luca VC, Nelson CA, Fremont DH (2013) Structure of the St. Louis encephalitis virus postfusion envelope trimer. J Virol 87:818–828
Ma H, Dang Y, Wu Y, Jia G, Anaya E, Zhang J, Abraham S, Choi JG, Shi G, Qi L, Manjunath N, Wu H (2015) A CRISPR-Based Screen Identifies Genes Essential for West-Nile-Virus-Induced Cell Death. Cell Rep 12:673–683
Marceau CD, Puschnik AS, Majzoub K, Ooi YS, Brewer SM, Fuchs G, Swaminathan K, Mata MA, Elias JE, Sarnow P, Carette JE (2016) Genetic dissection of Flaviviridae host factors through genome-scale CRISPR screens. Nature 535:159–163
Mateo R, Nagamine CM, Spagnolo J, Méndez E, Rahe M, Gale M, Yuan J, Kirkegaard K (2013) Inhibition of cellular autophagy deranges dengue virion maturation. J Virol 87(3):1312–1321
Meertens L, Carnec X, Lecoin MP, Ramdasi R, Guivel-Benhassine F, Lew E, Lemke G, Schwartz O, Amara A (2012) The TIM and TAM families of phosphatidylserine receptors mediate dengue virus entry. Cell Host Microbe 12:544–557

Miller JL, de Wet BJ, Martinez-Pomares L, Radcliffe CM, Dwek RA, Rudd PM, Gordon S (2008) The mannose receptor mediates dengue virus infection of macrophages. PLoS Pathog 4:e17
Miner JJ, Sene A, Richner JM, Smith AM, Santeford A, Ban N, Weger-Lucarelli J, Manzella F, Ruckert C, Govero J, Noguchi KK, Ebel GD, Diamond MS, Apte RS (2016) Zika Virus Infection in Mice Causes Panuveitis with Shedding of Virus in Tears. Cell Rep 16:3208–3218
Moller-Tank S, Maury W (2014) Phosphatidylserine receptors: enhancers of enveloped virus entry and infection. Virology 468–470:565–580
Mondotte JA, Lozach PY, Amara A, Gamarnik AV (2007) Essential role of dengue virus envelope protein N glycosylation at asparagine-67 during viral propagation. J Virol 81:7136–7148
Mukhopadhyay S, Kim BS, Chipman PR, Rossmann MG, Kuhn RJ (2003) Structure of West Nile virus. Science 302:248
Muylaert IR, Chambers TJ, Galler R, Rice CM (1996) Mutagenesis of the N-linked glycosylation sites of the yellow fever virus NS1 protein: effects on virus replication and mouse neurovirulence. Virology 222:159–168
Nagy PD, Strating JR, van Kuppeveld FJ (2016) Building Viral Replication Organelles: Close Encounters of the Membrane Types. PLoS Pathog 12:e1005912
Navarro-Sanchez E, Altmeyer R, Amara A, Schwartz O, Fieschi F, Virelizier JL, Arenzana-Seisdedos F, Despres P (2003) Dendritic-cell-specific ICAM3-grabbing non-integrin is essential for the productive infection of human dendritic cells by mosquito-cell-derived dengue viruses. EMBO Rep 4:723–728
Nour AM, Li Y, Wolenski J, Modis Y (2013) Viral membrane fusion and nucleocapsid delivery into the cytoplasm are distinct events in some flaviviruses. PLoS Pathog 9:e1003585
Okamoto K, Kinoshita H, del Carmen Parquet M, Raekiansyah M, Kimura D, Yui K, Islam MA, Hasebe F, Morita K (2012) Dengue virus strain DEN2 16681 utilizes a specific glycochain of syndecan-2 proteoglycan as a receptor. J Gen Virol 93(4):761–770
Padwad YS, Mishra KP, Jain M, Chanda S, Ganju L (2010) Dengue virus infection activates cellular chaperone Hsp70 in THP-1 cells: downregulation of Hsp70 by siRNA revealed decreased viral replication. Viral Immunol 23:557–565
Pena J, Harris E (2012) Early dengue virus protein synthesis induces extensive rearrangement of the endoplasmic reticulum independent of the UPR and SREBP-2 pathway. PLoS ONE 7: e38202
Perera-Lecoin M, Meertens L, Carnec X, Amara A (2014) Flavivirus entry receptors: an update. Viruses 6:69–88
Perera R, Riley C, Isaac G, Hopf-Jannasch AS, Moore RJ, Weitz KW, Pasa-Tolic L, Metz TO, Adamec J, Kuhn RJ (2012) Dengue virus infection perturbs lipid homeostasis in infected mosquito cells. PLoS Pathog 8:e1002584
Perreira JM, Aker AM, Savidis G, Chin CR, McDougall WM, Portmann JM, Meraner P, Smith MC, Rahman M, Baker RE, Gauthier A, Franti M, Brass AL (2015) RNASEK Is a V-ATPase-Associated Factor Required for Endocytosis and the Replication of Rhinovirus, Influenza A Virus, and Dengue Virus. Cell Rep 12:850–863
Pokidysheva E, Zhang Y, Battisti AJ, Bator-Kelly CM, Chipman PR, Xiao C, Gregorio GG, Hendrickson WA, Kuhn RJ, Rossmann MG (2006) Cryo-EM reconstruction of dengue virus in complex with the carbohydrate recognition domain of DC-SIGN. Cell 124:485–493
Polacek C, Friebe P, Harris E (2009) Poly(A)-binding protein binds to the non-polyadenylated 3' untranslated region of dengue virus and modulates translation efficiency. J Gen Virol 90:687–692
Reyes-del Valle J, Chávez-Salinas S, Medina F, del Angel RM (2005) Heat shock protein 90 and heat shock protein 70 are components of dengue virus receptor complex in human cells. J Virol 79(8):4557–4567
Richard M, Boulin T, Robert VJ, Richmond JE, Bessereau JL (2013) Biosynthesis of ionotropic acetylcholine receptors requires the evolutionarily conserved ER membrane complex. Proc Natl Acad Sci U S A 110:E1055–E1063
Roehrig JT, Butrapet S, Liss NM, Bennett SL, Luy BE, Childers T, Boroughs KL, Stovall JL, Calvert AE, Blair CD, Huang CY (2013) Mutation of the dengue virus type 2 envelope protein

heparan sulfate binding sites or the domain III lateral ridge blocks replication in Vero cells prior to membrane fusion. Virology 441:114–125

Romero-Brey I, Bartenschlager R (2014) Membranous replication factories induced by plus-strand RNA viruses. Viruses 6:2826–2857

Samsa MM, Mondotte JA, Iglesias NG, Assuncao-Miranda I, Barbosa-Lima G, da Poian AT, Bozza PT, Gamarnik AV (2009) Dengue virus capsid protein usurps lipid droplets for viral particle formation. PLoS Pathog 5:e1000632

Satoh T, Ohba A, Liu Z, Inagaki T, Satoh AK (2015) dPob/EMC is essential for biosynthesis of rhodopsin and other multi-pass membrane proteins in Drosophila photoreceptors. Elife 4: e06306

Savidis G, McDougall WM, Meraner P, Perreira JM, Portmann JM, Trincucci G, John SP, Aker AM, Renzette N, Robbins DR, Guo Z, Green S, Kowalik TF, Brass AL (2016) Identification of Zika Virus and Dengue Virus Dependency Factors using Functional Genomics. Cell Rep 16:232–246

Seitz HM, Camenisch TD, Lemke G, Earp HS, Matsushima GK (2007) Macrophages and dendritic cells use different Axl/Mertk/Tyro3 receptors in clearance of apoptotic cells. J Immunol 178:5635–5642

Sessions OM, Barrows NJ, Souza-Neto JA, Robinson TJ, Hershey CL, Rodgers MA, Ramirez JL, Dimopoulos G, Yang PL, Pearson JL, Garcia-Blanco MA (2009) Discovery of insect and human dengue virus host factors. Nature 458:1047–1050

Sirohi D, Chen Z, Sun L, Klose T, Pierson TC, Rossmann MG, Kuhn RJ (2016) The 3.8 A resolution cryo-EM structure of Zika virus. Science 352:467–470

Smit JM, Moesker B, Rodenhuis-Zybert I, Wilschut J (2011) Flavivirus cell entry and membrane fusion. Viruses 3:160–171

Somnuke P, Hauhart RE, Atkinson JP, Diamond MS, Avirutnan P (2011) N-linked glycosylation of dengue virus NS1 protein modulates secretion, cell-surface expression, hexamer stability, and interactions with human complement. Virology 413:253–264

Stadler K, Allison SL, Schalich J, Heinz FX (1997) Proteolytic activation of tick-borne encephalitis virus by furin. J Virol 71:8475–8481

Tabata K, Arimoto M, Arakawa M, Nara A, Saito K, Omori H, Arai A, Ishikawa T, Konishi E, Suzuki R, Matsuura Y, Morita E (2016a) Unique Requirement for ESCRT Factors in Flavivirus Particle Formation on the Endoplasmic Reticulum. Cell Rep 16:2339–2347

Tabata T, Petitt M, Puerta-Guardo H, Michlmayr D, Wang C, Fang-Hoover J, Harris E, Pereira L (2016b) Zika Virus Targets Different Primary Human Placental Cells, Suggesting Two Routes for Vertical Transmission. Cell Host Microbe 20:155–166

Taguwa S, Maringer K, Li X, Bernal-Rubio D, Rauch JN, Gestwicki JE, Andino R, Fernandez-Sesma A, Frydman J (2015) Defining Hsp70 Subnetworks in Dengue Virus Replication Reveals Key Vulnerability in Flavivirus Infection. Cell 163:1108–1123

Tassaneetrithep B, Burgess TH, Granelli-Piperno A, Trumpfheller C, Finke J, Sun W, Eller MA, Pattanapanyasat K, Sarasombath S, Birx DL, Steinman RM, Schlesinger S, Marovich MA (2003) DC-SIGN (CD209) mediates dengue virus infection of human dendritic cells. J Exp Med 197:823–829

Thepparit C, Smith DR (2004) Serotype-specific entry of dengue virus into liver cells: identification of the 37-kilodalton/67-kilodalton high-affinity laminin receptor as a dengue virus serotype 1 receptor. J Virol 78:12647–12656

Tio PH, Jong WW, Cardosa MJ (2005) Two dimensional VOPBA reveals laminin receptor (LAMR1) interaction with dengue virus serotypes 1, 2 and 3. Virol J 2:25

Tung YT, Wu MF, Wang GJ, Hsieh SL (2014) Nanostructured electrochemical biosensor for th0065 detection of the weak binding between the dengue virus and the CLEC5A receptor. Nanomedicine 10:1335–1341

Uchida L, Espada-Murao LA, Takamatsu Y, Okamoto K, Hayasaka D, Yu F, Nabeshima T, Buerano CC, Morita K (2014) The dengue virus conceals double-stranded RNA in the intracellular membrane to escape from an interferon response. Sci Rep 4:7395 van der Schaar HM, Rust MJ, Chen C, Van der Ende-Metselaar H, Wilschut J, Zhuang X, Smit JM (2008) Dissecting the cell entry pathway of dengue virus by single-particle tracking in living cells. PLoS Pathog 4:e1000244

Vashist S, Anantpadma M, Sharma H, Vrati S (2009) La protein binds the predicted loop structures in the 3' non-coding region of Japanese encephalitis virus genome: role in virus replication. J Gen Virol 90:1343–1352

Walter P, Blobel G (1981a) Translocation of proteins across the endoplasmic reticulum III. Signal recognition protein (SRP) causes signal sequence-dependent and site-specific arrest of chain elongation that is released by microsomal membranes. J Cell Biol 91:557–561

Walter P, Blobel G (1981b) Translocation of proteins across the endoplasmic reticulum. II. Signal recognition protein (SRP) mediates the selective binding to microsomal membranes of in-vitro-assembled polysomes synthesizing secretory protein. J Cell Biol 91:551–556

Walter P, Ibrahimi I, Blobel G (1981) Translocation of proteins across the endoplasmic reticulum. I. Signal recognition protein (SRP) binds to in-vitro-assembled polysomes synthesizing secretory protein. J Cell Biol 91:545–550

Walter P, Johnson AE (1994) Signal sequence recognition and protein targeting to the endoplasmic reticulum membrane. Annu Rev Cell Biol 10:87–119

Wang P, Hu K, Luo S, Zhang M, Deng X, Li C, Jin W, Hu B, He S, Li M, Du T, Xiao G, Zhang B, Liu Y, Hu Q (2016) DC-SIGN as an attachment factor mediates Japanese encephalitis virus infection of human dendritic cells via interaction with a single high-mannose residue of viral E glycoprotein. Virology 488:108–119

Ward AM, Bidet K, Yinglin A, Ler SG, Hogue K, Blackstock W, Gunaratne J, Garcia-Blanco MA (2011) Quantitative mass spectrometry of DENV-2 RNA-interacting proteins reveals that the DEAD-box RNA helicase DDX6 binds the DB1 and DB2 3' UTR structures. RNA Biol 8:1173–1186

Ward AM, Calvert ME, Read LR, Kang S, Levitt BE, Dimopoulos G, Bradrick SS, Gunaratne J, Garcia-Blanco MA (2016) The Golgi associated ERI3 is a Flavivirus host factor. Sci Rep 6:34379

Weaver SC, Costa F, Garcia-Blanco MA, Ko AI, Ribeiro GS, Saade G, Shi PY, Vasilakis N (2016) Zika virus: history, emergence, biology, and prospects for control. Antiviral research. 30(130):69–80

Wells MF, Salick MR, Wiskow O, Ho DJ, Worringer KA, Ihry RJ, Kommineni S, Bilican B, Klim JR, Hill EJ, Kane LT, Ye C, Kaykas A, Eggan K (2016) Genetic Ablation of AXL Does Not Protect Human Neural Progenitor Cells and Cerebral Organoids from Zika Virus Infection. Cell Stem Cell 19:703–708

Welsch S, Miller S, Romero-Brey I, Merz A, Bleck CK, Walther P, Fuller SD, Antony C, Krijnse-Locker J, Bartenschlager R (2009) Composition and three-dimensional architecture of the dengue virus replication and assembly sites. Cell Host Microbe 5:365–375

Wilder-Smith A, Byass P (2016) The elusive global burden of dengue. Lancet Infect Dis 16 (6):629–631

Xu Z, Hobman TC (2012) The helicase activity of DDX56 is required for its role in assembly of infectious West Nile virus particles. Virology 433:226–235

Ye J, Chen Z, Zhang B, Miao H, Zohaib A, Xu Q, Chen H, Cao S (2013) Heat shock protein 70 is associated with replicase complex of Japanese encephalitis virus and positively regulates viral genome replication. PLoS ONE 8:e75188

Yu IM, Zhang W, Holdaway HA, Li L, Kostyuchenko VA, Chipman PR, Kuhn RJ, Rossmann MG, Chen J (2008) Structure of the immature dengue virus at low pH primes proteolytic maturation. Science 319:1834–1837

Zaitseva E, Yang ST, Melikov K, Pourmal S, Chernomordik LV (2010) Dengue virus ensures its fusion in late endosomes using compartment-specific lipids. PLoS Pathog 6:e1001131

Zeisel MB, Felmlee DJ, Baumert TF (2013) Hepatitis C virus entry. Curr Top Microbiol Immunol 369:87–112

Zhang R, Miner JJ, Gorman MJ, Rausch K, Ramage H, White JP, Zuiani A, Zhang P, Fernandez E, Zhang Q, Dowd KA, Pierson TC, Cherry S, Diamond MS (2016) A CRISPR screen defines a signal peptide processing pathway required by flaviviruses. Nature 535: 164–168

Zhang X, Ge P, Yu X, Brannan JM, Bi G, Zhang Q, Schein S, Zhou ZH (2013) Cryo-EM structure of the mature dengue virus at 3.5-A resolution. Nat Struct Mol Biol 20:105–110

Zhou X, Liao WJ, Liao JM, Liao P, Lu H (2015) Ribosomal proteins: functions beyond the ribosome. J Mol Cell Biol 7:92–104

Zhouravleva G, Frolova L, le Goff X, le Guellec R, Inge-Vechtomov S, Kisselev L, Philippe M (1995) Termination of translation in eukaryotes is governed by two interacting polypeptide chain release factors, eRF1 and eRF3. EMBO J 14:4065–4072

Zybert IA, Van der Ende-Metselaar H, Wilschut J, Smit JM (2008) Functional importance of dengue virus maturation: infectious properties of immature virions. J Gen Virol 89:3047–3051

Role of Innate Genes in HIV Replication

Kerstin Schott, Maximilian Riess and Renate König

Abstract Cells use an elaborate innate immune surveillance and defense system against virus infections. Here, we discuss recent studies that reveal how HIV-1 is sensed by the innate immune system. Furthermore, we present mechanisms on the counteraction of HIV-1. We will provide an overview how HIV-1 actively utilizes host cellular factors to avoid sensing. Additionally, we will summarize effectors of the innate response that provide an antiviral cellular state. HIV-1 has evolved passive mechanism to avoid restriction and to regulate the innate response. We review in detail two prominent examples of these cellular factors: (i) NLRX1, a negative regulator of the innate response that HIV-1 actively usurps to block cytosolic innate sensing; (ii) SAMHD1, a restriction factor blocking the virus at the reverse transcription step that HIV-1 passively avoids to escape sensing.

Contents

Kerstin Schott and Maximilian Riess contributed equally.

K. Schott · M. Riess · R. König (✉)
Host-Pathogen Interactions, Paul-Ehrlich-Institute, 63225 Langen, Germany
e-mail: renate.koenig@pei.de

R. König
Immunity and Pathogenesis Program, Sanford Burnham Prebys Medical Discovery Institute, La Jolla, CA 92037, USA

R. König
German Center for Infection Research (DZIF), 63225 Langen, Germany

Current Topics in Microbiology and Immunology (2018) 419:69–112
DOI 10.1007/82_2017_29

Published Online: 08 July 2017

1 Introduction

Cells use an elaborate surveillance system to detect incoming viruses so they can respond quickly to foreign invaders. Induction of these intracellular innate immune defense mechanisms strongly relies on a variety of cellular pattern recognition receptors (PRRs) to detect viral proteins or nucleic acid structures, called pathogen-associated patterns (PAMPs). The recognition of these patterns results in the activation of inflammatory responses and the interferon (IFN) system. IFN mediates its antiviral effects by inducing the expression of antiviral interferon-stimulated genes (ISGs). The PRRs include membranous Toll-like receptors (TLRs), cytoplasmic RIG-I-like receptors (RLRs), NOD-like receptors (NLRs), and DNA sensors that signal through the adapter protein stimulator of interferon genes (STING) [reviewed in Takeuchi and Akira (2010), Chen et al. (2016)]. They survey almost every cellular compartment and alert neighboring cells, building up a protective antiviral state.

Interestingly, even so human immunodeficiency virus (HIV)-1 is highly sensitive to the action of IFN and can be inhibited by antiviral ISGs, the virus is a poor inducer of IFN [reviewed in Doyle et al. (2015), McMichael et al. (2010)]. For many years, it has been unknown how the virus evades the detection of the innate immune system. More than a decade ago, the discovery of the first restriction factors, and within recent years, the identification of intracellular DNA sensors, paved the way to unravel the conundrum of HIV-1 sensing and restriction.

Here, in this chapter, we provide a brief summary of how HIV-1 is sensed by the innate immune system. Moreover, we provide recent discoveries that cytosolic sensing is a major contributor in monocyte-derived dendritic cells (MDDCs) and other myeloid cells mediated by the cyclic GMP-AMP synthase (cGAS) together with polyglutamine binding protein 1 (PQBP1). We will provide an overview how HIV-1 utilizes host cellular factors to avoid sensing by either degrading excess PAMPs or shielding its capsid and cargo from premature uncoating and sensing in the cytosol. Additionally, we will summarize known restriction factors providing, as effectors of the innate response, an antiviral cellular state. We review in detail two prominent examples of cellular factors that recently got into research focus. The first one, NLR family member X1 (NLRX1), a member of the NOD-like receptor family, is a dependency factor that HIV-1 actively usurps to block cytosolic innate sensing. Second, it seems that HIV-1 passively avoids infection of myeloid cells that contain enzymatically active SAM and HD domain containing deoxynucleoside triphosphate triphosphohydrolase 1 (SAMHD1), a restriction factor that blocks reverse transcription of the virus, to prevent innate immune responses.

2 Sensors of the Innate Response

2.1 TLR Sensors

The presence of "non-host" nucleic acid structures upon viral infection suggests that nucleic acid sensing PRRs are important components of the innate immune surveillance machinery. The first and acute cell responders to viral infections are plasmacytoid dendritic cells (pDCs) (Swiecki and Colonna 2015), that express TLR7/8 and TLR9 in their endosomal compartments responding to single-stranded RNA (ssRNA) and CpG-DNA, respectively and mediate an IFNα response mainly through interferon regulatory factor 7 (IRF7) and nuclear factor kappa B (NF-κB) (Takeuchi and Akira 2010). In fact, pDCs respond rapidly to HIV-1 (Yonezawa et al. 2003), this response is endosome-dependent (Schmidt et al. 2005) and seems to be more potent when HIV transfer occurs through cell-to-cell contact rather than by endocytosis of HIV virions (Lepelley et al. 2011). HIV virions contain two copies of single-stranded RNA that are non-covalently linked and can build a self-complementary loop sequence resulting in short double-stranded RNA (dsRNA) regions (Paillart et al. 1996). It has been suggested that HIV-1 ssRNA represents a physiological ligand to stimulate TLR7/8 (Heil et al. 2004; Meier et al. 2007). There is a substantial body of evidence indicating that TLR7/8 signaling pathways regulate HIV-1 infection in pDCs (Beignon et al. 2005; Lepelley et al. 2011), though TLR9 may play a role as well (Mandl et al. 2008; Beignon et al. 2005). Besides "non-host" nucleic acids, viral proteins, such as HIV-1 envelope gp120, have been reported to trigger proinflammatory responses through TLR2 and TLR4 (Nazli et al. 2013). The importance of TLR signaling in the outcome of HIV-1 infections is reflected by reported associations between specific polymorphisms of TLR4, TLR7, and TLR9 genes and HIV disease progression and/or higher viral loads (Oh et al. 2009; Pine et al. 2009).

2.2 Cytosolic Sensors

The HIV life cycle suggests that "non-host" nucleic structures or viral proteins could be detected in the cytosol as well. HIV-1 fuses with the plasma membrane and releases its capsid into the cytoplasm. The capsid disintegrates and could potentially expose viral RNA and reverse transcribed DNA before the latter is transported to the nucleus and integrated into the host genome. During these processes, viral nucleic acids and reverse transcription intermediates (RTIs) such as ssRNA, RNA:DNA hybrids, ssDNA, and dsDNA could potentially serve as PAMPs for cytosolic sensors.

Notably, studies demonstrated that IRF3-dependent pathways participate in HIV recognition (Lepelley et al. 2011; Doehle et al. 2009; Manel et al. 2010). IRF3 is a master regulator of cytosolic responses known to regulate expression of IFNβ and a subset of ISGs independent of the action of IFN (Hiscott 2007; Honda et al. 2006). Interestingly, HIV seems to counteract the IRF3 pathway (Doehle et al. 2009, 2012; Okumura et al. 2008; Hotter et al. 2013) though this has been controversially reported and discussed (Rustagi and Gale 2014; Hotter et al. 2013). Still, these data hint toward the importance of cytoplasmic sensors.

Viral RNA can be sensed by cytosolic DexD/H box RNA helicases RIG-I and melanoma differentiation-associated protein 5 (MDA5) that recognize shorter dsRNA structures or ssRNA containing 5'triphosphates and longer dsRNA structures, respectively, and signal through the adapter mitochondrial antiviral-signaling protein (MAVS) (Schlee 2013). Some reports suggest the involvement of the RIG-I/MAVS pathway in controlling HIV-1 replication (Wang et al. 2013; Gupta et al. 2016). RIG-I has been demonstrated to recognize secondary structured HIV-derived RNA to trigger a MAVS and IRF3-dependent response (Berg et al. 2012). HIV-1 counters the response by sequestration of RIG-I with a HIV protease-dependent mechanism (Solis et al. 2011). These observations have not been confirmed so far. Newest evidence suggests that the RNA helicase DDX3 serves as a sensor for abortive HIV-1 RNA in MDDCs via the adaptor protein MAVS (Gringhuis et al. 2017).

In recent years, it became clear that cytosolic DNA can activate IRF3-dependent innate immune responses (Stetson and Medzhitov 2006). A plethora of novel DNA sensors have been identified since then (Dempsey and Bowie 2015). For the first time, in 2010, a cell-intrinsic recognition of HIV was reported in MDDCs resulting in activation of IRF3 (Manel et al. 2010) suggesting a novel cytosolic sensor. Intriguingly, a screening approach identified a number of host factors involved in retroviral sensing including STING, TANK-binding kinase 1 (TBK1), and IRF3 (Lee et al. 2013). Not long after, the cyclic GMP-AMP synthase (cGAS) was identified as a novel cytosolic DNA sensor (Sun et al. 2013) that was capable of activating the adapter protein STING and IRF3 through the synthesis of the second messenger cyclic GMP-AMP (cGAMP) (Wu et al. 2013). In fact, Gao et al. could prove that cGAS can sense HIV-1 and other retroviruses in myeloid cells (Gao et al. 2013). Interestingly, they reported that inhibitors of reverse transcriptase, but not integrase, abrogated IFNβ induction by the virus suggesting that cytosolic RTIs might be the PAMPs for recognition through cGAS, but a direct association with the viral PAMP has not been demonstrated. Very recently, a targeted RNAi screen in MDDCs revealed PQBP1 as a proximal sensor for HIV-1 (Yoh et al. 2015) (Fig. 1). The authors demonstrated that PQBP1 directly binds to reverse transcribed

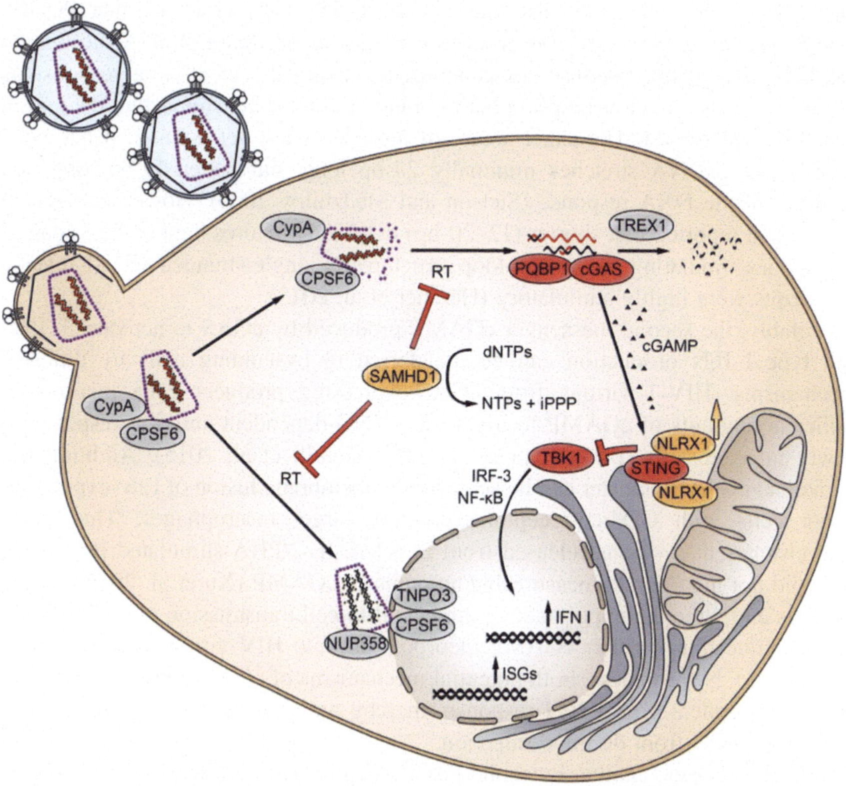

Fig. 1 HIV-1 exploits cellular factors to escape innate immune sensing. After fusion with a target cell, HIV-1 capsid is released into the cytoplasm and a timely and spatially highly organized process is initiated. Host cell factors like cyclophilins (CypA) cloak the viral capsid and aid in preserving its structural integrity on its travel to the nucleus. Reverse transcription takes place *en route* or at the nuclear pore complex, where the reverse transcribed genome is translocated into the nucleus with help of several nuclear pore complex proteins including NUP358 and CPSF6 together with TNPO3 involved in orchestrating this process. These processes are susceptible to perturbations resulting in unproductive infections. Destabilization of the viral capsid within the cytoplasm due to viral capsid mutations may result in premature loss of capsid structural integrity and exposure of viral PAMPS (reverse transcription intermediates, RTIs) to cellular cytoplasmic PRRs. Sensing of the viral PAMP by PQBP1 and cGAS initiates production of second messenger cGAMP by cGAS, which activates STING triggering TBK1 to induce downstream pathways NF-κB and IRF-3. These initiate production of mainly IFNβ followed by expression of IFN stimulated genes (ISGs) and antiviral restriction factors. Cellular proteins are exploited by the virus to prevent recognition by innate PRRs: SAMHD1 not only acts as an innate antiviral restriction factor, but it also supports the virus in staying under the radar of innate immunity. TREX1 is a cytoplasmic nuclease that degrades excess cytoplasmic DNA. Cytoplasmic or mitochondria-associated ER membranes localized NLRX1 is a negative regulator of the central innate sensing node STING. Upregulated NLRX1 activity aids the virus in attenuating antiviral innate responses

HIV-1 DNA and physically associates with cGAS. They conclude that PQBP1 would function as a specific co-sensor to retroviruses upstream of cGAS to initiate the IRF3-dependent response. In concordance with Gao et al., the response is dependent on reverse transcription but not integration, further supporting the notion that RTIs are sensed. The exact nature of the RTI is not yet defined. It has been known that dsDNA stretches minimally 24 bp long, but better 45 bp long, are inducers of the DNA response (Stetson and Medzhitov 2006). However, a recent publication reported that shorter (12–20 bp) dsDNA structures flanked by unpaired guanosines simulating short stem-loop structures in single-stranded HIV-1 reverse transcripts were highly stimulatory (Herzner et al. 2015).

Notably, the second messenger cGAMP produced by cGAS to activate STING and type I IFN production can be transferred to bystanding cells by different mechanisms. HIV-1 virions from cGAS-expressing producer cells incorporate sufficient amounts of cGAMP to trigger a STING-dependent antiviral response in newly infected cells (Bridgeman et al. 2015; Gentili et al. 2015). Additionally, horizontal cGAMP transfer can be achieved by membrane fusion of Env-expressing donor cells with CD4/co-receptor-expressing target macrophages. This study demonstrates that virions released from infected, IL-2/PHA-stimulated primary T cells did not incorporate measureable amounts of cGAMP (Xu et al. 2016). These findings underline the importance of direct cell-to-cell transmission of cGAMP and raise the question whether cGAMP incorporation into HIV virions might be relevant in vivo. Nevertheless, both potential mechanisms of cGAMP transfer induce a STING-dependent type I IFN response, thereby antiviral state that could protect HIV target cells from de novo infection.

In macrophages, another cytosolic DNA receptor was reported to sense HIV-1. The interferon gamma inducible protein 16 (IFI16) was shown to bind to stem-loop containing ssDNA derived from HIV-1 proviral DNA and initiated a STING-TBK1-IRF3 pathway (Jakobsen et al. 2013, 2015). Intriguingly, very recent evidence suggests that IFI16 regulates DNA sensing in macrophages and keratinocytes by promoting the activity of STING and efficient cGAMP production (Almine et al. 2017; Jønsson et al. 2017). Intriguingly, IFI16 is a host DNA sensor that initiates pyroptosis in abortively infected CD4 T cells through an inflammatory response triggered mainly by IL-1β (Monroe et al. 2014; Doitsh et al. 2014). This response requires cell-to-cell transmission and was observed only in resting T cells derived from lymphoid tissues (Galloway et al. 2015; Munoz-Arias et al. 2015). Interestingly, the situation seems to differ in activated infected T cells: the activation of DNA-dependent protein kinase (DNA-PK) during sensing of the viral integration process plays a central role in CD4 T cell depletion (Cooper et al. 2013).

3 HIV-1 Utilizes or Avoids Cellular Factors to Escape Sensing

3.1 Cloaking of HIV-1 Cargo

Only in recent years, it became clear that HIV avoids certain host factors, while on the other hand utilizes several cellular factors to evade sensing followed by innate immune signaling resulting in an antiviral state. Still, it stays controversial to what time point during infection and under which cellular prerequisites sensing of HIV-1 does occur in myeloid cells (Landau 2014).

It is common knowledge that conventional DCs (cDCs) per se do not respond to HIV infection with production of type I IFNα (Smed-Sorensen et al. 2005) resulting in incomplete cDC maturation (Luban 2012), probably due to inefficient infection. In order to establish efficient replication in MDDCs and other myeloid cells, so that —in consequence—sensing of the viral PAMPs could occur, the addition of the lentiviral accessory protein Vpx is required as a means to overcome the restriction posed by SAMHD1 (Manel et al. 2010; Laguette et al. 2011; Hrecka et al. 2011; Berger et al. 2011). We will review SAMHD1 and its function in detail in the upcoming paragraph.

For that reason, Gao et al. used virus-like particles (VLPs) containing Vpx to trigger innate immune responses in their experiments to identify cGAS as a sensor for HIV-1 (Gao et al. 2013). Other labs reported that HIV-1 is not sensed even after removal of SAMHD1. They report that capsid mutations that increase the affinity for the peptidylprolyl isomerase A (CypA) allow for sensing. In that case, the HIV capsid is a determinant of innate sensing of viral DNA after integration (Lahaye et al. 2013). In a third scenario, HIV-1 is not sensed by cGAS because capsid and the cargo is cloaked and protected by CypA and the cleavage and polyadenylation specific factor 6 (CPSF6) (Rasaiyaah et al. 2013).

HIV utilizes several host cell factors to promote timely capsid uncoating and initiation of reverse transcription (RT) and efficient nuclear import [reviewed in (Le Sage et al. 2014; Campbell and Hope 2015)]. These include CypA (Braaten et al. 1996), Pin1 (Misumi et al. 2010), PDZD8 (Henning et al. 2010), CPSF6 (Lee et al. 2010; Hori et al. 2013), TNPO3, Nup358, and Nup153 (Brass et al. 2008; Konig et al. 2008; Zhou et al. 2008; Bushman et al. 2009).

Perturbation of the uncoating process by polymorphisms in CypA binding affinity to HIV-1 capsid (An et al. 2007; Rits et al. 2008) or by cytoplasmic located CPSF6 (Lee et al. 2010; Hori et al. 2013; Iaco et al. 2013) induces premature capsid disassembly which may lead to exposure of viral RT products to cytosolic sensors (Lahaye et al. 2013; Rasaiyaah et al. 2013). In fact, it was shown that capsid mutants N74D and P90A that are impaired for interaction with CPSF6 and CypA or Nup358, respectively, lead to type I IFN responses (Rasaiyaah et al. 2013).

3.2 Exploitation of a Cellular Nuclease

Moreover, it is thought that incomplete or aberrant RT products in unproductive infections due to the error-prone RT process might be removed by the three prime repair exonuclease 1 (TREX1). It has been shown that TREX1 as a negative regulator of the IFN-stimulatory DNA (ISD) response prevents autoimmunity (Stetson et al. 2008) by binding and degrading unwanted self-DNA in the cytosol. TREX1 was found to digest ssDNA derived from endogenous retroelements and in consequence prevented triggering of cytosolic sensing (Stetson et al. 2008). Notably, the absence of TREX1 provides a cellular state where cytosolic HIV-1 RT products accumulate and lead to IRF3 activation (Yan et al. 2010a; Lee et al. 2013). Additionally, TREX1 seems to prevent autointegration as part of the ER-associated SET complex (Yan et al. 2009).

In conclusion, the steps of HIV-1 following entry into the target cells require safe and "unrecognized" travel of the viral capsid and the cargo through the cellular cytosol, spatially and timely tuned reverse transcription, translocation into the nucleus and integration in the host genome. Infection is a highly organized process, which turns out to be quite susceptible to perturbation.

3.3 Active Exploitation of NLRX1

Another recently described factor that HIV-1 actively exploits to its advantage is NLRX1. In the next chapter, we will first review the general known functions of NRLX1. Next, we will provide a detailed review on its involvement in the innate signaling response toward HIV-1.

3.3.1 NLRX1—A Regulator of Innate Signaling

NLRX1 is a member of the NLR family, a group of cytosolic PRRs which are involved in sensing microbial and danger signals to induce innate immune activation and inflammation. All protein members of this family include a nucleotide-binding domain (NBD) and leucine-rich repeats (LRR) but while most also contain an N-terminal CARD or PYD domain, these are absent in NLRX1. Instead it is the only family member with an N-terminal transit peptide directing it to mitochondria (Tattoli et al. 2008; Moore et al. 2008). Its exact localizations may be dynamic or dependent on the inner mitochondrial transmembrane potential (Arnoult et al. 2009) as NLRX1 has been found at both, the mitochondrial outer membrane (MOM) (Moore et al. 2008) and within the mitochondrial matrix (Arnoult et al. 2009; Rhee et al. 2013; Sasaki et al. 2013). A crystal structure of the C-terminal fragment of NLRX1 revealed that the protein forms dimers that oligomerize into hexamers via extensive interdomain and intersubunit interactions (Hong et al. 2012).

Intensive research has linked NLRX1 to various functions in innate immunity. At the MOM, NLRX1 is acting as a negative regulator attenuating the innate immune response to viral infections, including influenza virus, that are sensed through the RIG-I/MAVS (also known as IPS-1, VISA, and CARDIF) axis by directly competing for MAVS binding (Moore et al. 2008; Allen et al. 2011; Xia et al. 2011). This perception was challenged by two independently generated NLRX1-deficient mouse models (Rebsamen et al. 2011; Soares et al. 2013). However, it is not clear why the different mouse models came to opposite conclusions. Of note, the observed effect on RIG-I/MAVS axis was not present in bone marrow-derived macrophages (BMDMs) (Allen et al. 2011; Jaworska et al. 2014) or alveolar macrophages (Jaworska et al. 2014) and pDCs (Allen et al. 2011), arguing for a specialized role of NLRX1 in these immune cells.

Besides inhibiting RIG-I/MAVS-dependent type I IFN induction, NLRX1 emerged as essential factor to promote autophagy. Lei et al. identified a mitochondrial interaction network including NLRX1, MAVS, TUFM, and autophagy induction proteins Atg5-Atg12 and ATG16L1 (Lei et al. 2012). Both, attenuation of IFN production as well as promotion of autophagy synergistically promoted viral infection (Lei et al. 2012). Additional evidence for the cooperative augmentation of autophagy by NLRX1 and TUFM was provided recently as knockdown of either factor in squamous cell carcinoma of the head and neck cells attenuated autophagy upon EGFR inhibition (Lei et al. 2016).

It seems that the control of MAVS signaling leading to downregulation of type I IFN and the induction of autophagy are intertwined processes mediated by a group of proteins with dual functions at the mitochondrial outer membrane.

Resembling its negative influence on the MAVS-dependent type I IFN induction, NLRX1 also attenuates NF-κB signaling through MAVS-independent LPS-induced TLR4 signaling via TRAF6 (Allen et al. 2011; Xia et al. 2011), but did not affect NF-κB activity in response to TNFα treatment (Rebsamen et al. 2011; Allen et al. 2011; Xia et al. 2011) or TCR or BCR activation (Xia et al. 2011). Besides this direct negative regulation of NF-κB signaling, NLRX1 may also have a positive effect on NF-κB activity by modulating reactive oxygen species (ROS) production (Tattoli et al. 2008). NLRX1 synergistically enhances ROS production in vitro upon bacterial (Tattoli et al. 2008; Abdul-Sater et al. 2010)/viral (Unger et al. 2014) infection and TNFα (Tattoli et al. 2008; Hong et al. 2012)/polyI:C (Hong et al. 2012) treatment.

Given that NLRX1 might be localized in the mitochondrial matrix, it seems difficult to explain how this NLR may sense a microbial associated molecular pattern (Arnoult et al. 2009). Abdul-Sater et al. provide an elegant explanation suggesting that NLRX1 augments ROS production in a secondary step after primary ROS produced by membrane-bound NADPH oxidases NOX/DUOX diffuses into mitochondria, triggering NLRX1-dependent ROS production (Abdul-Sater et al. 2010). Such a link of NLRX1 to ROS production is further backed by NLRX1 interaction with the mitochondrial respiratory chain protein UQCRC2 (Arnoult et al. 2009; Rebsamen et al. 2011). Similarly, viral infection by rhinovirus induced both NOX- and NLRX1-dependent ROS production in vitro (Unger et al. 2014).

Interestingly, the LRR containing part of NLRX1 can interact with dsRNA over ssRNA ligands, but not dsDNA or ssDNA, as demonstrated with the purified C-terminal fragment of NLRX1 (Hong et al. 2012). Binding of dsRNA analog polyI:C was abolished by a specific mutation in NLRX1 coincident with the loss of ROS induction upon polyI:C treatment in vitro (Hong et al. 2012), suggesting that RNA binding to NLRX1 may also be a mechanism to trigger NLRX1-dependent ROS production. Tattoli et al. suggested that an NLRX1-enhanced ROS production may translate to NF-κB and JNK activation (Tattoli et al. 2008; Gloire et al. 2006). This in vitro finding is questioned by in vivo models, though (Rebsamen et al. 2011; Allen et al. 2011; Xia et al. 2011). Recently, new results were added to this discussion demonstrating that NLRX1 regulates basal ROS production only in transformed, but not primary mouse embryonic fibroblasts (MEFs) (Soares et al. 2014), which is in accordance to the in vivo experiments (Rebsamen et al. 2011). Clearly, the influence of NLRX1 on ROS production and positive or negative effects on NF-κB activity is strongly dependent on the investigated stimulus. Thus, it will be interesting to see whether previous results will hold true in untransformed cell models. Further confirmation of NLRX1-RNA interaction or a solved structure of the interaction would be much needed for physiological verification.

Recently, NLRX1 was linked to apoptosis and cancer. It exhibited neuroprotective capacities in a mouse model for multiple sclerosis (Eitas et al. 2014) and directed cells toward an apoptotic pathway protecting the cells from necrosis-like cell death (Imbeault et al. 2014). It seems that NLRX1 protects cells against extrinsic apoptosis signals and sensitizes them to intrinsic apoptosis (Soares et al. 2014; Singh et al. 2015). Interestingly, this effect was only found in transformed, but not in primary cells suggesting a link between NLRX1 and tumorigenesis (Soares et al. 2014). NLRX1 was highly downregulated in transformed versus untransformed MEFs (Soares et al. 2014), in human colon tumors (Koblansky et al. 2016) and in chronic obstructive pulmonary disease, inversely correlating with disease severity (Kang et al. 2015). Fittingly, key cancer-promoting pathways STAT3, NF-κB, MAPK, and IL-6 were augmented in colons of NLRX1$^{-/-}$ mice (Koblansky et al. 2016) and NF-κB signaling was also significantly upregulated in mouse NLRX1$^{-/-}$ cells of a splenic histiocytic sarcoma (Coutermarsh-Ott et al. 2016b). Indeed, tumor suppressor functions of NLRX1 were observed in several tumor models (Soares et al. 2014; Koblansky et al. 2016; Tattoli et al. 2016; Singh et al. 2015; Coutermarsh-Ott et al. 2016b; Li et al. 2016). Intriguingly, NLRX1 was upregulated in a pulmonary metastasis model concurrent with downregulation of RIG-I, MDA-5, LGP-2 and MAVS, which can be reverted by antitumor poly I:C application (Ma et al. 2016), indicating an important role for NLRX1 suppression of RIG-I/MAVS signaling in control of tumor spread (Ma et al. 2016).

3.3.2 NLRX1—A Negative Regulator of HIV-1 Sensing

Recently, NLRX1 has been discovered as a negative regulator of the central DNA-sensing pathway node STING to hamper activation of TBK1/IRF3/NF-κB - axis leading to reduced expression of IFNs and inflammatory cytokines (Guo et al. 2016). Mechanistically, NLRX1 seems to sequester STING, thus preventing activation of TBK1 and downstream signaling (Guo et al. 2016). This function seems to be actively exploited by HIV-1 and simian immunodeficiency virus (SIV) (Guo et al. 2016; Barouch et al. 2016). Knockdown of NLRX1 reduces retroviral infection in primary human cells and a mouse model accompanied by increased production of innate immune signaling effectors (Guo et al. 2016). An elegant study by Barouch and colleagues demonstrated that mucosal infection of rhesus monkeys with SIVmac251 triggered an early proinflammatory response that intriguingly lacks the expression of antiviral restriction factors. NLRX1 was upregulated very early in the first 24 h upon SIV infection in the infected tissues. Its expression correlated inversely with the expression of antiviral restriction factors (Barouch et al. 2016). These correlations imply an inhibitory effect of NLRX1 on innate immune signaling allowing for more efficient viral replication, similar to the findings by Guo et al. in the case of HIV-1. Interestingly, another study demonstrated upregulation of NLRX1 in the lamia propria leukocytes also at late times after SIV infection (Mohan et al. 2012). Furthermore, the TGF-β pathway signaling was activated after SIV infection which suppressed successful initiation of adaptive immune responses (Barouch et al. 2016). In summary, SIV triggers host responses in the first 24 h that suppress antiviral innate and adaptive responses and facilitate early systemic dissemination (Barouch et al. 2016). Importantly, TGF-β is also involved in the pathogenesis of HIV-1 in human patients as CD4 and CD8 T cell counts are reduced with increased TGF-β levels (Wiercinska-Drapalo et al. 2004) e.g. the antiviral cytolytic CD4 T cell response towards HIV-1 is strongly affected by TGF-β (Lewis et al. 2016) thus, by extension, findings of SIV infection may also translate to HIV-1 infection (Mar and Schoggins 2016).

A recent study analyzed expression of various signaling molecules of the inflammasome and mitochondrial signaling in peripheral blood mononuclear cells (PBMCs). They compared HIV patients on successful combination antiretroviral therapy (cART) for at least ten months to healthy donors. Interestingly, the only molecule that was differentially regulated was NLRX1 (Nasi et al. 2015). NLRX1 was found to be downregulated in the patient samples suggesting that it might be modulated by the virus to avoid triggering of cell death through NLRX1 at this late stage in infection (Nasi et al. 2015). Still, the consequences of NLRX1 regulation on viral infection are not yet conclusively demonstrated. A study comparing CD4 T cells of elite controllers (ECs) to HIV negative and HIV-infected patients on successful ART could not reveal a differential expression pattern of NLRX1 (Vigneault et al. 2011). Opposing findings may result from limited sample sizes or analyzed stages of infection, nevertheless, a comprehensive study of all infection phases of substantial cohort sizes would allow a more detailed picture if how HIV-1 affects regulation of various genes during infection, including NLRX1.

3.3.3 Localization of NLRX1 and Signal Components

Although in vitro and in vivo data clearly point to the effect of NLRX1 on STING-mediated signaling, it seems unclear how the spatial separation of these proteins in the cell can be explained. In the section below, we will discuss current knowledge. We would like to point out that the localization of neither NLRX1 nor STING excludes their functional connection.

The exact localization of NLRX1 is under debate, while several reports state NLRX1 is localized to the mitochondrial matrix (Arnoult et al. 2009; Rhee et al. 2013; Sasaki et al. 2013), it has also been detected on the MOM (Moore et al. 2008) and in the cytoplasm (Unger et al. 2014).

NLRX1 contains an N-terminal mitochondrial targeting signal (Moore et al. 2008; Tattoli et al. 2008), which is cleaved upon translocation to the inner side of the mitochondrial inner membrane (MIM) (Arnoult et al. 2009). The mitochondrial matrix pool of NLRX1 may represent its final localization (Arnoult et al. 2009; Rhee et al. 2013; Sasaki et al. 2013), but is readily replaced with newly synthesized NLRX1 (Arnoult et al. 2009), opening many possibilities for NLRX1 localizations on the route. Translocation of NLRX1 to the mitochondrial matrix is dependent on the inner mitochondrial transmembrane potential (Arnoult et al. 2009). Perturbation of this by e.g. stress or viral infection could influence NLRX1 translocation. Cytoplasmic NLRX1 interacts with MAVS (Moore et al. 2008; Arnoult et al. 2009; Sasaki et al. 2013; Li et al. 2016), but not mitochondrial matrix-localized NLRX1s (Rebsamen et al. 2011; Sasaki et al. 2013). Sequestration of NLRX1 in supramolecular complexes (Lei et al. 2012; Moore et al. 2008) at the MOM (Moore et al. 2008; Guo et al. 2016) possibly through MAVS (Moore et al. 2008) or TUFM (Lei et al. 2012) interaction and release upon RIG-I stimulation (Lei et al. 2012) could prevent NLRX1 from translocation to the mitochondrial matrix. Sufficient activation could induce NLRX1 translocation to the mitochondrial matrix, where it may support antimicrobial defense by additional ROS stimulation (Tattoli et al. 2008; Abdul-Sater et al. 2010; Hong et al. 2012; Unger et al. 2014). Speculations about a cytoplasmic pool of NLRX1 have been repeatedly expressed (Xia et al. 2011; Hong et al. 2012; Sasaki et al. 2013) and translocation from cytoplasm to mitochondria has been demonstrated upon rhinovirus infection in primary differentiated airway epithelial cells (Unger et al. 2014). As discussed earlier, the picture of NLRX1 localization and function might be more complex than we anticipate today (Coutermarsh-Ott et al. 2016a). Given the marked difference in NLRX1 protein abundance as well as functional differences between transformed and untransformed cell, it appears very plausible that also the localization of NLRX1 might differ between those. Additional knowledge about the temporal and spatial organization and localization of NLRX1 and careful interpretation of in vitro and in vivo data would be advantageous.

STING is primarily localized to the endoplasmic reticulum (Ishikawa and Barber 2008), which makes contact to various other organelles including mitochondria, giving rise to so-called mitochondria-associated ER membranes (MAM). These facilitate lipid and calcium exchange and are sites for autophagosome formation and

NLRP3 inflammasome formation trough MAVS [reviewed in Raturi and Simmen (2013); Marchi et al. (2014); Zhang and Hu (2016)]. Importantly, MAVS (Horner et al. 2011) and STING (Ishikawa and Barber 2008; Marchi et al. 2014) both localize to MAMs in the mitochondrial and ER sides, respectively. Upon detection of RNA virus infection RIG-I translocates to MAMs to interact with MAVS (Horner et al. 2011) where STING is able to interact with this complex (Ishikawa and Barber 2008). Both, a potential cytoplasmic NLRX1 pool as well as contacts of MOM localized NLRX1 to STING via mitochondria-associated ER membranes would be plausible explanations to successful NLRX1-STING interaction, probably in macromolecular complexes at the membranes. Furthermore, STING is subject of extensive re-localization upon activation, as it is shuttled to the Golgi network and some yet undefined signaling vesicles (Chow et al. 2015). Autophagy proteins play important roles in translocation processes of STING (Saitoh et al. 2009; Konno et al. 2013; Chow et al. 2015). Additionally, RIG-I/MAVS signaling is connected to autophagy induction (Jounai et al. 2007) and, recently, NLRX1 emerged as a positive regulator of autophagy (Lei et al. 2012). Therefore, a functional connection of all proteins to autophagy processes would be interesting to explore.

3.3.4 Possible Impact of NLRX1 on Latent Reservoirs

NLRX1 has become a versatile player in the regulation of innate immune responses and the recent studies underline the in vivo relevance as NLRX1 is clearly manipulated by HIV/SIV virus to suppress antiviral innate immunity and support viral infection. It is conceivable that NLRX1 affects HIV infection at further steps beyond initial dissemination. One major obstacle of HIV eradication is the latent reservoir of the virus. We hypothesize that NLRX1 may have functional connections to the reservoir. First, NLRX1 dampens the innate immune response upon early infection (Guo et al. 2016; Barouch et al. 2016), a time point at which latent reservoirs are already initiated if not established (Lindbäck et al. 2000; Hogan et al. 2012; Chun et al. 1998; Whitney et al. 2014), thus a patient would probably benefit from a strong early antiviral innate response by inhibition of NLRX1 (Towers and Noursadeghi 2014). Second, NLRX1 attenuates signaling activities which affect latency: (i) NLRX1 dampens RIG-I/MAVS signaling (Moore et al. 2008; Allen et al. 2011; Xia et al. 2011), which can be beneficial in proviral transcriptional activation and killing of latently infected cells (Li et al. 2016); (ii) NLRX1 may directly attenuate the NF-κB pathway (Allen et al. 2011; Xia et al. 2011), which is enhancing HIV transcription (Nabel and Baltimore 1987; Duh et al. 1989; Böhnlein et al. 1988; Tong-Starksen et al. 1987) and reactivating the latent reservoir. Thus, attenuation of both pathways by NLRX1 may drive and sustain latency. The positive effect of NLRX1 on ROS production upon external stimuli (Tattoli et al. 2008; Abdul-Sater et al. 2010; Hong et al. 2012; Unger et al. 2014) under certain conditions (Rebsamen et al. 2011; Allen et al. 2011; Xia et al. 2011) may have a positive effect on NF-κB (Gloire et al. 2006; Yang et al. 2009) and the transcriptional status of latent HIV reservoirs.

4 Restriction Factors as Effectors of the Innate Response

Host target cells express a diverse set of antivirally active proteins, mostly activated by the innate immune response. Restriction factors can execute immediate blocks at different steps of the HIV replication cycle. Therefore, they are an important component of the innate immune response against HIV infection. All restriction factors against viral infection share common features (Duggal and Emerman 2012; Malim and Bieniasz 2012; Doyle et al. 2015): (i) germline-encoded, constitutive or IFN-inducible expression; (ii) they act dominantly against viral infections in a cell-autonomous manner; (iii) responsible for species-specific suppression, therefore determining the viral host range; (iv) partly counteracted by viral (accessory) proteins, namely Nef, Vif, Vpr, Vpu, and Vpx; (v) showing evidence of positive genetic selection, as sign of an evolutionary "arms race" between host survival/viral replication (Table 1).

Well studied and only recently discovered examples of HIV-mediated evasion of restriction factors involve nearly every step of the HIV replication cycle (see Table 1). Furthermore, some restriction factors can also act as innate sensors, namely tetherin and TRIM5α. If tetherin is present, newly formed HIV-1 virions are retained at the plasma membrane of infected cells (Neil et al. 2008; van Damme et al. 2008). Related to this, tetherin can induce proinflammatory responses by activating NF-κB (Galao et al. 2012, 2014). As countermeasure, Vpu induces tetherin downregulation from the cell surface (Neil et al. 2008; van Damme et al. 2008). TRIM5α induces AP-1 and NF-κB-dependent factors and acts as a PRR for the HIV-1 capsid lattice (Pertel et al. 2011). In general, IFN treatment leads to HIV-1 replication blocks at early and late stages in different cell types (Cheney and McKnight 2010; Goujon and Malim 2010), through the induction of known and yet to be identified antiviral ISGs (Doyle et al. 2015).

5 SAMHD1—HIV-1 is Passively Avoiding Restriction

Here, we highlight the HIV-1 restriction factor SAMHD1 to demonstrate how cell-autonomous antiviral factors influence sensing of HIV PAMPs and thereby actively shape the innate/adaptive immune response to HIV-1 infection.

5.1 Discovery of SAMHD1

The accessory protein Vpx, which originated from duplication of the *vpr* gene (Tristem et al. 1990), is encoded by the primate lentiviruses of the HIV-2/SIV_{smm}

Table 1 Innate Genes in HIV-1 Replication

	Role in HIV replication	Regulation by type I/II IFN	Utilization and/or counteraction by HIV	Positive selection	Reference
Innate sensors					
cGAS	senses HIV-1 RTIs	type I/II IFNs (Ma et al. 2015)	cloaking of RTIs by capsid up to integration (Lahaye et al. 2013)	yes (Hancks et al. 2015)	Gao et al. (2013)
IFI16	senses stem-loop containing HIV-1 ssDNA in macrophages	type I/II IFNs (Dawson and Trapani 1995; Trapani et al. 1992)		yes (McLaren et al. 2015; Cagliani et al. 2014)	Jakobsen et al. (2013)
PQBP1	binds reverse transcribed HIV-1 DNA and interacts/cooperates with cGAS in MDDCs	unknown			Yoh et al. (2015)
RIG-I (DDX58)	(potentially) senses dimeric/monomeric HIV RNA	type I/II IFNs (Imaizumi et al. 2004; Sumpter et al. 2005)	HIV-1: PR (Solis et al. 2011)	yes (Vasseur et al. 2011)	Berg et al. (2012), Solis et al. (2011), Wang et al. (2013)
TLR2/4	sense gp120 in female genital epithelial cells	type I/II IFN (Faure et al. 2001; Miettinen et al. 2001)			Nazli et al. (2013)
TLR7/8	senses genomic HIV ssRNA	type I/II IFN (Miettinen et al. 2001)		yes, for specific residues (Ortiz et al. 2009)	Beignon et al. (2005), Heil et al. (2004), Lepelley et al. (2011)
TLR9	senses CpG DNA potentially present in HIV-1 virions	unknown			Mandl et al. (2008), Beignon et al. (2005)

(continued)

Table 1 (continued)

	Role in HIV replication	Regulation by type I/II IFN	Utilization and/or counteraction by HIV	Positive selection	Reference
Innate sensors and restriction factors					
Tetherin (BST2, CD317)	inhibits release of newly formed virions in infected cells; activates NF-κB/initiates proinflammatory cytokine production	type I/II IFNs (Neil et al. 2008; Cobos Jimenez et al. 2012)	HIV-1: Vpu (group M, N) or Nef (group O) (Sauter et al. 2009); HIV-2: Env (Le Tortorec and Neil 2009)	yes (Lim et al. 2010; McNatt et al. 2009)	Neil et al. (2008), van Damme et al. (2008), Galao et al. (2012)
TRIM5α	binds to surface of viral capsid, thereby inducing its dissociation; activates TAK1/NF-κB upon capsid recognition	type I/II IFNs (Asaoka et al. 2005)	capsid mutation (Stremlau et al. 2004)	yes (Sawyer et al. 2005)	Pertel et al. (2011), Stremlau et al. (2004)
Dependency factors interfering with sensing					
ADAR1/2	(potentially) stimulate HIV-1 replication by increasing viral protein production and specific HIV-1 RNA editing	ADAR1: type I/II IFNs (Patterson et al. 1995); ADAR2: not by type I/II IFNs (Samuel 2011)		yes, for specific residues (Forni et al. 2015)	Doria et al. (2009), (2011), Phuphuakrat et al. (2008), Biswas et al. (2012)
CPSF6	interacts with CA; targets HIV-1 integration into transcriptionally active euchromatin	not by type I IFN (Bulli et al. 2016)	cloaking of RTIs (Rasaiyaah et al. 2013)	no (Lee et al. 2012)	Sowd et al. (2016), Rasheedi et al. (2016), Lee et al. (2010)
CypA	interacts with CA; promotes RT; needed for HIV-1 infection of certain human cell types/lines	type I/II IFN (Cobos Jimenez et al. 2012)	cloaking of RTIs (Rasaiyaah et al. 2013)	no (Ribeiro et al. 2005)	Luban et al. (1993), Towers et al. (2003), Iaco and Luban 2014)
NLRX1	facilitates HIV-1 infection/spread by inhibiting STING-mediated IFN response	unknown	HIV-1: upregulates NLRX1, thereby inhibiting IFN induction (Barouch et al. 2016)		Barouch et al. (2016), Guo et al. (2016)

(continued)

Table 1 (continued)

	Role in HIV replication	Regulation by type I/II IFN	Utilization and/or counteraction by HIV	Positive selection	Reference
RanBP2 (Nup358)	interacts with CA; involved in nuclear import of HIV-1 PICs	unknown	cloaking of RTIs (Rasaiyaah et al. 2013)	yes (Meyerson et al. 2014)	Zhang et al. (2010), Ocwieja et al. (2011), Schaller et al. (2011)
RNAse H2 complex	promotes HIV-1 replication through unknown mechanism	unknown			Genovesio et al. (2011)
SUN2	co-factor for CypA-mediated positive effects on RT/infection of $CD4^+$ T cells	ISG library (Schoggins et al. 2011); not by type I/II IFNs (Donahue et al. 2016)			Lahaye et al. (2016)
SLX4 complex	recruited by Vpr for degradation of RTIs to avoid sensing/IFN response	unknown	HIV-1: Vpr prematurely activates SLX4 complex (Laguette et al. 2014)		Laguette et al. (2014)
TNPO3 (Transportin-3)	interacts with CA; involved in nuclear import of HIV-1 PICs	unknown		no (Meyerson et al. 2014)	Brass et al. (2008), Konig et al. (2008), Zhou et al. (2008)
TREX-1	degrades excess RTIs in the cytoplasm to avoid sensing/IFN response	type I/II IFNs (Serra et al. 2011; Cobos Jimenez et al. 2012)			Yan et al. (2010b)
Restriction factors					
90 K	reduces particle infectivity by interfering with maturation and Env incorporation into HIV-1 virions	type I/II IFNs (Marth et al. 1994)			Lodermeyer et al. (2013)

(continued)

Table 1 (continued)

	Role in HIV replication	Regulation by type I/II IFN	Utilization and/or counteraction by HIV	Positive selection	Reference
APOBECs (B, D, F, G, H)	deaminate cytidines to uracils, leading to hypermutated/defective HIV genomes; interfere with cDNA elongation step during RT	A3G: type I IFN (Wang et al. 2008); A3F+H: type I IFN (Koning et al. 2009)	HIV-1/HIV-2: Vif (Sheehy et al. 2002; Smith et al. 2014)	yes (some) (OhAinle et al. 2006; Sawyer et al. 2004)	Bishop et al. (2008), Sheehy et al. (2002)
RNF115 (BCA2, Rabring 7)	co-factor of tetherin-mediated restriction; reduces cellular Gag levels by promoting its ubiquitination and lysosomal degradation	unknown			Miyakawa et al. (2009), Nityanandam and Serra-Moreno (2014)
GBP5	interferes with HIV-1 Env processing and incorporation	type I/II IFNs (Krapp et al. 2016)	vpu mutations/increase in Env expression (Krapp et al. 2016)	yes (McLaren et al. 2015)	Krapp et al. (2016), McLaren et al. (2015)
HERC5	inhibits early stage of HIV-1 Gag particle assembly and interferes with nuclear export of Rev/RRE-dependent RNA	type I IFN (Wong et al. 2006; Dastur et al. 2006)		yes (Woods et al. 2014)	Woods et al. (2011), (2014)
IFITM1-3	interfere with virus entry and/or Gag production	type I/II IFNs (Friedman et al. 1984; Lewin et al. 1991)	Vpu/Env mutation (Ding et al. 2014)	no, rather negative selection (Wilkins et al. 2016)	Lu et al. (2011)
ISG15	inhibits Tsg101 (ESCRT-I)-Gag interaction, thereby interfering with efficient HIV-1 virion budding	(mainly by) type I IFN (Korant et al. 1984; Farrell et al. 1979)			Okumura et al. (2006), Pincetic et al. (2010)

(continued)

Table 1 (continued)

	Role in HIV replication	Regulation by type I/II IFN	Utilization and/or counteraction by HIV	Positive selection	Reference
KAP1 (TRIM28)	deacetylates HIV-1 IN in complex with HDAC1, thereby represses integration	unknown			Allouch et al. (2011)
MARCH8	blocks Env incorporation into virions	(slightly by) type I IFN (Tada et al. 2015)			Tada et al. (2015)
MOV10	packaged into virions, inhibits proteolytic Gag processing and RT in infected target cells	ISG library (Schoggins et al. 2011)			Furtak et al. (2010), Burdick et al. (2010), Wang et al. (2010)
MxB (Mx2)	acts after RT at late post-entry step, potentially by inhibiting CA-dependent nuclear import and/or DNA integration	type I/II IFN (Aebi et al. 1989)	capsid mutation (Liu et al. 2013, 2015)	yes (Busnadiego et al. 2014)	Goujon et al. (2013), Kane et al. (2013), Liu et al. (2013)
PAF1 complex	inhibits early steps of HIV life cycle (RT to integration)	unknown			Liu et al. (2011)
RSAD2 (Viperin)	(potentially) inhibits assembly/release of de novo synthesized HIV-1 virions	type I/II IFNs (Chin and Cresswell 2001)		yes, but not driven by lentiviruses (Lim et al. 2012b)	Nasr et al. (2012), Lim et al. (2012b)
SAMHD1	inhibits HIV-1 replication in non-cycling cells at RT step by dNTP depletion and/or degradation of HIV-1 RNA	type I/II IFNs (Li et al. 2000; Berger et al. 2011; Riess et al. 2017)	HIV-2: Vpx (Berger et al. 2011; Hrecka et al. 2011; Laguette et al. 2011)	yes (Laguette et al. 2012; Lim et al. 2012a)	Berger et al. (2011), Hrecka et al. (2011), Laguette et al. (2011)

(continued)

Table 1 (continued)

	Role in HIV replication	Regulation by type I/II IFN	Utilization and/or counteraction by HIV	Positive selection	Reference
SERINC3/5	interfere with efficient virus fusion to target cell and potentially early post-entry steps	not by type I IFN (Rosa et al. 2015; Usami et al. 2015)	HIV-1: Nef (Rosa et al. 2015; Usami et al. 2015)	no (Murrell et al. 2016)	Rosa et al. (2015), Usami et al. (2015)
SLFN11	inhibits expression of viral proteins by binding tRNAs and altering codon usage	type I IFN (Li et al. 2012)			Li et al. (2012)

lineage, not by HIV-1. In vitro, Vpx is necessary to efficiently infect non-cycling myeloid cells (including macrophages and MDDCs) (Yu et al. 1991; Goujon et al. 2008). Vpx, which is incorporated into virions (Kappes et al. 1988), was proposed to induce the proteasomal degradation of a myeloid-specific restriction factor acting at an early step of the HIV-1 replication cycle. Indeed, Vpx interaction with the DDB1-Cul4A-associated-factor-1 (VprBP/DCAF1) substrate receptor for the Cullin-4A (Cul4A) E3 ubiquitin ligase and the proteasomal pathway are needed for Vpx-mediated promotion of infection (Goujon et al. 2007; Sharova et al. 2008; Srivastava et al. 2008).

Affinity purification of Vpx-associated proteins or Vpx-recruited substrates to the CRL4-DCAF1 complex with subsequent mass spectrometry (MS) analysis were used to identify SAMHD1 as the unknown HIV-1 restriction factor being active in non-cycling myeloid cells (Berger et al. 2011; Hrecka et al. 2011; Laguette et al. 2011) and resting $CD4^+$ T cells (Baldauf et al. 2012; Descours et al. 2012). Vpx loads SAMHD1 through DCAF1 onto the Cul4A E3 ubiquitin ligase complex, leading to its polyubiquitination and subsequent proteasomal degradation (Ahn et al. 2012).

5.2 *SAMHD1 Enzymatic Functions*

SAMHD1 contains an N-terminal nuclear localization sequence (NLS), therefore being mainly present in the nucleus (Brandariz-Nunez et al. 2012; Hofmann et al. 2012). As a dGTP/GTP-stimulated triphosphohydrolase (= dNTPase), SAMHD1 is able to efficiently degrade dNTPs via its HD domain to the respective deoxynucleoside and inorganic triphosphate (Amie et al. 2013; Goldstone et al. 2011; Powell et al. 2011). For full dNTPase and restrictive activity, SAMHD1 tetramerization is required (Yan et al. 2013). SAMHD1-mediated restriction was proposed to work through depletion of dNTPs. To complete reverse transcription (RT) of its RNA genome into double-stranded DNA, HIV-1 is depending on a sufficient supply of cellular dNTPs. With SAMHD1 being present, dNTP pools are reduced below concentrations that support the RT step of HIV-1 (Lahouassa et al. 2012). Indeed, adding exogenous deoxynucleosides (dNs) to MDMs or resting $CD4^+$ T cells improves HIV-1 infection (Baldauf et al. 2012; Lahouassa et al. 2012).

Although expressed in both activated and resting $CD4^+$ T cells, SAMHD1 is only able to block HIV-1 infection in the latter, indicative of SAMHD1 restriction activity being controlled by post-translational modifications. Indeed, in cycling cells SAMHD1 is phosphorylated at threonine (T) 592 by cyclin-dependent kinase (CDKs)/cyclin A2, while T592 phosphorylation is absent when cells are in a non-cycling/resting state (Cribier et al. 2013; Welbourn et al. 2013; White et al. 2013). Cell cycle-related control of T592 phosphorylation is also reflected by the phosphatase acting on SAMHD1, when cells enter G_1/G_0 phase: Specific protein phosphatase 2A (PP2A) holoenzymes, containing the regulatory PP2A B55α

subunit, remove SAMHD1 T592 phosphorylation during mitotic exit (Schott et al., in review)—indicating a tight control of dephosphorylation during cell cycle progression. Interestingly, innate stimuli influence SAMHD1 phosphorylation at T592 differently: Type I/II IFN treatment of activated $CD4^+$ T cells and MDMs reduces T592 phosphorylation level (Cribier et al. 2013), potentially to strengthen the IFN-induced antiviral state by activating SAMHD1's antiviral activity. On the other hand, stimulation and activation of $CD4^+$ T cells using anti-CD3/CD28 antibodies, IL-2 or IL-7 results in increase of SAMHD1 T592 phosphorylation, therefore SAMHD1 inactivation (Coiras et al. 2016; Cribier et al. 2013; White et al. 2013).

Mutation of T592 to a phosphomimetic residue (= T592D/E) abolishes SAMHD1-mediated restriction, surprisingly, without an apparent effect on its dNTPase activity in vitro or in SAMHD1 mutant-expressing U937 cells (Welbourn et al. 2013; White et al. 2013). However, recent biochemical/structural studies suggest that a change to phosphomimetic residues (T592D/E) can directly destabilize active, tetrameric SAMHD1 (Tang et al. 2015; Yan et al. 2015). More specifically, T592 phosphorylation could interfere with the formation of "long-lived" SAMHD1 tetramers at conditions of low dNTP concentrations present in non-cycling cells (Arnold et al. 2015). But it remains puzzling that T592 phosphorylation is not perturbing SAMHD1 tetramerization equilibrium itself, but rather kinetics of tetramer disassembly (Bhattacharya et al. 2016). Therefore, the strong effect of T592 phosphorylation on restriction observed in vivo still needs to be further clarified. Further research is needed to mechanistically understand how T592 phosphorylation is regulating SAMHD1's dNTPase and antiviral activity, particularly in the cellular context.

Another mechanism for SAMHD1-mediated restriction came into play only recently. Besides a potent dNTPase, SAMHD1 is able to bind single-stranded nucleic acids, with a preference of RNA over DNA (Goncalves et al. 2012; Tungler et al. 2013). Interestingly, SAMHD1 oligomers induced by nucleic acid binding seem to be distinct from dNTPase-competent tetramers (Seamon et al. 2015). SAMHD1 was further proposed to exert an 3'-5'RNA exonuclease (= RNase) activity (Beloglazova et al. 2013; Ryoo et al. 2014), which could be inactivated by a single point mutation (Q548A). RNase-defective and dNTPase-competent SAMHD1 Q548A was unable to restrict HIV-1, suggesting that direct degradation of the incoming HIV-1 RNA genome by SAMHD1 prevents replication. Additionally, T592 phosphorylation was shown to negatively regulate SAMHD1's RNase activity (Ryoo et al. 2014). Subsequent studies could confirm single-stranded nucleic acid binding by SAMHD1, but not an intrinsic nuclease activity (Antonucci et al. 2016; Seamon et al. 2015). Different protein purification methods or assay/buffer conditions used for in vitro nuclease assays could explain the varying results: For instance, SAMHD1 was proposed to be a phosphorolytic RNase depending on inorganic phosphate (Ryoo et al. 2016)—a finding which could not be reproduced until now (Seamon et al. 2016).

Nevertheless, since reverse transcription is probably initiated in the cytoplasm, it is challenging to apprehend how nuclear SAMHD1, although to a certain extent being present in the cytoplasm (Baldauf et al. 2012; Ryoo et al. 2014), could be able to reach and degrade incoming viral genomic RNA, especially since viral nucleic acids are shielded by the viral capsid to avoid innate immune sensing. Additionally, reversibility of SAMHD1-mediated restriction (Hofmann et al. 2013) is challenging to explain in the light of a proposed degradation of the HIV-1 RNA genome. Therefore, further studies are needed, for instance involving comprehensive investigation of SAMHD1 mutants, to clarify the contribution of SAMHD1's dNTPase and/or RNase activity to HIV-1 restriction.

5.3 SAMHD1 Involved in Innate Immunity

Previously, SAMHD1 has been recognized as an important part of the innate immune system in another context: Mutations in the *SAMHD1* gene are associated with the rare autoimmune disease Aicardi-Goutières syndrome (AGS) (Rice et al. 2009), which is associated with chronically elevated levels of the antiviral cytokine IFNα in patients (Lebon et al. 1988). Therefore, SAMHD1 was proposed to act as a negative regulator of the innate immune response (Rice et al. 2009). Underlining its involvement in the innate immune response, SAMHD1 expression is induced in a cell type-dependent manner by type I/II IFNs, for instance in monocytes (Berger et al. 2011; Riess et al. 2017), microglia (Jin et al. 2016), different cell lines and primary hepatocytes (Sommer et al. 2016; St. Gelais et al. 2012). Mechanistically, increasing SAMHD1 expression levels after IFN treatment correlate inversely with reduced levels of specific microRNAs (miRs) (Jin et al. 2016; Riess et al. 2017). Interestingly, SAMHD1 levels are differentially regulated by miRs in distinct myeloid cell types: In monocytes, miR-181a and -30a are down-regulated after IFN treatment, concurrent with increased SAMHD1 protein expression. miR levels and SAMHD1 expression levels stay unchanged in macrophages and MDDCs (Riess et al. 2017).

In line with SAMHD1's potential regulatory role *Samhd1*$^{-/-}$ mice spontaneously produce IFN and exhibit constitutively upregulated ISGs (Behrendt et al. 2013; Rehwinkel et al. 2013). In this context, it was proposed that SAMHD1 could be involved in the control of endogenous retroelements, either by degrading them or preventing their synthesis through low dNTP levels (Zhao et al. 2013). By now, however, the exact source and identity of endogenous nucleic acids that potentially accumulate and induce an overshooting IFN response in SAMHD1-deficient AGS patients are unknown.

Intriguingly, restriction of HIV replication through SAMHD1 is directly influencing the outcome of innate and virus-specific adaptive immune responses: HIV-1 restriction imposed by SAMHD1 limits virus-induced IFN production, activation of myeloid cells and lentivirus-specific CD8^{+} T cell responses (Ayinde et al. 2015; Maelfait et al. 2016). Only in the absence of SAMHD1 HIV-1 RTIs are efficiently produced and can be sensed through the cGAS/STING pathway (Maelfait et al. 2016). Therefore, SAMHD1 restriction prevents innate sensing and curtails the subsequent adaptive immune response. Related to this, PBMCs from AGS patients lacking functional SAMHD1 are more susceptible to and produce increased amounts of pro-inflammatory cytokines after HIV-1 infection (Berger et al. 2011). How SAMHD1 expression and/or activity correlate with increased HIV control in patients, as seen in elite controllers (ECs), was shown recently in more detail: HIV-1 challenge of conventional DCs (cDCs) from ECs lead to an increased type I IFN production and upregulation of ISGs. Additionally, HIV-1 RT products accumulated disproportionally in cDCs from ECs compared to chronically-infected and HAART-treated patients. Remarkably, SAMHD1 mRNA expression was not significantly upregulated upon HIV-1 exposure in cDCs from ECs, whereas a significant induction of mRNA levels were observed in cDCs of other patient groups (Martin-Gayo et al. 2015). Taken together, these findings fit the observation that cDCs from ECs mount a better innate as well as HIV-1-specific $CD8^+$ T cell response (Martin-Gayo et al. 2015), a process that might otherwise be obstructed by SAMHD1 upregulation. Moreover, treatment of $CD4^+$ T cells with a combination of five cytokines that are specifically upregulated in ECs suppressed *SAMHD1* mRNA expression (Jacobs et al. 2017).

Therefore, it is tempting to speculate that HIV-1 did not evolve strategies for counteracting SAMHD1 to reduce infection of immune-competent myeloid cells and consequently escape immune sensing (Puigdomenech et al. 2013). By avoiding cGAS/STING-mediated RTI sensing, HIV-1 could be able to sidestep an efficient innate as well as adaptive immune response. In contrast, HIV-2 infection of DCs induced IFN production and DC activation (Lahaye et al. 2013). Supporting this notion, HIV-2-infected patients display are more efficient control of the virus by the immune system and progress less frequently as well as delayed to AIDS (Nyamweya et al. 2013; Rowland-Jones and Whittle 2007). While SAMHD1 is limiting the adaptive response during HIV-1 infection, strong HIV-2-specific $CD8^+$ T cell responses could be observed that were at times even cross-reactive to HIV-1 (Rowland-Jones et al. 1995). Additionally, HIV-1 disease progression was slowed down, which was reflected through higher $CD4^+$ T cell counts and delayed AIDS onset, by a preceding HIV-2 infection in patients (Esbjornsson et al. 2012).

However, viral dissemination and disease progression are severely impaired in macaques infected with Vpx-deficient SIVsmmPBj. By facilitating the infection of intraepithelial lymphocytes and macrophages in vivo, Vpx enables the virus to amplify at the initial sites of viral attack (Hirsch et al. 1998). Recent studies underline the importance of Vpx to establish and maintain infection in vivo (Belshan et al. 2012; Shingai et al. 2015). Additionally, infected macrophages were almost absent in vivo after infection without Vpx and, related to this, dissemination to/loss of $CD4^+$ T cells in the gut-associated lymphoid tissue (GALT) was minimized (Westmoreland et al. 2014). In contradiction, a recent publication proposed that Vpx may not have any influence on the infectivity of myeloid cells in various monkey models. They propose that myeloid cells may not be a major source of SIV in vivo (Calantone et al. 2014).

Despite its importance during SIV infections, the *vpx* gene was lost during the viral transfer from Old World monkeys to hominids. SIVcpz as well as HIV-1 do not encode *vpx*, but a reconstructed *vif* gene to efficiently antagonize hominid APOBEC3C (A3C) proteins. For cross-species transmission, Vif adaption to chimpanzee A3Cs seemed to be more important than retained SAMHD1 antagonism through Vpx (Etienne et al. 2013). Two possible explanations could help to solve this conundrum: On the one hand, HIV-1 and/or its viral precursors had to evolve different strategies to establish initial as well as ongoing infection to overcome lack of Vpx, especially in resting $CD4^+$ T cells and macrophages at the sites of initial infection. For instance, HIV-1 reverse transcriptase displays a higher affinity for dNTPs compared to its HIV-2 counterpart (Lenzi et al. 2014). On the other hand, avoiding an efficient innate response at the site of infection could be sufficient for HIV-1 to ensure initial viral acquisition. Here, intact SAMHD1-mediated restriction would prevent efficient replication, thereby sensing (Ahn et al. 2012; Maelfait et al. 2016; Puigdomenech et al. 2013), in immune-competent myeloid cells. Regardless, further elucidating the mechanism of how HIV-1 can overcome the lack of Vpx in vivo would be of great interest.

6 Conclusion

The understanding of the complex interplay of innate sensing is still works in progress. More than a decade ago, it was believed that HIV-1 is a more or less stealth virus that would not induce any innate responses. Only, with the discovery of the first restriction factors, such as APOBEC3G in 2002, and the recognition that these factors not only underlie positive selection and a constant arms race with the virus, but also that these factors are effectors of an innate immune response, the HIV field appreciated the possibility of a not yet defined cryptic recognition by PRRs. Second, the major research discoveries in the field of DNA sensing made it possible to ask the right questions for HIV-1 sensing. This drove important discoveries that cGAS together with PQBP1 and STING as well as IFI16-dependent pathways are recognizing HIV-1. Furthermore, this led to the insights of many other cellular factors that play important roles in protecting the viral core against sensing. Intriguingly, HIV-1 seems to be able to shape its own fate, it actively alters the outcome of sensing. Moreover, during evolution, HIV-1 may have lost the ability to replicate in immune-competent cells. This evolutionary trick may have allowed HIV-1 to escape the most prominent sensing-competent cells. A better understanding of this complex immune response will be helpful to develop better therapeutic strategies and to improve the efficiency of HIV vaccine candidates.

Acknowledgements We thank Dr. Manja Burggraf for critical reading and helpful discussions.

References

Abdul-Sater AA, Said-Sadier N, Lam VM, Singh B, Pettengill MA, Soares F, Tattoli I, Lipinski S, Girardin SE, Rosenstiel P, Ojcius DM (2010) Enhancement of reactive oxygen species production and chlamydial infection by the mitochondrial Nod-like family member NLRX1. J Biol Chem 285:41637–41645. doi:10.1074/jbc.M110.137885

Aebi M, Fah J, Hurt N, Samuel CE, Thomis D, Bazzigher L, Pavlovic J, Haller O, Staeheli P (1989) cDNA structures and regulation of two interferon-induced human Mx proteins. Mol Cell Biol 9:5062–5072

Ahn J, Hao C, Yan J, DeLucia M, Mehrens J, Wang C, Gronenborn AM, Skowronski J (2012) HIV/simian immunodeficiency virus (SIV) accessory virulence factor Vpx loads the host cell restriction factor SAMHD1 onto the E3 ubiquitin ligase complex CRL4DCAF1. J Biol Chem 287:12550–12558. doi:10.1074/jbc.M112.340711

Allen IC, Moore CB, Schneider M, Lei Y, Davis BK, Scull MA, Gris D, Roney KE, Zimmermann AG, Bowzard JB, Ranjan P, Monroe KM, Pickles RJ, Sambhara S, Ting JPY (2011) NLRX1 protein attenuates inflammatory responses to infection by interfering with the RIG-I-MAVS and TRAF6-NF-kappaB signaling pathways. Immunity 34:854–865. doi:10.1016/j.immuni.2011.03.026

Allouch A, Di Primio C, Alpi E, Lusic M, Arosio D, Giacca M, Cereseto A (2011) The TRIM family protein KAP1 inhibits HIV-1 integration. Cell Host Microbe 9:484–495. doi:10.1016/j.chom.2011.05.004

Almine JF, O'Hare CAJ, Dunphy G, Haga IR, Naik RJ, Atrih A, Connolly DJ, Taylor J, Kelsall IR, Bowie AG, Beard PM, Unterholzner L (2017) IFI16 and cGAS cooperate in the activation of STING during DNA sensing in human keratinocytes. Nat Commun 8:14392. doi:10.1038/ncomms14392

Amie SM, Bambara RA, Kim B (2013) GTP is the primary activator of the anti-HIV restriction factor SAMHD1. J Biol Chem 288:25001–25006. doi:10.1074/jbc.C113.493619

An P, Wang LH, Hutcheson-Dilks H, Nelson G, Donfield S, Goedert JJ, Rinaldo CR, Buchbinder S, Kirk GD, O'Brien SJ, Winkler CA (2007) Regulatory polymorphisms in the cyclophilin A gene, PPIA, accelerate progression to AIDS. PLoS Pathog 3:e88. doi:10.1371/journal.ppat.0030088

Antonucci JM, St Gelais C, de Silva S, Yount JS, Tang C, Ji X, Shepard C, Xiong Y, Kim B, Wu L (2016) SAMHD1-mediated HIV-1 restriction in cells does not involve ribonuclease activity. Nat Med 22:1072–1074. doi:10.1038/nm.4163

Arnold LH, Groom HCT, Kunzelmann S, Schwefel D, Caswell SJ, Ordonez P, Mann MC, Rueschenbaum S, Goldstone DC, Pennell S, Howell SA, Stoye JP, Webb M, Taylor IA, Bishop KN (2015) Phospho-dependent regulation of SAMHD1 oligomerisation couples catalysis and restriction. PLoS Pathog 11:e1005194. doi:10.1371/journal.ppat.1005194

Arnoult D, Soares F, Tattoli I, Castanier C, Philpott DJ, Girardin SE (2009) An N-terminal addressing sequence targets NLRX1 to the mitochondrial matrix. J Cell Sci 122:3161–3168. doi:10.1242/jcs.051193

Asaoka K, Ikeda K, Hishinuma T, Horie-Inoue K, Takeda S, Inoue S (2005) A retrovirus restriction factor TRIM5alpha is transcriptionally regulated by interferons. Biochem Biophys Res Commun 338:1950–1956. doi:10.1016/j.bbrc.2005.10.173

Ayinde D, Bruel T, Cardinaud S, Porrot F, Prado JG, Moris A, Schwartz O (2015) SAMHD1 limits HIV-1 antigen presentation by monocyte-derived dendritic cells. J Virol 89:6994–7006. doi:10.1128/JVI.00069-15

Baldauf H-M, Pan X, Erikson E, Schmidt S, Daddacha W, Burggraf M, Schenkova K, Ambiel I, Wabnitz G, Gramberg T, Panitz S, Flory E, Landau NR, Sertel S, Rutsch F, Lasitschka F, Kim B, Konig R, Fackler OT, Keppler OT (2012) SAMHD1 restricts HIV-1 infection in resting CD4(+) T cells. Nat Med 18:1682–1687. doi:10.1038/nm.2964

Barouch DH, Ghneim K, Bosche WJ, Li Y, Berkemeier B, Hull M, Bhattacharyya S, Cameron M, Liu J, Smith K, Borducchi E, Cabral C, Peter L, Brinkman A, Shetty M, Li H, Gittens C, Baker C, Wagner W, Lewis MG, Colantonio A, Kang H-J, Li W, Lifson JD, Piatak M Jr, Sekaly R-P (2016) Rapid inflammasome activation following mucosal SIV infection of rhesus monkeys. Cell 165:656–667. doi:10.1016/j.cell.2016.03.021

Behrendt R, Schumann T, Gerbaulet A, Nguyen LA, Schubert N, Alexopoulou D, Berka U, Lienenklaus S, Peschke K, Gibbert K, Wittmann S, Lindemann D, Weiss S, Dahl A, Naumann R, Dittmer U, Kim B, Mueller W, Gramberg T, Roers A (2013) Mouse SAMHD1 has antiretroviral activity and suppresses a spontaneous cell-intrinsic antiviral response. Cell Rep 4:689–696. doi:10.1016/j.celrep.2013.07.037

Beignon A-S, McKenna K, Skoberne M, Manches O, DaSilva I, Kavanagh DG, Larsson M, Gorelick RJ, Lifson JD, Bhardwaj N (2005) Endocytosis of HIV-1 activates plasmacytoid dendritic cells via Toll-like receptor-viral RNA interactions. J Clin Investig 115:3265–3275. doi:10.1172/JCI26032

Beloglazova N, Flick R, Tchigvintsev A, Brown G, Popovic A, Nocek B, Yakunin AF (2013) Nuclease activity of the human SAMHD1 protein implicated in the Aicardi-Goutieres syndrome and HIV-1 restriction. J Biol Chem 288:8101–8110. doi:10.1074/jbc.M112.431148

Belshan M, Kimata JT, Brown C, Cheng X, McCulley A, Larsen A, Thippeshappa R, Hodara V, Giavedoni L, Hirsch V, Ratner L (2012) Vpx is critical for SIVmne infection of pigtail macaques. Retrovirology 9:32. doi:10.1186/1742-4690-9-32

Berg RK, Melchjorsen J, Rintahaka J, Diget E, Soby S, Horan KA, Gorelick RJ, Matikainen S, Larsen CS, Ostergaard L, Paludan SR, Mogensen TH (2012) Genomic HIV RNA induces innate immune responses through RIG-I-dependent sensing of secondary-structured RNA. PLoS ONE 7:e29291. doi:10.1371/journal.pone.0029291

Berger A, Sommer AFR, Zwarg J, Hamdorf M, Welzel K, Esly N, Panitz S, Reuter A, Ramos I, Jatiani A, Mulder LCF, Fernandez-Sesma A, Rutsch F, Simon V, Konig R, Flory E (2011) SAMHD1-deficient CD14+ cells from individuals with Aicardi-Goutieres syndrome are highly susceptible to HIV-1 infection. PLoS Pathog 7:e1002425. doi:10.1371/journal.ppat.1002425

Bhattacharya A, Wang Z, White T, Buffone C, Nguyen LA, Shepard CN, Kim B, Demeler B, Diaz-Griffero F, Ivanov DN (2016) Effects of T592 phosphomimetic mutations on tetramer stability and dNTPase activity of SAMHD1 can not explain the retroviral restriction defect. Sci Rep 6:31353. doi:10.1038/srep31353

Bishop KN, Verma M, Kim E-Y, Wolinsky SM, Malim MH (2008) APOBEC3G inhibits elongation of HIV-1 reverse transcripts. PLoS Pathog 4:e1000231. doi:10.1371/journal.ppat.1000231

Biswas N, Wang T, Ding M, Tumne A, Chen Y, Wang Q, Gupta P (2012) ADAR1 is a novel multi targeted anti-HIV-1 cellular protein. Virology 422:265–277. doi:10.1016/j.virol.2011.10.024

Böhnlein E, Lowenthal JW, Siekevitz M, Ballard DW, Franza B, Greene WC (1988) The same inducible nuclear proteins regulates mitogen activation of both the interleukin-2 receptor-alpha gene and type 1 HIV. Cell 53:827–836. doi:10.1016/0092-8674(88)90099-2

Braaten D, Franke EK, Luban J (1996) Cyclophilin A is required for an early step in the life cycle of human immunodeficiency virus type 1 before the initiation of reverse transcription. J Virol 70:3551–3560

Brandariz-Nunez A, Valle-Casuso JC, White TE, Laguette N, Benkirane M, Brojatsch J, Diaz-Griffero F (2012) Role of SAMHD1 nuclear localization in restriction of HIV-1 and SIVmac. Retrovirology 9:49. doi:10.1186/1742-4690-9-49

Brass AL, Dykxhoorn DM, Benita Y, Yan N, Engelman A, Xavier RJ, Lieberman J, Elledge SJ (2008) Identification of host proteins required for HIV infection through a functional genomic screen. Science (New York, N.Y.) 319:921–926. doi:10.1126/science.1152725

Bridgeman A, Maelfait J, Davenne T, Partridge T, Peng Y, Mayer A, Dong T, Kaever V, Borrow P, Rehwinkel J (2015) Viruses transfer the antiviral second messenger cGAMP between cells. Science (New York, N.Y.) 349:1228–1232. doi:10.1126/science.aab3632

Bulli L, Apolonia L, Kutzner J, Pollpeter D, Goujon C, Herold N, Schwarz S-M, Giernat Y, Keppler OT, Malim MH, Schaller T (2016) Complex interplay between HIV-1 capsid and MX2-independent alpha interferon-induced antiviral factors. J Virol 90:7469–7480. doi:10.1128/JVI.00458-16

Burdick R, Smith JL, Chaipan C, Friew Y, Chen J, Venkatachari NJ, Delviks-Frankenberry KA, Hu W-S, Pathak VK (2010) P body-associated protein Mov10 inhibits HIV-1 replication at multiple stages. J Virol 84:10241–10253. doi:10.1128/JVI.00585-10

Bushman FD, Malani N, Fernandes J, D'Orso I, Cagney G, Diamond TL, Zhou H, Hazuda DJ, Espeseth AS, Konig R, Bandyopadhyay S, Ideker T, Goff SP, Krogan NJ, Frankel AD,

Young JAT, Chanda SK (2009) Host cell factors in HIV replication: meta-analysis of genome-wide studies. PLoS Pathog 5:e1000437. doi:10.1371/journal.ppat.1000437

Busnadiego I, Kane M, Rihn SJ, Preugschas HF, Hughes J, Blanco-Melo D, Strouvelle VP, Zang TM, Willett BJ, Boutell C, Bieniasz PD, Wilson SJ (2014) Host and viral determinants of Mx2 antiretroviral activity. J Virol 88:7738–7752. doi:10.1128/JVI.00214-14

Cagliani R, Forni D, Biasin M, Comabella M, Guerini FR, Riva S, Pozzoli U, Agliardi C, Caputo D, Malhotra S, Montalban X, Bresolin N, Clerici M, Sironi M (2014) Ancient and recent selective pressures shaped genetic diversity at AIM2-like nucleic acid sensors. Genome Biol Evol 6:830–845. doi:10.1093/gbe/evu066

Calantone N, Wu F, Klase Z, Deleage C, Perkins M, Matsuda K, Thompson EA, Ortiz AM, Vinton CL, Ourmanov I, Lore K, Douek DC, Estes JD, Hirsch VM, Brenchley JM (2014) Tissue myeloid cells in SIV-infected primates acquire viral DNA through phagocytosis of infected T cells. Immunity 41:493–502. doi:10.1016/j.immuni.2014.08.014

Campbell EM, Hope TJ (2015) HIV-1 capsid: the multifaceted key player in HIV-1 infection. Nat Rev Microbiol 13:471–483. doi:10.1038/nrmicro3503

Chen Q, Sun L, Chen ZJ (2016) Regulation and function of the cGAS-STING pathway of cytosolic DNA sensing. Nat Immunol 17:1142–1149. doi:10.1038/ni.3558

Cheney KM, McKnight A (2010) Interferon-alpha mediates restriction of human immunodeficiency virus type-1 replication in primary human macrophages at an early stage of replication. PLoS ONE 5:e13521. doi:10.1371/journal.pone.0013521

Chin KC, Cresswell P (2001) Viperin (cig5), an IFN-inducible antiviral protein directly induced by human cytomegalovirus. Proc Natl Acad Sci USA 98:15125–15130. doi:10.1073/pnas.011593298

Chow J, Franz KM, Kagan JC (2015) PRRs are watching you: localization of innate sensing and signaling regulators. Virology 479–480:104–109. doi:10.1016/j.virol.2015.02.051

Chun T-W, Engel D, Berrey MM, Shea T, Corey L, Fauci AS (1998) Early establishment of a pool of latently infected, resting CD4+ T cells during primary HIV-1 infection. Proc Natl Acad Sci 95:8869–8873. doi:10.1073/pnas.95.15.8869

Cobos Jimenez V, Booiman T, de Taeye SW, van Dort KA, Rits MAN, Hamann J, Kootstra NA (2012) Differential expression of HIV-1 interfering factors in monocyte-derived macrophages stimulated with polarizing cytokines or interferons. Sci Rep 2:763. doi:10.1038/srep00763

Coiras M, Bermejo M, Descours B, Mateos E, Garcia-Perez J, Lopez-Huertas M-R, Lederman MM, Benkirane M, Alcami J (2016) IL-7 Induces SAMHD1 Phosphorylation in CD4+ T Lymphocytes, Improving Early Steps of HIV-1 Life Cycle. Cell Rep 14:2100–2107. doi:10.1016/j.celrep.2016.02.022

Cooper A, Garcia M, Petrovas C, Yamamoto T, Koup RA, Nabel GJ (2013) HIV-1 causes CD4 cell death through DNA-dependent protein kinase during viral integration. Nature 498:376–379. doi:10.1038/nature12274

Coutermarsh-Ott S, Eden K, Allen IC (2016a) Beyond the inflammasome: regulatory NOD-like receptor modulation of the host immune response following virus exposure. J Gen Virol 97:825–838. doi:10.1099/jgv.0.000401

Coutermarsh-Ott S, Simmons A, Capria V, LeRoith T, Wilson JE, Heid B, Philipson CW, Qin Q, Hontecillas-Magarzo R, Bassaganya-Riera J, Ting JP-Y, Dervisis N, Allen IC (2016b) NLRX1 suppresses tumorigenesis and attenuates histiocytic sarcoma through the negative regulation of NF-kappaB signaling. Oncotarget 7:33096–33110. doi:10.18632/oncotarget.8861

Cribier A, Descours B, Valadao ALC, Laguette N, Benkirane M (2013) Phosphorylation of SAMHD1 by cyclin A2/CDK1 regulates its restriction activity toward HIV-1. Cell Rep 3:1036–1043. doi:10.1016/j.celrep.2013.03.017

Dastur A, Beaudenon S, Kelley M, Krug RM, Huibregtse JM (2006) Herc5, an interferon-induced HECT E3 enzyme, is required for conjugation of ISG15 in human cells. J Biol Chem 281:4334–4338. doi:10.1074/jbc.M512830200

Dawson MJ, Trapani JA (1995) IFI 16 gene encodes a nuclear protein whose expression is induced by interferons in human myeloid leukaemia cell lines. J Cell Biochem 57:39–51. doi:10.1002/jcb.240570106

Dempsey A, Bowie AG (2015) Innate immune recognition of DNA: a recent history. Virology 479–480:146–152. doi:10.1016/j.virol.2015.03.013

Descours B, Cribier A, Chable-Bessia C, Ayinde D, Rice G, Crow Y, Yatim A, Schwartz O, Laguette N, Benkirane M (2012) SAMHD1 restricts HIV-1 reverse transcription in quiescent CD4(+) T-cells. Retrovirology 9:87. doi:10.1186/1742-4690-9-87

Ding S, Pan Q, Liu S-L, Liang C (2014) HIV-1 mutates to evade IFITM1 restriction. Virology 454–455:11–24. doi:10.1016/j.virol.2014.01.020

Doehle BP, Chang K, Fleming L, McNevin J, Hladik F, McElrath MJ, Gale M Jr (2012) Vpu-deficient HIV strains stimulate innate immune signaling responses in target cells. J Virol 86:8499–8506. doi:10.1128/JVI.00424-12

Doehle BP, Hladik F, McNevin JP, McElrath MJ, Gale M Jr (2009) Human immunodeficiency virus type 1 mediates global disruption of innate antiviral signaling and immune defenses within infected cells. J Virol 83:10395–10405. doi:10.1128/JVI.00849-09

Doitsh G, Galloway NLK, Geng X, Yang Z, Monroe KM, Zepeda O, Hunt PW, Hatano H, Sowinski S, Munoz-Arias I, Greene WC (2014) Cell death by pyroptosis drives CD4 T-cell depletion in HIV-1 infection. Nature 505:509–514. doi:10.1038/nature12940

Donahue DA, Amraoui S, Di Nunzio F, Kieffer C, Porrot F, Opp S, Diaz-Griffero F, Casartelli N, Schwartz O (2016) SUN2 overexpression deforms nuclear shape and inhibits HIV. J Virol 90:4199–4214. doi:10.1128/JVI.03202-15

Doria M, Neri F, Gallo A, Farace MG, Michienzi A (2009) Editing of HIV-1 RNA by the double-stranded RNA deaminase ADAR1 stimulates viral infection. Nucleic Acids Res 37:5848–5858. doi:10.1093/nar/gkp604

Doria M, Tomaselli S, Neri F, Ciafre SA, Farace MG, Michienzi A, Gallo A (2011) ADAR2 editing enzyme is a novel human immunodeficiency virus-1 proviral factor. J Gen Virol 92:1228–1232. doi:10.1099/vir.0.028043-0

Doyle T, Goujon C, Malim MH (2015) HIV-1 and interferons: who's interfering with whom? Nature reviews. Microbiology 13:403–413. doi:10.1038/nrmicro3449

Duggal NK, Emerman M (2012) Evolutionary conflicts between viruses and restriction factors shape immunity. Nat Rev Immunol 12:687–695. doi:10.1038/nri3295

Duh EJ, Maury WJ, Folks TM, Fauci AS, Rabson AB (1989) Tumor necrosis factor alpha activates human immunodeficiency virus type 1 through induction of nuclear factor binding to the NF-kappa B sites in the long terminal repeat. Proc Natl Acad Sci USA 86:5974–5978

Eitas TK, Chou W-C, Wen H, Gris D, Robbins GR, Brickey J, Oyama Y, Ting JP-Y (2014) The nucleotide-binding leucine-rich repeat (NLR) family member NLRX1 mediates protection against experimental autoimmune encephalomyelitis and represses macrophage/microglia-induced inflammation. J Biol Chem 289:4173–4179. doi:10.1074/jbc.M113.533034

Esbjornsson J, Mansson F, Kvist A, Isberg P-E, Nowroozalizadeh S, Biague AJ, da Silva ZJ, Jansson M, Fenyo EM, Norrgren H, Medstrand P (2012) Inhibition of HIV-1 disease progression by contemporaneous HIV-2 infection. N Engl J Med 367:224–232. doi:10.1056/NEJMoa1113244

Etienne L, Hahn BH, Sharp PM, Matsen FA, Emerman M (2013) Gene loss and adaptation to hominids underlie the ancient origin of HIV-1. Cell Host Microbe 14:85–92. doi:10.1016/j.chom.2013.06.002

Farrell PJ, Broeze RJ, Lengyel P (1979) Accumulation of an mRNA and protein in interferon-treated Ehrlich ascites tumour cells. Nature 279:523–525

Faure E, Thomas L, Xu H, Medvedev A, Equils O, Arditi M (2001) Bacterial lipopolysaccharide and IFN-gamma induce Toll-like receptor 2 and Toll-like receptor 4 expression in human endothelial cells: role of NF-kappa B activation. J Immunol (Baltimore, Md.: 1950) 166:2018–2024

Forni D, Mozzi A, Pontremoli C, Vertemara J, Pozzoli U, Biasin M, Bresolin N, Clerici M, Cagliani R, Sironi M (2015) Diverse selective regimes shape genetic diversity at ADAR genes and at their coding targets. RNA Biol 12:149–161. doi:10.1080/15476286.2015.1017215

Friedman RL, Manly SP, McMahon M, Kerr IM, Stark GR (1984) Transcriptional and posttranscriptional regulation of interferon-induced gene expression in human cells. Cell 38:745–755

Furtak V, Mulky A, Rawlings SA, Kozhaya L, Lee K, Kewalramani VN, Unutmaz D (2010) Perturbation of the P-body component Mov10 inhibits HIV-1 infectivity. PLoS ONE 5:e9081. doi:10.1371/journal.pone.0009081

Galao RP, Le Tortorec A, Pickering S, Kueck T, Neil SJD (2012) Innate sensing of HIV-1 assembly by tetherin induces NFkappaB-dependent proinflammatory responses. Cell Host Microbe 12:633–644. doi:10.1016/j.chom.2012.10.007

Galao RP, Pickering S, Curnock R, Neil SJD (2014) Retroviral retention activates a Syk-dependent HemITAM in human tetherin. Cell Host Microbe 16:291–303. doi:10.1016/j.chom.2014.08.005

Galloway NLK, Doitsh G, Monroe KM, Yang Z, Munoz-Arias I, Levy DN, Greene WC (2015) Cell-to-cell transmission of HIV-1 is required to trigger pyroptotic death of lymphoid-tissue-derived CD4 T cells. Cell Rep 12:1555–1563. doi:10.1016/j.celrep.2015.08.011

Gao D, Wu J, Wu Y-T, Du F, Aroh C, Yan N, Sun L, Chen ZJ (2013) Cyclic GMP-AMP synthase is an innate immune sensor of HIV and other retroviruses. Science (New York, N.Y.) 341:903–906. doi:10.1126/science.1240933

Genovesio A, Kwon Y-J, Windisch MP, Kim NY, Choi SY, Kim HC, Jung S, Mammano F, Perrin V, Boese AS, Casartelli N, Schwartz O, Nehrbass U, Emans N (2011) Automated genome-wide visual profiling of cellular proteins involved in HIV infection. J Biomol Screen 16:945–958. doi:10.1177/1087057111415521

Gentili M, Kowal J, Tkach M, Satoh T, Lahaye X, Conrad C, Boyron M, Lombard B, Durand S, Kroemer G, Loew D, Dalod M, Thery C, Manel N (2015) Transmission of innate immune signaling by packaging of cGAMP in viral particles. Science (New York, N.Y.) 349:1232–1236. doi:10.1126/science.aab3628

Gloire G, Legrand-Poels S, Piette J (2006) NF-kappaB activation by reactive oxygen species: fifteen years later. Biochem Pharmacol 72:1493–1505. doi:10.1016/j.bcp.2006.04.011

Goldstone DC, Ennis-Adeniran V, Hedden JJ, Groom HCT, Rice GI, Christodoulou E, Walker PA, Kelly G, Haire LF, Yap MW, de Carvalho Luiz, Pedro S, Stoye JP, Crow YJ, Taylor IA, Webb M (2011) HIV-1 restriction factor SAMHD1 is a deoxynucleoside triphosphate triphosphohydrolase. Nature 480:379–382. doi:10.1038/nature10623

Goncalves A, Karayel E, Rice GI, Bennett KL, Crow YJ, Superti-Furga G, Burckstummer T (2012) SAMHD1 is a nucleic-acid binding protein that is mislocalized due to aicardi-goutieres syndrome-associated mutations. Hum Mutat 33:1116–1122. doi:10.1002/humu.22087

Goujon C, Arfi V, Pertel T, Luban J, Lienard J, Rigal D, Darlix J-L, Cimarelli A (2008) Characterization of simian immunodeficiency virus SIVSM/human immunodeficiency virus type 2 Vpx function in human myeloid cells. J Virol 82:12335–12345. doi:10.1128/JVI.01181-08

Goujon C, Malim MH (2010) Characterization of the alpha interferon-induced postentry block to HIV-1 infection in primary human macrophages and T cells. J Virol 84:9254–9266. doi:10.1128/JVI.00854-10

Goujon C, Moncorge O, Bauby H, Doyle T, Ward CC, Schaller T, Hue S, Barclay WS, Schulz R, Malim MH (2013) Human MX2 is an interferon-induced post-entry inhibitor of HIV-1 infection. Nature 502:559–562. doi:10.1038/nature12542

Goujon C, Riviere L, Jarrosson-Wuilleme L, Bernaud J, Rigal D, Darlix J-L, Cimarelli A (2007) SIVSM/HIV-2 Vpx proteins promote retroviral escape from a proteasome-dependent restriction pathway present in human dendritic cells. Retrovirology 4:2. doi:10.1186/1742-4690-4-2

Gringhuis SI, Hertoghs N, Kaptein TM, Zijlstra-Willems EM, Sarrami-Fooroshani R, Sprokholt JK, van Teijlingen NH, Kootstra NA, Booiman T, van Dort KA, Ribeiro CMS, Drewniak A, Geijtenbeek TBH (2017) HIV-1 blocks the signaling adaptor MAVS to evade antiviral host defense after sensing of abortive HIV-1 RNA by the host helicase DDX3. Nat Immunol 18:225–235. doi:10.1038/ni.3647

Guo H, Konig R, Deng M, Riess M, Mo J, Zhang L, Petrucelli A, Yoh SM, Barefoot B, Samo M, Sempowski GD, Zhang A, Colberg-Poley AM, Feng H, Lemon SM, Liu Y, Zhang Y, Wen H, Zhang Z, Damania B, Tsao L-C, Wang Q, Su L, Duncan JA, Chanda SK, Ting JP-Y (2016) NLRX1 sequesters STING to negatively regulate the interferon response, thereby facilitating the replication of HIV-1 and DNA viruses. Cell Host Microbe 19:515–528. doi:10.1016/j.chom.2016.03.001

Gupta S, Termini JM, Issac B, Guirado E, Stone GW (2016) Constitutively active MAVS inhibits HIV-1 replication via Type I interferon secretion and induction of HIV-1 restriction factors. PLoS ONE 11:e0148929. doi:10.1371/journal.pone.0148929

Hancks DC, Hartley MK, Hagan C, Clark NL, Elde NC (2015) Overlapping patterns of rapid evolution in the nucleic acid sensors cGAS and OAS1 suggest a common mechanism of pathogen antagonism and escape. PLoS Genet 11:e1005203. doi:10.1371/journal.pgen.1005203

Heil F, Hemmi H, Hochrein H, Ampenberger F, Kirschning C, Akira S, Lipford G, Wagner H, Bauer S (2004) Species-specific recognition of single-stranded RNA via toll-like receptor 7 and 8. Science (New York, N.Y.) 303:1526–1529. doi:10.1126/science.1093620

Henning MS, Morham SG, Goff SP, Naghavi MH (2010) PDZD8 is a novel Gag-interacting factor that promotes retroviral infection. J Virol 84:8990–8995. doi:10.1128/JVI.00843-10

Herzner A-M, Hagmann CA, Goldeck M, Wolter S, Kubler K, Wittmann S, Gramberg T, Andreeva L, Hopfner K-P, Mertens C, Zillinger T, Jin T, Xiao TS, Bartok E, Coch C, Ackermann D, Hornung V, Ludwig J, Barchet W, Hartmann G, Schlee M (2015) Sequence-specific activation of the DNA sensor cGAS by Y-form DNA structures as found in primary HIV-1 cDNA. Nat Immunol 16:1025–1033. doi:10.1038/ni.3267

Hirsch VM, Sharkey ME, Brown CR, Brichacek B, Goldstein S, Wakefield J, Byrum R, Elkins WR, Hahn BH, Lifson JD, Stevenson M (1998) Vpx is required for dissemination and pathogenesis of SIV(SM) PBj: evidence of macrophage-dependent viral amplification. Nat Med 4:1401–1408. doi:10.1038/3992

Hiscott J (2007) Triggering the innate antiviral response through IRF-3 activation. J Biol Chem 282:15325–15329. doi:10.1074/jbc.R700002200

Hofmann H, Logue EC, Bloch N, Daddacha W, Polsky SB, Schultz ML, Kim B, Landau NR (2012) The Vpx lentiviral accessory protein targets SAMHD1 for degradation in the nucleus. J Virol 86:12552–12560. doi:10.1128/JVI.01657-12

Hofmann H, Norton TD, Schultz ML, Polsky SB, Sunseri N, Landau NR (2013) Inhibition of CUL4A Neddylation causes a reversible block to SAMHD1-mediated restriction of HIV-1. J Virol 87:11741–11750. doi:10.1128/JVI.02002-13

Hogan CM, Degruttola V, Sun X, Fiscus SA, Del Rio C, Hare CB, Markowitz M, Connick E, Macatangay B, Tashima KT, Kallungal B, Camp R, Morton T, Daar ES, Little S (2012) The setpoint study (ACTG A5217): effect of immediate versus deferred antiretroviral therapy on virologic set point in recently HIV-1-infected individuals. J Infect Dis 205:87–96. doi:10.1093/infdis/jir699

Honda K, Takaoka A, Taniguchi T (2006) Type I interferon corrected gene induction by the interferon regulatory factor family of transcription factors. Immunity 25:349–360. doi:10.1016/j.immuni.2006.08.009

Hong M, S-i Yoon, Wilson IA (2012) Structure and functional characterization of the RNA-binding element of the NLRX1 innate immune modulator. Immunity 36:337–347. doi:10.1016/j.immuni.2011.12.018

Hori T, Takeuchi H, Saito H, Sakuma R, Inagaki Y, Yamaoka S (2013) A carboxy-terminally truncated human CPSF6 lacking residues encoded by exon 6 inhibits HIV-1 cDNA synthesis and promotes capsid disassembly. J Virol 87:7726–7736. doi:10.1128/JVI.00124-13

Horner SM, Liu HM, Park HS, Briley J, Gale M Jr (2011) Mitochondrial-associated endoplasmic reticulum membranes (MAM) form innate immune synapses and are targeted by hepatitis C virus. Proc Natl Acad Sci USA 108:14590–14595. doi:10.1073/pnas.1110133108

Hotter D, Kirchhoff F, Sauter D (2013) HIV-1 Vpu does not degrade interferon regulatory factor 3. J Virol 87:7160–7165. doi:10.1128/JVI.00526-13

Hrecka K, Hao C, Gierszewska M, Swanson SK, Kesik-Brodacka M, Srivastava S, Florens L, Washburn MP, Skowronski J (2011) Vpx relieves inhibition of HIV-1 infection of macrophages mediated by the SAMHD1 protein. Nature 474:658–661. doi:10.1038/nature10195
de Iaco A, Luban J (2014) Cyclophilin A promotes HIV-1 reverse transcription but its effect on transduction correlates best with its effect on nuclear entry of viral cDNA. Retrovirology 11:11. doi:10.1186/1742-4690-11-11
de Iaco A, Santoni F, Vannier A, Guipponi M, Antonarakis S, Luban J (2013) TNPO3 protects HIV-1 replication from CPSF6-mediated capsid stabilization in the host cell cytoplasm. Retrovirology 10:20. doi:10.1186/1742-4690-10-20
Imaizumi T, Hatakeyama M, Yamashita K, Yoshida H, Ishikawa A, Taima K, Satoh K, Mori F, Wakabayashi K (2004) Interferon-gamma induces retinoic acid-inducible gene-I in endothelial cells. Endothelium: J Endothelial Cell Res 11:169–173. doi:10.1080/10623320490512156
Imbeault E, Mahvelati TM, Braun R, Gris P, Gris D (2014) Nlrx1 regulates neuronal cell death. Mol Brain 7:90. doi:10.1186/s13041-014-0090-x
Ishikawa H, Barber GN (2008) STING is an endoplasmic reticulum adaptor that facilitates innate immune signalling. Nature 455:674–678. doi:10.1038/nature07317
Jacobs ES, Keating SM, Abdel-Mohsen M, Gibb SL, Heitman JW, Inglis HC, Martin JN, Zhang J, Kaidarova Z, Deng X, Wu S, Anastos K, Crystal H, Villacres MC, Young M, Greenblatt RM, Landay AL, Gange SJ, Deeks SG, Golub ET, Pillai SK, Norris PJ (2017) Cytokines elevated in HIV elite controllers reduce HIV replication in vitro and modulate HIV restriction factor expression. J Virol. doi:10.1128/JVI.02051-16
Jakobsen MR, Bak RO, Andersen A, Berg RK, Jensen SB, Tengchuan J, Laustsen A, Hansen K, Ostergaard L, Fitzgerald KA, Xiao TS, Mikkelsen JG, Mogensen TH, Paludan SR (2013) IFI16 senses DNA forms of the lentiviral replication cycle and controls HIV-1 replication. Proc Natl Acad Sci USA 110:E4571–E4580. doi:10.1073/pnas.1311669110
Jakobsen MR, Olagnier D, Hiscott J (2015) Innate immune sensing of HIV-1 infection. Curr Opin HIV AIDS 10:96–102. doi:10.1097/COH.0000000000000129
Jaworska J, Coulombe F, Downey J, Tzelepis F, Shalaby K, Tattoli I, Berube J, Rousseau S, Martin JG, Girardin SE, McCullers JA, Divangahi M (2014) NLRX1 prevents mitochondrial induced apoptosis and enhances macrophage antiviral immunity by interacting with influenza virus PB1-F2 protein. Proc Natl Acad Sci USA 111:E2110–E2119. doi:10.1073/pnas.1322118111
Jin C, Peng X, Liu F, Cheng L, Xie T, Lu X, Wu H, Wu N (2016) Interferon-induced sterile alpha motif and histidine/aspartic acid domain-containing protein 1 expression in astrocytes and microglia is mediated by microRNA-181a. AIDS (London, England) 30:2053–2064. doi:10.1097/QAD.0000000000001166
Jounai N, Takeshita F, Kobiyama K, Sawano A, Miyawaki A, Xin K-Q, Ishii KJ, Kawai T, Akira S, Suzuki K, Okuda K (2007) The Atg5 Atg12 conjugate associates with innate antiviral immune responses. Proc Natl Acad Sci USA 104:14050–14055. doi:10.1073/pnas.0704014104
Jønsson KL, Laustsen A, Krapp C, Skipper KA, Thavachelvam K, Hotter D, Egedal JH, Kjolby M, Mohammadi P, Prabakaran T, Sørensen LK, Sun C, Jensen SB, Holm CK, Lebbink RJ, Johannsen M, Nyegaard M, Mikkelsen JG, Kirchhoff F, Paludan SR, Jakobsen MR (2017) IFI16 is required for DNA sensing in human macrophages by promoting production and function of cGAMP. Nat Commun 8:14391. doi:10.1038/ncomms14391
Kane M, Yadav SS, Bitzegeio J, Kutluay SB, Zang T, Wilson SJ, Schoggins JW, Rice CM, Yamashita M, Hatziioannou T, Bieniasz PD (2013) MX2 is an interferon-induced inhibitor of HIV-1 infection. Nature 502:563–566. doi:10.1038/nature12653
Kang M-J, Yoon CM, Kim BH, Lee C-M, Zhou Y, Sauler M, Homer R, Dhamija A, Boffa D, West AP, Shadel GS, Ting JP, Tedrow Jr, Kaminski N, Kim WJ, Lee CG, Oh Y-M, Elias JA (2015) Suppression of NLRX1 in chronic obstructive pulmonary disease. J Clin Investig 125:2458–2462. doi:10.1172/JCI71747
Kappes JC, Morrow CD, Lee SW, Jameson BA, Kent SB, Hood LE, Shaw GM, Hahn BH (1988) Identification of a novel retroviral gene unique to human immunodeficiency virus type 2 and simian immunodeficiency virus SIVMAC. J Virol 62:3501–3505

Koblansky AA, Truax AD, Liu R, Montgomery SA, Ding S, Wilson JE, Brickey WJ, Muhlbauer M, McFadden R-MT, Hu P, Li Z, Jobin C, Lund PK, Ting JP-Y (2016) The innate immune receptor NLRX1 functions as a tumor suppressor by reducing colon tumorigenesis and key tumor-promoting signals. Cell Rep 14:2562–2575. doi:10.1016/j.celrep.2016.02.064

Konig R, Zhou Y, Elleder D, Diamond TL, Bonamy GMC, Irelan JT, Chiang C-Y, Tu BP, de Jesus PD, Lilley CE, Seidel S, Opaluch AM, Caldwell JS, Weitzman MD, Kuhen KL, Bandyopadhyay S, Ideker T, Orth AP, Miraglia LJ, Bushman FD, Young JA, Chanda SK (2008) Global analysis of host-pathogen interactions that regulate early-stage HIV-1 replication. Cell 135:49–60. doi:10.1016/j.cell.2008.07.032

Koning FA, Newman ENC, Kim E-Y, Kunstman KJ, Wolinsky SM, Malim MH (2009) Defining APOBEC3 expression patterns in human tissues and hematopoietic cell subsets. J Virol 83:9474–9485. doi:10.1128/JVI.01089-09

Konno H, Konno K, Barber GN (2013) Cyclic dinucleotides trigger ULK1 (ATG1) phosphorylation of STING to prevent sustained innate immune signaling. Cell 155:688–698. doi:10.1016/j.cell.2013.09.049

Korant BD, Blomstrom DC, Jonak GJ, Knight E Jr (1984) Interferon-induced proteins. Purification and characterization of a 15,000-dalton protein from human and bovine cells induced by interferon. J Biol Chem 259:14835–14839

Krapp C, Hotter D, Gawanbacht A, McLaren PJ, Kluge SF, Sturzel CM, Mack K, Reith E, Engelhart S, Ciuffi A, Hornung V, Sauter D, Telenti A, Kirchhoff F (2016) Guanylate binding protein (GBP) 5 is an interferon-inducible inhibitor of HIV-1 infectivity. Cell Host Microbe 19:504–514. doi:10.1016/j.chom.2016.02.019

Laguette N, Bregnard C, Hue P, Basbous J, Yatim A, Larroque M, Kirchhoff F, Constantinou A, Sobhian B, Benkirane M (2014) Premature activation of the SLX4 complex by Vpr promotes G2/M arrest and escape from innate immune sensing. Cell 156:134–145. doi:10.1016/j.cell.2013.12.011

Laguette N, Rahm N, Sobhian B, Chable-Bessia C, Munch J, Snoeck J, Sauter D, Switzer WM, Heneine W, Kirchhoff F, Delsuc F, Telenti A, Benkirane M (2012) Evolutionary and functional analyses of the interaction between the myeloid restriction factor SAMHD1 and the lentiviral Vpx protein. Cell Host Microbe 11:205–217. doi:10.1016/j.chom.2012.01.007

Laguette N, Sobhian B, Casartelli N, Ringeard M, Chable-Bessia C, Segeral E, Yatim A, Emiliani S, Schwartz O, Benkirane M (2011) SAMHD1 is the dendritic- and myeloid-cell-specific HIV-1 restriction factor counteracted by Vpx. Nature 474:654–657. doi:10.1038/nature10117

Lahaye X, Satoh T, Gentili M, Cerboni S, Conrad C, Hurbain I, El Marjou A, Lacabaratz C, Lelievre J-D, Manel N (2013) The capsids of HIV-1 and HIV-2 determine immune detection of the viral cDNA by the innate sensor cGAS in dendritic cells. Immunity 39:1132–1142. doi:10.1016/j.immuni.2013.11.002

Lahaye X, Satoh T, Gentili M, Cerboni S, Silvin A, Conrad C, Ahmed-Belkacem A, Rodriguez EC, Guichou J-F, Bosquet N, Piel M, Le Grand R, King MC, Pawlotsky J-M, Manel N (2016) Nuclear envelope protein SUN2 promotes cyclophilin-A-dependent steps of HIV replication. Cell Rep. doi:10.1016/j.celrep.2016.03.074

Lahouassa H, Daddacha W, Hofmann H, Ayinde D, Logue EC, Dragin L, Bloch N, Maudet C, Bertrand M, Gramberg T, Pancino G, Priet S, Canard B, Laguette N, Benkirane M, Transy C, Landau NR, Kim B, Margottin-Goguet F (2012) SAMHD1 restricts the replication of human immunodeficiency virus type 1 by depleting the intracellular pool of deoxynucleoside triphosphates. Nat Immunol 13:223–228. doi:10.1038/ni.2236

Landau NR (2014) The innate immune response to HIV-1: to sense or not to sense. DNA Cell Biol 33:271–274. doi:10.1089/dna.2014.2378

Le Sage V, Mouland AJ, Valiente-Echeverria F (2014) Roles of HIV-1 capsid in viral replication and immune evasion. Virus Res 193:116–129. doi:10.1016/j.virusres.2014.07.010

Le Tortorec A, Neil SJD (2009) Antagonism to and intracellular sequestration of human tetherin by the human immunodeficiency virus type 2 envelope glycoprotein. J Virol 83:11966–11978. doi:10.1128/JVI.01515-09

Lebon P, Badoual J, Ponsot G, Goutieres F, Hemeury-Cukier F, Aicardi J (1988) Intrathecal synthesis of interferon-alpha in infants with progressive familial encephalopathy. J Neurol Sci 84:201–208
Lee K, Ambrose Z, Martin TD, Oztop I, Mulky A, Julias JG, Vandegraaff N, Baumann JG, Wang R, Yuen W, Takemura T, Shelton K, Taniuchi I, Li Y, Sodroski J, Littman DR, Coffin JM, Hughes SH, Unutmaz D, Engelman A, Kewalramani VN (2010) Flexible use of nuclear import pathways by HIV-1. Cell Host Microbe 7:221–233. doi:10.1016/j.chom.2010.02.007
Lee K, Mulky A, Yuen W, Martin TD, Meyerson NR, Choi L, Yu H, Sawyer SL, Kewalramani VN (2012) HIV-1 capsid-targeting domain of cleavage and polyadenylation specificity factor 6. J Virol 86:3851–3860. doi:10.1128/JVI.06607-11
Lee MN, Roy M, Ong S-E, Mertins P, Villani A-C, Li W, Dotiwala F, Sen J, Doench JG, Orzalli MH, Kramnik I, Knipe DM, Lieberman J, Carr SA, Hacohen N (2013) Identification of regulators of the innate immune response to cytosolic DNA and retroviral infection by an integrative approach. Nat Immunol 14:179–185. doi:10.1038/ni.2509
Lei Y, Kansy BA, Li J, Cong L, Liu Y, Trivedi S, Wen H, Ting JP-Y, Ouyang H, Ferris RL (2016) EGFR-targeted mAb therapy modulates autophagy in head and neck squamous cell carcinoma through NLRX1-TUFM protein complex. Oncogene 35:4698–4707. doi:10.1038/onc.2016.11
Lei Y, Wen H, Yu Y, Taxman DJ, Zhang L, Widman DG, Swanson KV, Wen K-W, Damania B, Moore CB, Giguere PM, Siderovski DP, Hiscott J, Razani B, Semenkovich CF, Chen X, Ting JP-Y (2012) The mitochondrial proteins NLRX1 and TUFM form a complex that regulates type I interferon and autophagy. Immunity 36:933–946. doi:10.1016/j.immuni.2012.03.025
Lenzi GM, Domaoal RA, Kim D-H, Schinazi RF, Kim B (2014) Kinetic variations between reverse transcriptases of viral protein X coding and noncoding lentiviruses. Retrovirology 11:111. doi:10.1186/s12977-014-0111-y
Lepelley A, Louis S, Sourisseau M, Law HKW, Pothlichet J, Schilte C, Chaperot L, Plumas J, Randall RE, Si-Tahar M, Mammano F, Albert ML, Schwartz O (2011) Innate sensing of HIV-infected cells. PLoS Pathog 7:e1001284. doi:10.1371/journal.ppat.1001284
Lewin AR, Reid LE, McMahon M, Stark GR, Kerr IM (1991) Molecular analysis of a human interferon-inducible gene family. Eur J Biochem 199:417–423
Lewis GM, Wehrens EJ, Labarta-Bajo L, Streeck H, Zuniga EI (2016) TGF-beta receptor maintains CD4 T helper cell identity during chronic viral infections. J Clin Investig 126:3799–3813. doi:10.1172/JCI87041
Li H, Zhang S, Li F, Qin L (2016) NLRX1 attenuates apoptosis and inflammatory responses in myocardial ischemia by inhibiting MAVS-dependent NLRP3 inflammasome activation. Mol Immunol 76:90–97. doi:10.1016/j.molimm.2016.06.013
Li M, Kao E, Gao X, Sandig H, Limmer K, Pavon-Eternod M, Jones TE, Landry S, Pan T, Weitzman MD, David M (2012) Codon-usage-based inhibition of HIV protein synthesis by human schlafen 11. Nature 491:125–128. doi:10.1038/nature11433
Li N, Zhang W, Cao X (2000) Identification of human homologue of mouse IFN-gamma induced protein from human dendritic cells. Immunol Lett 74:221–224
Lim ES, Fregoso OI, McCoy CO, Matsen FA, Malik HS, Emerman M (2012a) The ability of primate lentiviruses to degrade the monocyte restriction factor SAMHD1 preceded the birth of the viral accessory protein Vpx. Cell Host Microbe 11:194–204. doi:10.1016/j.chom.2012.01.004
Lim ES, Malik HS, Emerman M (2010) Ancient adaptive evolution of tetherin shaped the functions of Vpu and Nef in human immunodeficiency virus and primate lentiviruses. J Virol 84:7124–7134. doi:10.1128/JVI.00468-10
Lim ES, Wu LI, Malik HS, Emerman M (2012b) The function and evolution of the restriction factor Viperin in primates was not driven by lentiviruses. Retrovirology 9:55. doi:10.1186/1742-4690-9-55
Lindbäck S, Karlsson AC, Mittler J, Blaxhult A, Carlsson M, Briheim G, Sönnerborg A, Gaines H (2000) Viral dynamics in primary HIV-1 infection. AIDS 14:2283–2291. doi:10.1097/00002030-200010200-00009

Liu L, Oliveira NMM, Cheney KM, Pade C, Dreja H, Bergin A-MH, Borgdorff V, Beach DH, Bishop CL, Dittmar MT, McKnight A (2011) A whole genome screen for HIV restriction factors. Retrovirology 8:94. doi:10.1186/1742-4690-8-94

Liu Z, Pan Q, Ding S, Qian J, Xu F, Zhou J, Cen S, Guo F, Liang C (2013) The interferon-inducible MxB protein inhibits HIV-1 infection. Cell Host Microbe 14:398–410. doi:10.1016/j.chom.2013.08.015

Liu Z, Pan Q, Liang Z, Qiao W, Cen S, Liang C (2015) The highly polymorphic cyclophilin A-binding loop in HIV-1 capsid modulates viral resistance to MxB. Retrovirology 12:1. doi:10.1186/s12977-014-0129-1

Lodermeyer V, Suhr K, Schrott N, Kolbe C, Sturzel CM, Krnavek D, Munch J, Dietz C, Waldmann T, Kirchhoff F, Goffinet C (2013) 90 K, an interferon-stimulated gene product, reduces the infectivity of HIV-1. Retrovirology 10:111. doi:10.1186/1742-4690-10-111

Lu J, Pan Q, Rong L, He W, Liu S-L, Liang C (2011) The IFITM proteins inhibit HIV-1 infection. J Virol 85:2126–2137. doi:10.1128/JVI.01531-10

Luban J (2012) Innate immune sensing of HIV-1 by dendritic cells. Cell Host Microbe 12:408–418. doi:10.1016/j.chom.2012.10.002

Luban J, Bossolt KL, Franke EK, Kalpana GV, Goff SP (1993) Human immunodeficiency virus type 1 Gag protein binds to cyclophilins A and B. Cell 73:1067–1078

Ma B, Herzog EL, Moore M, Lee C-M, Na SH, Lee CG, Elias JA (2016) RIG-like helicase regulation of chitinase 3-like 1 axis and pulmonary metastasis. Sci Rep 6:26299. doi:10.1038/srep26299

Ma F, Li B, Liu S-y, Iyer SS, Yu Y, Wu A, Cheng G (2015) Positive feedback regulation of type I IFN production by the IFN-inducible DNA sensor cGAS. J Immunol (Baltimore, Md.: 1950) 194:1545–1554. doi:10.4049/jimmunol.1402066

Maelfait J, Bridgeman A, Benlahrech A, Cursi C, Rehwinkel J (2016) Restriction by SAMHD1 limits cGAS/STING-dependent innate and adaptive immune responses to HIV-1. Cell Rep 16:1492–1501. doi:10.1016/j.celrep.2016.07.002

Malim MH, Bieniasz PD (2012) HIV restriction factors and mechanisms of evasion. Cold Spring Harb Perspect Med 2:a006940. doi:10.1101/cshperspect.a006940

Mandl JN, Barry AP, Vanderford TH, Kozyr N, Chavan R, Klucking S, Barrat FJ, Coffman RL, Staprans SI, Feinberg MB (2008) Divergent TLR7 and TLR9 signaling and type I interferon production distinguish pathogenic and nonpathogenic AIDS virus infections. Nat Med 14:1077–1087. doi:10.1038/nm.1871

Manel N, Hogstad B, Wang Y, Levy DE, Unutmaz D, Littman DR (2010) A cryptic sensor for HIV-1 activates antiviral innate immunity in dendritic cells. Nature 467:214–217. doi:10.1038/nature09337

Mar KB, Schoggins JW (2016) NLRX1 Helps HIV Avoid a STING Operation. Cell Host Microbe 19:430–431. doi:10.1016/j.chom.2016.03.011

Marchi S, Patergnani S, Pinton P (2014) The endoplasmic reticulum-mitochondria connection: one touch, multiple functions. Biochem Biophys Acta 1837:461–469. doi:10.1016/j.bbabio.2013.10.015

Marth C, Dreps A, Natoli C, Zeimet AG, Lang T, Widschwendter M, Daxenbichler G, Ullrich A, Iacobelli S (1994) Effects of type-I and -II interferons on 90 K antigen expression in ovarian carcinoma cells. Int J Cancer 59:808–813

Martin-Gayo E, Buzon MJ, Ouyang Z, Hickman T, Cronin J, Pimenova D, Walker BD, Lichterfeld M, Yu XG (2015) Potent cell-intrinsic immune responses in dendritic cells facilitate HIV-1-specific T cell immunity in HIV-1 elite controllers. PLoS Pathog 11:e1004930. doi:10.1371/journal.ppat.1004930

McLaren PJ, Gawanbacht A, Pyndiah N, Krapp C, Hotter D, Kluge SF, Gotz N, Heilmann J, Mack K, Sauter D, Thompson D, Perreaud J, Rausell A, Munoz M, Ciuffi A, Kirchhoff F, Telenti A (2015) Identification of potential HIV restriction factors by combining evolutionary genomic signatures with functional analyses. Retrovirology 12:41. doi:10.1186/s12977-015-0165-5

McMichael AJ, Borrow P, Tomaras GD, Goonetilleke N, Haynes BF (2010) The immune response during acute HIV-1 infection: clues for vaccine development. Nat Rev Immunol 10:11–23. doi:10.1038/nri2674

McNatt MW, Zang T, Hatziioannou T, Bartlett M, Fofana IB, Johnson WE, Neil SJD, Bieniasz PD (2009) Species-specific activity of HIV-1 Vpu and positive selection of tetherin transmembrane domain variants. PLoS Pathog 5:e1000300. doi:10.1371/journal.ppat.1000300

Meier A, Alter G, Frahm N, Sidhu H, Li B, Bagchi A, Teigen N, Streeck H, Stellbrink H-J, Hellman J, van Lunzen J, Altfeld M (2007) MyD88-dependent immune activation mediated by human immunodeficiency virus type 1-encoded Toll-like receptor ligands. J Virol 81:8180–8191. doi:10.1128/JVI.00421-07

Meyerson NR, Rowley PA, Swan CH, Le DT, Wilkerson GK, Sawyer SL (2014) Positive selection of primate genes that promote HIV-1 replication. Virology 454–455:291–298. doi:10.1016/j.virol.2014.02.029

Miettinen M, Sareneva T, Julkunen I, Matikainen S (2001) IFNs activate toll-like receptor gene expression in viral infections. Genes Immun 2:349–355. doi:10.1038/sj.gene.6363791

Misumi S, Inoue M, Dochi T, Kishimoto N, Hasegawa N, Takamune N, Shoji S (2010) Uncoating of human immunodeficiency virus type 1 requires prolyl isomerase Pin1. J Biol Chem 285:25185–25195. doi:10.1074/jbc.M110.114256

Miyakawa K, Ryo A, Murakami T, Ohba K, Yamaoka S, Fukuda M, Guatelli J, Yamamoto N (2009) BCA2/Rabring7 promotes tetherin-dependent HIV-1 restriction. PLoS Pathog 5: e1000700. doi:10.1371/journal.ppat.1000700

Mohan M, Kaushal D, Aye PP, Alvarez X, Veazey RS, Lackner AA (2012) Focused examination of the intestinal lamina propria yields greater molecular insight into mechanisms underlying SIV induced immune dysfunction. PLoS ONE 7:e34561. doi:10.1371/journal.pone.0034561

Monroe KM, Yang Z, Johnson JR, Geng X, Doitsh G, Krogan NJ, Greene WC (2014) IFI16 DNA sensor is required for death of lymphoid CD4 T cells abortively infected with HIV. Science (New York, N.Y.) 343:428–432. doi:10.1126/science.1243640

Moore CB, Bergstralh DT, Duncan JA, Lei Y, Morrison TE, Zimmermann AG, Accavitti-Loper MA, Madden VJ, Sun L, Ye Z, Lich JD, Heise MT, Chen Z, Ting JP-Y (2008) NLRX1 is a regulator of mitochondrial antiviral immunity. Nature 451:573–577. doi:10.1038/nature06501

Munoz-Arias I, Doitsh G, Yang Z, Sowinski S, Ruelas D, Greene WC (2015) Blood-derived CD4 T cells naturally resist pyroptosis during abortive HIV-1 infection. Cell Host Microbe 18:463–470. doi:10.1016/j.chom.2015.09.010

Murrell B, Vollbrecht T, Guatelli J, Wertheim JO (2016) The evolutionary histories of antiretroviral proteins SERINC3 and SERINC5 do not support an evolutionary arms race in primates. J Virol 90:8085–8089. doi:10.1128/JVI.00972-16

Nabel G, Baltimore D (1987) An inducible transcription factor activates expression of human immunodeficiency virus in T cells. Nature 326:711–713. doi:10.1038/326711a0

Nasi M, Biasi S de, Bianchini E, Digaetano M, Pinti M, Gibellini L, Pecorini S, Carnevale G, Guaraldi G, Borghi V, Mussini C, Cossarizza A (2015) Analysis of inflammasomes and antiviral sensing components reveals decreased expression of NLRX1 in HIV-positive patients assuming efficient antiretroviral therapy. AIDS (London, England) 29:1937–1941. doi:10.1097/QAD.0000000000000830

Nasr N, Maddocks S, Turville SG, Harman AN, Woolger N, Helbig KJ, Wilkinson J, Bye CR, Wright TK, Rambukwelle D, Donaghy H, Beard MR, Cunningham AL (2012) HIV-1 infection of human macrophages directly induces viperin which inhibits viral production. Blood 120:778–788. doi:10.1182/blood-2012-01-407395

Nazli A, Kafka JK, Ferreira VH, Anipindi V, Mueller K, Osborne BJ, Dizzell S, Chauvin S, Mian MF, Ouellet M, Tremblay MJ, Mossman KL, Ashkar AA, Kovacs C, Bowdish DME, Snider DP, Kaul R, Kaushic C (2013) HIV-1 gp120 induces TLR2- and TLR4-mediated innate immune activation in human female genital epithelium. J Immunol (Baltimore, Md.: 1950) 191:4246–4258. doi:10.4049/jimmunol.1301482

Neil SJD, Zang T, Bieniasz PD (2008) Tetherin inhibits retrovirus release and is antagonized by HIV-1 Vpu. Nature 451:425–430. doi:10.1038/nature06553

Nityanandam R, Serra-Moreno R (2014) BCA2/Rabring7 targets HIV-1 Gag for lysosomal degradation in a tetherin-independent manner. PLoS Pathog 10:e1004151. doi:10.1371/journal.ppat.1004151

Nyamweya S, Hegedus A, Jaye A, Rowland-Jones S, Flanagan KL, Macallan DC (2013) Comparing HIV-1 and HIV-2 infection: Lessons for viral immunopathogenesis. Rev Med Virol 23:221–240. doi:10.1002/rmv.1739

Ocwieja KE, Brady TL, Ronen K, Huegel A, Roth SL, Schaller T, James LC, Towers GJ, Young JAT, Chanda SK, Konig R, Malani N, Berry CC, Bushman FD (2011) HIV integration targeting: a pathway involving Transportin-3 and the nuclear pore protein RanBP2. PLoS Pathog 7:e1001313. doi:10.1371/journal.ppat.1001313

Oh D-Y, Baumann K, Hamouda O, Eckert JK, Neumann K, Kucherer C, Bartmeyer B, Poggensee G, Oh N, Pruss A, Jessen H, Schumann RR (2009) A frequent functional toll-like receptor 7 polymorphism is associated with accelerated HIV-1 disease progression. AIDS (London, England) 23:297–307. doi:10.1097/QAD.0b013e32831fb540

OhAinle M, Kerns JA, Malik HS, Emerman M (2006) Adaptive evolution and antiviral activity of the conserved mammalian cytidine deaminase APOBEC3H. J Virol 80:3853–3862. doi:10.1128/JVI.80.8.3853-3862.2006

Okumura A, Alce T, Lubyova B, Ezelle H, Strebel K, Pitha PM (2008) HIV-1 accessory proteins VPR and Vif modulate antiviral response by targeting IRF-3 for degradation. Virology 373:85–97. doi:10.1016/j.virol.2007.10.042

Okumura A, Lu G, Pitha-Rowe I, Pitha PM (2006) Innate antiviral response targets HIV-1 release by the induction of ubiquitin-like protein ISG15. Proc Natl Acad Sci USA 103:1440–1445. doi:10.1073/pnas.0510518103

Ortiz M, Guex N, Patin E, Martin O, Xenarios I, Ciuffi A, Quintana-Murci L, Telenti A (2009) Evolutionary trajectories of primate genes involved in HIV pathogenesis. Mol Biol Evol 26:2865–2875. doi:10.1093/molbev/msp197

Paillart JC, Marquet R, Skripkin E, Ehresmann C, Ehresmann B (1996) Dimerization of retroviral genomic RNAs: structural and functional implications. Biochimie 78:639–653

Patterson JB, Thomis DC, Hans SL, Samuel CE (1995) Mechanism of interferon action: double-stranded RNA-specific adenosine deaminase from human cells is inducible by alpha and gamma interferons. Virology 210:508–511. doi:10.1006/viro.1995.1370

Pertel T, Hausmann S, Morger D, Zuger S, Guerra J, Lascano J, Reinhard C, Santoni FA, Uchil PD, Chatel L, Bisiaux A, Albert ML, Strambio-De-Castillia C, Mothes W, Pizzato M, Grutter MG, Luban J (2011) TRIM5 is an innate immune sensor for the retrovirus capsid lattice. Nature 472:361–365. doi:10.1038/nature09976

Phuphuakrat A, Kraiwong R, Boonarkart C, Lauhakirti D, Lee T-H, Auewarakul P (2008) Double-stranded RNA adenosine deaminases enhance expression of human immunodeficiency virus type 1 proteins. J Virol 82:10864–10872. doi:10.1128/JVI.00238-08

Pincetic A, Kuang Z, Seo EJ, Leis J (2010) The interferon-induced gene ISG15 blocks retrovirus release from cells late in the budding process. J Virol 84:4725–4736. doi:10.1128/JVI.02478-09

Pine SO, McElrath MJ, Bochud PY (2009) Polymorphisms in toll-like receptor 4 and toll-like receptor 9 influence viral load in a seroincident cohort of HIV-1-infected individuals. AIDS (London, England) 23:2387–2395. doi:10.1097/QAD.0b013e328330b489

Powell RD, Holland PJ, Hollis T, Perrino FW (2011) Aicardi-Goutieres syndrome gene and HIV-1 restriction factor SAMHD1 is a dGTP-regulated deoxynucleotide triphosphohydrolase. J Biol Chem 286:43596–43600. doi:10.1074/jbc.C111.317628

Puigdomenech I, Casartelli N, Porrot F, Schwartz O (2013) SAMHD1 restricts HIV-1 cell-to-cell transmission and limits immune detection in monocyte-derived dendritic cells. J Virol 87:2846–2856. doi:10.1128/JVI.02514-12

Rasaiyaah J, Tan CP, Fletcher AJ, Price AJ, Blondeau C, Hilditch L, Jacques DA, Selwood DL, James LC, Noursadeghi M, Towers GJ (2013) HIV-1 evades innate immune recognition through specific cofactor recruitment. Nature 503:402–405. doi:10.1038/nature12769

Rasheedi S, Shun M-C, Serrao E, Sowd GA, Qian J, Hao C, Dasgupta T, Engelman AN, Skowronski J (2016) The cleavage and polyadenylation specificity factor 6 (CPSF6) subunit of the capsid-recruited pre-messenger RNA cleavage factor I (CFIm) complex mediates HIV-1 integration into genes. J Biol Chem 291:11809–11819. doi:10.1074/jbc.M116.721647

Raturi A, Simmen T (2013) Where the endoplasmic reticulum and the mitochondrion tie the knot: the mitochondria-associated membrane (MAM). Biochem Biophys Acta 1833:213–224. doi:10.1016/j.bbamcr.2012.04.013

Rebsamen M, Vazquez J, Tardivel A, Guarda G, Curran J, Tschopp J (2011) NLRX1/NOD5 deficiency does not affect MAVS signalling. Cell Death Differ 18:1387. doi:10.1038/cdd.2011.64

Rehwinkel J, Maelfait J, Bridgeman A, Rigby R, Hayward B, Liberatore RA, Bieniasz PD, Towers GJ, Moita LF, Crow YJ, Bonthron DT, Reis e Sousa C (2013) SAMHD1-dependent retroviral control and escape in mice. EMBO J 32:2454–2462. doi:10.1038/emboj.2013.163

Rhee H-W, Zou P, Udeshi ND, Martell JD, Mootha VK, Carr SA, Ting AY (2013) Proteomic mapping of mitochondria in living cells via spatially restricted enzymatic tagging. Science (New York, N.Y.) 339:1328–1331. doi:10.1126/science.1230593

Ribeiro IP, Menezes AN, Moreira MAM, Bonvicino CR, Seuanez HN, Soares MA (2005) Evolution of cyclophilin A and TRIMCyp retrotransposition in new world primates. J Virol 79:14998–15003. doi:10.1128/JVI.79.23.14998-15003.2005

Rice GI, Bond J, Asipu A et al (2009) Mutations involved in Aicardi-Goutieres syndrome implicate SAMHD1 as regulator of the innate immune response. Nat Genet 41:829–832. doi:10.1038/ng.373

Riess M, Fuchs NV, Idica A, Hamdorf M, Flory E, Pedersen IM, Konig R (2017) Interferons induce expression of SAMHD1 in monocytes through down-regulation of miR-181a and miR-30a. J Biol Chem 292:264–277. doi:10.1074/jbc.M116.752584

Rits MAN, van Dort KA, Kootstra NA (2008) Polymorphisms in the regulatory region of the Cyclophilin A gene influence the susceptibility for HIV-1 infection. PLoS ONE 3:e3975. doi:10.1371/journal.pone.0003975

Rosa A, Chande A, Ziglio S, de Sanctis V, Bertorelli R, Goh SL, McCauley SM, Nowosielska A, Antonarakis SE, Luban J, Santoni FA, Pizzato M (2015) HIV-1 Nef promotes infection by excluding SERINC5 from virion incorporation. Nature 526:212–217. doi:10.1038/nature15399

Rowland-Jones S, Sutton J, Ariyoshi K, Dong T, Gotch F, McAdam S, Whitby D, Sabally S, Gallimore A, Corrah T (1995) HIV-specific cytotoxic T-cells in HIV-exposed but uninfected Gambian women. Nat Med 1:59–64

Rowland-Jones SL, Whittle HC (2007) Out of Africa: what can we learn from HIV-2 about protective immunity to HIV-1? Nat Immunol 8:329–331. doi:10.1038/ni0407-329

Rustagi A, Gale M Jr (2014) Innate antiviral immune signaling, viral evasion and modulation by HIV-1. J Mol Biol 426:1161–1177. doi:10.1016/j.jmb.2013.12.003

Ryoo J, Choi J, Oh C, Kim S, Seo M, Kim S-Y, Seo D, Kim J, White TE, Brandariz-Nunez A, Diaz-Griffero F, Yun C-H, Hollenbaugh JA, Kim B, Baek D, Ahn K (2014) The ribonuclease activity of SAMHD1 is required for HIV-1 restriction. Nat Med 20:936–941. doi:10.1038/nm.3626

Ryoo J, Hwang S-Y, Choi J, Oh C, Ahn K (2016) SAMHD1, the Aicardi-Goutieres syndrome gene and retroviral restriction factor, is a phosphorolytic ribonuclease rather than a hydrolytic ribonuclease. Biochem Biophys Res Commun 477:977–981. doi:10.1016/j.bbrc.2016.07.013

Saitoh T, Fujita N, Hayashi T, Takahara K, Satoh T, Lee H, Matsunaga K, Kageyama S, Omori H, Noda T, Yamamoto N, Kawai T, Ishii K, Takeuchi O, Yoshimori T, Akira S (2009) Atg9a controls dsDNA-driven dynamic translocation of STING and the innate immune response. Proc Natl Acad Sci USA 106:20842–20846. doi:10.1073/pnas.0911267106

Samuel CE (2011) Adenosine deaminases acting on RNA (ADARs) are both antiviral and proviral. Virology 411:180–193. doi:10.1016/j.virol.2010.12.004

Sasaki O, Yoshizumi T, Kuboyama M, Ishihara T, Suzuki E, S-i Kawabata, Koshiba T (2013) A structural perspective of the MAVS-regulatory mechanism on the mitochondrial outer membrane using bioluminescence resonance energy transfer. Biochem Biophys Acta 1833:1017–1027. doi:10.1016/j.bbamcr.2013.01.010

Sauter D, Schindler M, Specht A, Landford WN, Munch J, Kim K-A, Votteler J, Schubert U, Bibollet-Ruche F, Keele BF, Takehisa J, Ogando Y, Ochsenbauer C, Kappes JC, Ayouba A,

Peeters M, Learn GH, Shaw G, Sharp PM, Bieniasz P, Hahn BH, Hatziioannou T, Kirchhoff F (2009) Tetherin-driven adaptation of Vpu and Nef function and the evolution of pandemic and nonpandemic HIV-1 strains. Cell Host Microbe 6:409–421. doi:10.1016/j.chom.2009.10.004
Sawyer SL, Emerman M, Malik HS (2004) Ancient adaptive evolution of the primate antiviral DNA-editing enzyme APOBEC3G. PLoS Biol 2:E275. doi:10.1371/journal.pbio.0020275
Sawyer SL, Wu LI, Emerman M, Malik HS (2005) Positive selection of primate TRIM5alpha identifies a critical species-specific retroviral restriction domain. Proc Natl Acad Sci USA 102:2832–2837. doi:10.1073/pnas.0409853102
Schaller T, Ocwieja KE, Rasaiyaah J, Price AJ, Brady TL, Roth SL, Hue S, Fletcher AJ, Lee K, Kewalramani VN, Noursadeghi M, Jenner RG, James LC, Bushman FD, Towers GJ (2011) HIV-1 capsid-cyclophilin interactions determine nuclear import pathway, integration targeting and replication efficiency. PLoS Pathog 7:e1002439. doi:10.1371/journal.ppat.1002439
Schlee M (2013) Master sensors of pathogenic RNA—RIG-I like receptors. Immunobiology 218:1322–1335. doi:10.1016/j.imbio.2013.06.007
Schmidt B, Ashlock BM, Foster H, Fujimura SH, Levy JA (2005) HIV-infected cells are major inducers of plasmacytoid dendritic cell interferon production, maturation, and migration. Virology 343:256–266. doi:10.1016/j.virol.2005.09.059
Schoggins JW, Wilson SJ, Panis M, Murphy MY, Jones CT, Bieniasz P, Rice CM (2011) A diverse range of gene products are effectors of the type I interferon antiviral response. Nature 472:481–485. doi:10.1038/nature09907
Seamon KJ, Sun Z, Shlyakhtenko LS, Lyubchenko YL, Stivers JT (2015) SAMHD1 is a single-stranded nucleic acid binding protein with no active site-associated nuclease activity. Nucleic Acids Res 43:6486–6499. doi:10.1093/nar/gkv633
Seamon KJ, Bumpus NN, Stivers JT (2016) Single-stranded nucleic acids bind to the tetramer interface of SAMHD1 and prevent formation of the catalytic homotetramer. Biochemistry 55:6087–6099. doi:10.1021/acs.biochem.6b00986
Serra M, Forcales S-V, Pereira-Lopes S, Lloberas J, Celada A (2011) Characterization of Trex1 induction by IFN-gamma in murine macrophages. J Immunol (Baltimore, Md.: 1950) 186:2299–2308. doi:10.4049/jimmunol.1002364
Sharova N, Wu Y, Zhu X, Stranska R, Kaushik R, Sharkey M, Stevenson M (2008) Primate lentiviral Vpx commandeers DDB1 to counteract a macrophage restriction. PLoS Pathog 4: e1000057. doi:10.1371/journal.ppat.1000057
Sheehy AM, Gaddis NC, Choi JD, Malim MH (2002) Isolation of a human gene that inhibits HIV-1 infection and is suppressed by the viral Vif protein. Nature 418:646–650. doi:10.1038/nature00939
Shingai M, Welbourn S, Brenchley JM, Acharya P, Miyagi E, Plishka RJ, Buckler-White A, Kwong PD, Nishimura Y, Strebel K, Martin MA (2015) The expression of functional Vpx during pathogenic SIVmac infections of rhesus macaques suppresses SAMHD1 in CD4+ memory T cells. PLoS Pathog 11:e1004928. doi:10.1371/journal.ppat.1004928
Singh K, Poteryakhina A, Zheltukhin A, Bhatelia K, Prajapati P, Sripada L, Tomar D, Singh R, Singh AK, Chumakov PM, Singh R (2015) NLRX1 acts as tumor suppressor by regulating TNF-alpha induced apoptosis and metabolism in cancer cells. Biochem Biophys Acta 1853:1073–1086. doi:10.1016/j.bbamcr.2015.01.016
Smed-Sorensen A, Lore K, Vasudevan J, Louder MK, Andersson J, Mascola JR, Spetz A-L, Koup RA (2005) Differential susceptibility to human immunodeficiency virus type 1 infection of myeloid and plasmacytoid dendritic cells. J Virol 79:8861–8869. doi:10.1128/JVI.79.14.8861-8869.2005
Smith JL, Izumi T, Borbet TC, Hagedorn AN, Pathak VK (2014) HIV-1 and HIV-2 Vif interact with human APOBEC3 proteins using completely different determinants. J Virol 88:9893–9908. doi:10.1128/JVI.01318-14
Soares F, Tattoli I, Rahman MA, Robertson SJ, Belcheva A, Liu D, Streutker C, Winer S, Winer DA, Martin A, Philpott DJ, Arnoult D, Girardin SE (2014) The mitochondrial protein NLRX1 controls the balance between extrinsic and intrinsic apoptosis. J Biol Chem 289:19317–19330. doi:10.1074/jbc.M114.550111

Soares F, Tattoli I, Wortzman ME, Arnoult D, Philpott DJ, Girardin SE (2013) NLRX1 does not inhibit MAVS-dependent antiviral signalling. Innate Immun 19:438–448. doi:10.1177/1753425912467383

Solis M, Nakhaei P, Jalalirad M, Lacoste J, Douville R, Arguello M, Zhao T, Laughrea M, Wainberg MA, Hiscott J (2011) RIG-I-mediated antiviral signaling is inhibited in HIV-1 infection by a protease-mediated sequestration of RIG-I. J Virol 85:1224–1236. doi:10.1128/JVI.01635-10

Sommer AFR, Riviere L, Qu B, Schott K, Riess M, Ni Y, Shepard C, Schnellbacher E, Finkernagel M, Himmelsbach K, Welzel K, Kettern N, Donnerhak C, Munk C, Flory E, Liese J, Kim B, Urban S, Konig R (2016) Restrictive influence of SAMHD1 on hepatitis B virus life cycle. Sci Rep 6:26616. doi:10.1038/srep26616

Sowd GA, Serrao E, Wang H, Wang W, Fadel HJ, Poeschla EM, Engelman AN (2016) A critical role for alternative polyadenylation factor CPSF6 in targeting HIV-1 integration to transcriptionally active chromatin. Proc Natl Acad Sci USA 113:E1054–E1063. doi:10.1073/pnas.1524213113

Srivastava S, Swanson SK, Manel N, Florens L, Washburn MP, Skowronski J (2008) Lentiviral Vpx accessory factor targets VprBP/DCAF1 substrate adaptor for cullin 4 E3 ubiquitin ligase to enable macrophage infection. PLoS Pathog 4:e1000059. doi:10.1371/journal.ppat.1000059

St Gelais C, de Silva S, Amie SM, Coleman CM, Hoy H, Hollenbaugh JA, Kim B, Wu L (2012) SAMHD1 restricts HIV-1 infection in dendritic cells (DCs) by dNTP depletion, but its expression in DCs and primary CD4+ T-lymphocytes cannot be upregulated by interferons. Retrovirology 9:105. doi:10.1186/1742-4690-9-105

Stetson DB, Ko JS, Heidmann T, Medzhitov R (2008) Trex1 prevents cell-intrinsic initiation of autoimmunity. Cell 134:587–598. doi:10.1016/j.cell.2008.06.032

Stetson DB, Medzhitov R (2006) Recognition of cytosolic DNA activates an IRF3-dependent innate immune response. Immunity 24:93–103. doi:10.1016/j.immuni.2005.12.003

Stremlau M, Owens CM, Perron MJ, Kiessling M, Autissier P, Sodroski J (2004) The cytoplasmic body component TRIM5alpha restricts HIV-1 infection in old world monkeys. Nature 427:848–853. doi:10.1038/nature02343

Sumpter R, Jr, Loo Y-M, Foy E, Li K, Yoneyama M, Fujita T, Lemon SM, Gale Jr M (2005) Regulating intracellular antiviral defense and permissiveness to hepatitis C virus RNA replication through a cellular RNA helicase, RIG-I. J Virol 79:2689–2699. doi:10.1128/JVI.79.5.2689-2699.2005

Sun L, Wu J, Du F, Chen X, Chen ZJ (2013) Cyclic GMP-AMP synthase is a cytosolic DNA sensor that activates the type I interferon pathway. Science (New York, N.Y.) 339:786–791. doi:10.1126/science.1232458

Swiecki M, Colonna M (2015) The multifaceted biology of plasmacytoid dendritic cells. Nat Rev Immunol 15:471–485. doi:10.1038/nri3865

Tada T, Zhang Y, Koyama T, Tobiume M, Tsunetsugu-Yokota Y, Yamaoka S, Fujita H, Tokunaga K (2015) MARCH8 inhibits HIV-1 infection by reducing virion incorporation of envelope glycoproteins. Nat Med 21:1502–1507. doi:10.1038/nm.3956

Takeuchi O, Akira S (2010) Pattern recognition receptors and inflammation. Cell 140:805–820. doi:10.1016/j.cell.2010.01.022

Tang C, Ji X, Wu L, Xiong Y (2015) Impaired dNTPase activity of SAMHD1 by phosphomimetic mutation of Thr-592. J Biol Chem 290:26352–26359. doi:10.1074/jbc.M115.677435

Tattoli I, Carneiro LA, Jehanno M, Magalhaes JG, Shu Y, Philpott DJ, Arnoult D, Girardin SE (2008) NLRX1 is a mitochondrial NOD-like receptor that amplifies NF-kappaB and JNK pathways by inducing reactive oxygen species production. EMBO Rep 9:293–300. doi:10.1038/sj.embor.7401161

Tattoli I, Killackey SA, Foerster EG, Molinaro R, Maisonneuve C, Rahman MA, Winer S, Winer DA, Streutker CJ, Philpott DJ, Girardin SE (2016) NLRX1 acts as an epithelial-intrinsic tumor suppressor through the modulation of TNF-mediated proliferation. Cell Rep 14:2576–2586. doi:10.1016/j.celrep.2016.02.065

Tong-Starksen SE, Luciw PA, Peterlin BM (1987) Human immunodeficiency virus long terminal repeat responds to T-cell activation signals. Proc Natl Acad Sci USA 84:6845–6849

Towers GJ, Hatziioannou T, Cowan S, Goff SP, Luban J, Bieniasz PD (2003) Cyclophilin A modulates the sensitivity of HIV-1 to host restriction factors. Nat Med 9:1138–1143. doi:10.1038/nm910

Towers GJ, Noursadeghi M (2014) Interactions between HIV-1 and the cell-autonomous innate immune system. Cell Host Microbe 16:10–18. doi:10.1016/j.chom.2014.06.009

Trapani JA, Browne KA, Dawson MJ, Ramsay RG, Eddy RL, Show TB, White PC, Dupont B (1992) A novel gene constitutively expressed in human lymphoid cells is inducible with interferon-gamma in myeloid cells. Immunogenetics 36:369–376

Tristem M, Marshall C, Karpas A, Petrik J, Hill F (1990) Origin of vpx in lentiviruses. Nature 347:341–342. doi:10.1038/347341b0

Tungler V, Staroske W, Kind B, Dobrick M, Kretschmer S, Schmidt F, Krug C, Lorenz M, Chara O, Schwille P, Lee-Kirsch MA (2013) Single-stranded nucleic acids promote SAMHD1 complex formation. J Mol Med (Berlin, Germany) 91:759–770. doi:10.1007/s00109-013-0995-3

Unger BL, Ganesan S, Comstock AT, Faris AN, Hershenson MB, Sajjan US (2014) Nod-like receptor X-1 is required for rhinovirus-induced barrier dysfunction in airway epithelial cells. J Virol 88:3705–3718. doi:10.1128/JVI.03039-13

Usami Y, Wu Y, Gottlinger HG (2015) SERINC3 and SERINC5 restrict HIV-1 infectivity and are counteracted by Nef. Nature 526:218–223. doi:10.1038/nature15400

van Damme N, Goff D, Katsura C, Jorgenson RL, Mitchell R, Johnson MC, Stephens EB, Guatelli J (2008) The interferon-induced protein BST-2 restricts HIV-1 release and is downregulated from the cell surface by the viral Vpu protein. Cell Host Microbe 3:245–252. doi:10.1016/j.chom.2008.03.001

Vasseur E, Patin E, Laval G, Pajon S, Fornarino S, Crouau-Roy B, Quintana-Murci L (2011) The selective footprints of viral pressures at the human RIG-I-like receptor family. Hum Mol Genet 20:4462–4474. doi:10.1093/hmg/ddr377

Vigneault F, Woods M, Buzon MJ, Li C, Pereyra F, Crosby SD, Rychert J, Church G, Martinez-Picado J, Rosenberg ES, Telenti A, Yu XG, Lichterfeld M (2011) Transcriptional profiling of CD4 T cells identifies distinct subgroups of HIV-1 elite controllers. J Virol 85:3015–3019. doi:10.1128/JVI.01846-10

Wang F-X, Huang J, Zhang H, Ma X, Zhang H (2008) APOBEC3G upregulation by alpha interferon restricts human immunodeficiency virus type 1 infection in human peripheral plasmacytoid dendritic cells. J Gen Virol 89:722–730. doi:10.1099/vir.0.83530-0

Wang X, Han Y, Dang Y, Fu W, Zhou T, Ptak RG, Zheng Y-H (2010) Moloney leukemia virus 10 (MOV10) protein inhibits retrovirus replication. J Biol Chem 285:14346–14355. doi:10.1074/jbc.M110.109314

Wang Y, Wang X, Li J, Zhou Y, Ho W (2013) RIG-I activation inhibits HIV replication in macrophages. J Leukoc Biol 94:337–341. doi:10.1189/jlb.0313158

Welbourn S, Dutta SM, Semmes OJ, Strebel K (2013) Restriction of virus infection but not catalytic dNTPase activity is regulated by phosphorylation of SAMHD1. J Virol 87:11516–11524. doi:10.1128/JVI.01642-13

Westmoreland SV, Converse AP, Hrecka K, Hurley M, Knight H, Piatak M, Lifson J, Mansfield KG, Skowronski J, Desrosiers RC (2014) SIV vpx is essential for macrophage infection but not for development of AIDS. PLoS ONE 9:e84463. doi:10.1371/journal.pone.0084463

White TE, Brandariz-Nunez A, Valle-Casuso JC, Amie S, Nguyen LA, Kim B, Tuzova M, Diaz-Griffero F (2013) The retroviral restriction ability of SAMHD1, but not its deoxynucleotide triphosphohydrolase activity, is regulated by phosphorylation. Cell Host Microbe 13:441–451. doi:10.1016/j.chom.2013.03.005

Whitney JB, Hill AL, Sanisetty S, Penaloza-MacMaster P, Liu J, Shetty M, Parenteau L, Cabral C, Shields J, Blackmore S, Smith JY, Brinkman AL, Peter LE, Mathew SI, Smith KM, Borducchi EN, Rosenbloom DIS, Lewis MG, Hattersley J, Li B, Hesselgesser J, Geleziunas R,

Robb ML, Kim JH, Michael NL, Barouch DH (2014) Rapid seeding of the viral reservoir prior to SIV viraemia in rhesus monkeys. Nature 512:74–77. doi:10.1038/nature13594

Wiercinska-Drapalo A, Flisiak R, Jaroszewicz J, Prokopowicz D (2004) Increased plasma transforming growth factor-beta1 is associated with disease progression in HIV-1-infected patients. Viral Immunol 17:109–113. doi:10.1089/088282404322875502

Wilkins J, Zheng Y-M, Yu J, Liang C, Liu S-L (2016) Nonhuman Primate IFITM Proteins Are Potent Inhibitors of HIV and SIV. PLoS ONE 11:e0156739. doi:10.1371/journal.pone.0156739

Wong JJY, Pung YF, Sze NS-K, Chin K-C (2006) HERC5 is an IFN-induced HECT-type E3 protein ligase that mediates type I IFN-induced ISGylation of protein targets. Proc Natl Acad Sci USA 103:10735–10740. doi:10.1073/pnas.0600397103

Woods MW, Kelly JN, Hattlmann CJ, Tong JGK, Xu LS, Coleman MD, Quest GR, Smiley JR, Barr SD (2011) Human HERC5 restricts an early stage of HIV-1 assembly by a mechanism correlating with the ISGylation of Gag. Retrovirology 8:95. doi:10.1186/1742-4690-8-95

Woods MW, Tong JG, Tom SK, Szabo PA, Cavanagh PC, Dikeakos JD, Haeryfar SMM, Barr SD (2014) Interferon-induced HERC5 is evolving under positive selection and inhibits HIV-1 particle production by a novel mechanism targeting Rev/RRE-dependent RNA nuclear export. Retrovirology 11:27. doi:10.1186/1742-4690-11-27

Wu J, Sun L, Chen X, Du F, Shi H, Chen C, Chen ZJ (2013) Cyclic GMP-AMP is an endogenous second messenger in innate immune signaling by cytosolic DNA. Science (New York, N.Y.) 339:826–830. doi:10.1126/science.1229963

Xia X, Cui J, Wang HY, Zhu L, Matsueda S, Wang Q, Yang X, Hong J, Songyang Z, Chen ZJ, Wang R-F (2011) NLRX1 negatively regulates TLR-induced NF-kappaB signaling by targeting TRAF6 and IKK. Immunity 34:843–853. doi:10.1016/j.immuni.2011.02.022

Xu S, Ducroux A, Ponnurangam A, Vieyres G, Franz S, Musken M, Zillinger T, Malassa A, Ewald E, Hornung V, Barchet W, Haussler S, Pietschmann T, Goffinet C (2016) cGAS-mediated innate immunity spreads intercellularly through HIV-1 Env-induced membrane fusion sites. Cell Host Microbe 20:443–457. doi:10.1016/j.chom.2016.09.003

Yan J, Hao C, DeLucia M, Swanson S, Florens L, Washburn MP, Ahn J, Skowronski J (2015) CyclinA2-Cyclin-dependent kinase regulates SAMHD1 protein phosphohydrolase domain. J Biol Chem 290:13279–13292. doi:10.1074/jbc.M115.646588

Yan J, Kaur S, DeLucia M, Hao C, Mehrens J, Wang C, Golczak M, Palczewski K, Gronenborn AM, Ahn J, Skowronski J (2013) Tetramerization of SAMHD1 is required for biological activity and inhibition of HIV infection. J Biol Chem 288:10406–10417. doi:10.1074/jbc.M112.443796

Yan N, Cherepanov P, Daigle JE, Engelman A, Lieberman J (2009) The SET complex acts as a barrier to autointegration of HIV-1. PLoS Pathog 5:e1000327. doi:10.1371/journal.ppat.1000327

Yan N, Regalado-Magdos AD, Stiggelbout B, Lee-Kirsch MA, Lieberman J (2010a) The cytosolic exonuclease TREX1 inhibits the innate immune response to human immunodeficiency virus type 1. Nat Immunol 11:1005–1013. doi:10.1038/ni.1941

Yan N, Regalado-Magdos AD, Stiggelbout B, Lee-Kirsch MA, Lieberman J (2010b) The cytosolic exonuclease TREX1 inhibits the innate immune response to human immunodeficiency virus type 1. Nat Immunol 11:1005–1013. doi:10.1038/ni.1941

Yang H-C, Xing S, Shan L, O'Connell K, Dinoso J, Shen A, Zhou Y, Shrum CK, Han Y, Liu JO, Zhang H, Margolick JB, Siliciano RF (2009) Small-molecule screening using a human primary cell model of HIV latency identifies compounds that reverse latency without cellular activation. J Clin Investig 119:3473–3486. doi:10.1172/JCI39199

Yoh SM, Schneider M, Seifried J, Soonthornvacharin S, Akleh RE, Olivieri KC, de Jesus PD, Ruan C, de Castro E, Ruiz PA, Germanaud D, Des Portes V, Garcia-Sastre A, Konig R, Chanda SK (2015) PQBP1 is a proximal sensor of the cGAS-dependent innate response to HIV-1. Cell 161:1293–1305. doi:10.1016/j.cell.2015.04.050

Yonezawa A, Morita R, Takaori-Kondo A, Kadowaki N, Kitawaki T, Hori T, Uchiyama T (2003) Natural alpha interferon-producing cells respond to human immunodeficiency virus type 1 with alpha interferon production and maturation into dendritic cells. J Virol 77:3777–3784
Yu XF, Yu QC, Essex M, Lee TH (1991) The vpx gene of simian immunodeficiency virus facilitates efficient viral replication in fresh lymphocytes and macrophage. J Virol 65:5088–5091
Zhang H, Hu J (2016) Shaping the endoplasmic reticulum into a social network. Trends Cell Biol. doi:10.1016/j.tcb.2016.06.002
Zhang R, Mehla R, Chauhan A (2010) Perturbation of host nuclear membrane component RanBP2 impairs the nuclear import of human immunodeficiency virus -1 preintegration complex (DNA). PLoS ONE 5:e15620. doi:10.1371/journal.pone.0015620
Zhao K, Du J, Han X, Goodier JL, Li P, Zhou X, Wei W, Evans SL, Li L, Zhang W, Cheung LE, Wang G, Kazazian HH Jr, Yu X-F (2013) Modulation of LINE-1 and Alu/SVA retrotransposition by Aicardi-Goutieres syndrome-related SAMHD1. Cell Rep 4:1108–1115. doi:10.1016/j.celrep.2013.08.019
Zhou H, Xu M, Huang Q, Gates AT, Zhang XD, Castle JC, Stec E, Ferrer M, Strulovici B, Hazuda DJ, Espeseth AS (2008) Genome-scale RNAi screen for host factors required for HIV replication. Cell Host Microbe 4:495–504. doi:10.1016/j.chom.2008.10.004

Host Factors Involved in Ebola Virus Replication

Angela L. Rasmussen

Abstract Ebola virus (EBOV) is a highly pathogenic emerging virus that represents a serious threat to global public health and a major priority for biodefense. The 2014 West African outbreak demonstrated the potential of EBOV to cause an epidemic affecting thousands of people. The severity of disease and high case fatality rate of EBOV is largely due to the host response elicited by the virus. EBOV infection hijacks a number of host pathways to carry out replication and stimulate potent inflammatory responses, while simultaneously subverting the host antiviral immune response. Together, these events trigger a complex, systemic, often lethal febrile disease characterized by high levels of inflammatory cytokines, acute hepatitis and liver dysfunction, immune antagonism, gastrointestinal distress, and, in some cases, hemorrhage caused by coagulopathy and vascular leakage. This review presents current knowledge about the particular host responses induced and disrupted by EBOV infection and how these contribute to virus replication, immune evasion, pathogenesis, and disease outcome.

Contents

A.L. Rasmussen (✉)
Center for Infection and Immunity, Columbia University Mailman School of Public Health, New York, NY 10032, USA
e-mail: alr2105@cumc.columbia.edu

Current Topics in Microbiology and Immunology (2018) 419:113–150
DOI 10.1007/82_2017_27

Published Online: 15 July 2017

1 Introduction

Ebola virus (EBOV) has commanded the attention of the world since its emergence four decades ago in the village of Yambuku in Zaire (now the Democratic Republic of Congo). The first outbreak of what is now known as Ebola virus disease (EVD) caused severe hemorrhagic fever with a mortality rate approaching 90% (WHO 1978b). Despite the sporadic nature of most EBOV outbreaks, the severity of symptoms and the high mortality rate triggered serious concerns about larger epidemics. In 2014, many of those fears were realized when a variant strain of EBOV emerged unexpectedly in Guinea and spread rapidly through West Africa (Baize et al. 2014). By June 2016, there were 28,616 cases of EVD with 11,310 deaths (WHO 2016), devastating parts of Sierra Leone, Liberia, and Guinea, and demonstrating the potential threat to global public health.

Scientific study of EBOV has typically focused on the function of viral proteins, although it is clear that the host response to infection is a major determinant of pathogenicity. Individual patients develop a diverse variety of disease manifestations ranging in severity from asymptomatic or mild febrile illness to severe, lethal hemorrhagic fever. EBOV infection elicits a potent global transcriptional response in multiple cell types, and severe EVD is linked to uncontrolled, systemic inflammation in vivo. A number of host proteins also play integral roles in virus entry, replication, and egress (Fig. Disease presentation, pathogenesis, and clinical outcome all depend on host background and response to infection as much or more as the viral strain causing the infection.

2 Host Factors Required for Cell Entry and Fusion

Cell entry is one of the most studied processes in the EBOV replication cycle. However, for decades, the EBOV receptor remained unknown, as filoviruses utilize an unconventional entry strategy that does not require a host receptor on the surface of the cell. Recently, great advances have been made in understanding the particular host machinery required to mediate virus entry into the cell and then into the cytoplasm. However, many aspects of this process remain enigmatic, particularly regarding host factors that trigger fusion of the viral envelope with late endosomal membranes.

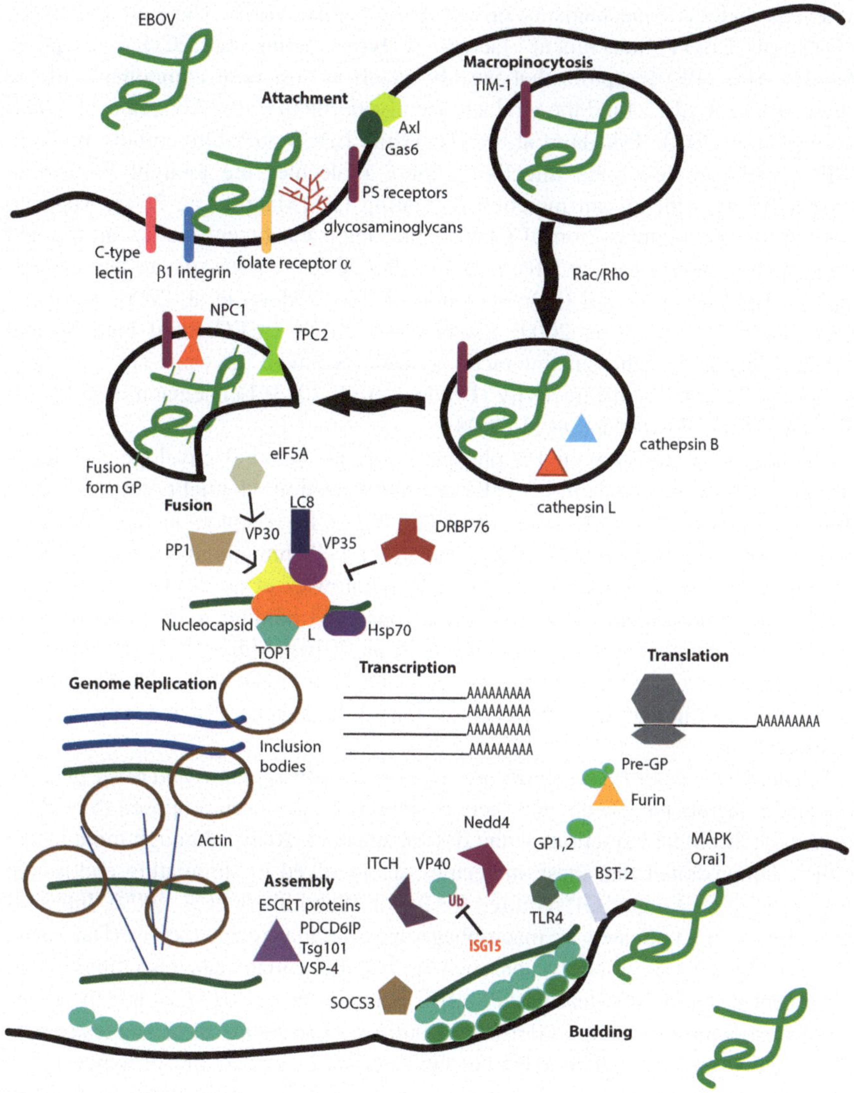

Fig.1 A number of host proteins also play integral roles in virus entry, replication, and egress

2.1 *Attachment Factors*

A host cell surface receptor capable of independently mediating entry has yet to be conclusively identified. Instead of binding a single host cell surface receptor molecule or molecular complex to mediate internalization, EBOV virions interact nonspecifically with various cell surface molecules and are subsequently taken up by several endocytic mechanisms, primarily macropinocytosis. There are two major classes of EBOV attachment factors, C-type lectins (CLECs) and phosphatidylserine (PS) receptors that weakly attach to structural components of the virion surface. CLECs bind the viral surface glycoprotein (GP) (Alvarez et al. 2002; Brudner et al. 2013; Takada et al. 2004). While EBOV GP differentially binds to CLECs and can determine infectivity, these molecules are thought to interact nonspecifically with glycan moieties decorating GP (Marzi et al. 2007; Matsuno et al. 2010). The composition of CLECs and other attachment factors on the cell surface partly determines the infectivity of different *Ebolavirus* species and cellular tropism (Brudner et al. 2013; Dahlmann et al. 2015; Marzi et al. 2006; Simmons et al. 2003a; Takada et al. 2004; Usami et al. 2011). CLECs bind both N- and O-linked glycans, although interactions with N-linked glycans are generally stronger determinants of infectivity (Brudner et al. 2013; Lennemann et al. 2014; Lin et al. 2003; Powlesland et al. 2008).

PS receptors interact with PS phospholipids in the viral envelope and are a GP-independent mechanism of cellular internalization (Jemielity et al. 2013; Moller-Tank et al. 2013; Morizono and Chen 2014). PS receptors include members of the Tyro3/Axl/Mer (TAM) family and the T cell immunoglobulin and mucin domain (TIM) family (Hunt et al. 2011; Kondratowicz et al. 2011). These molecules mediate attachment and entry via a conserved binding pocket that directly binds PS and triggers phagocytosis (Rhein et al. 2016). Although TIM-1 can bind both GP and envelope PS and mediate entry through either molecule (Yuan et al. 2015), interaction between PS receptors and GP is generally not required for internalization.

Neither CLECs nor PS receptors are required for EBOV entry, and other putative attachment factors have been identified. β1-integrins have been proposed to mediate attachment through interactions with GP (Takada et al. 2000), although more recent studies demonstrated that these molecules are involved in stimulating endosomal proteolytic processing of GP required for fusion (Schornberg et al. 2009). Integrins have also been implicated as macrophage-specific attachment factors (Dahlmann et al. 2015). Glycosaminoglycans such as heparin sulfate can also function as attachment receptors (O'Hearn et al. 2015; Salvador et al. 2013), as has the folate receptor α (Simmons et al. 2003b). Cell culture experiments have suggested that EBOV entry is restricted to adherent cells, possibly due to the presence of an adhesion-specific co-receptor (Dube et al. 2010). Soluble innate immune effectors such as mannose-binding lectin (Ji et al. 2005) and ficolin-1 (Favier et al. 2016) have both been shown to enhance EBOV infection by augmenting entry. No individual molecule or molecular family has been shown to be required for virus attachment and entry. However, both CLECs and TAM family members have been shown to have immunomodulatory activity that can alter susceptibility, which may explain cell type-specific differences in susceptibility (Bhattacharyya et al. 2013; Zhao et al. 2016).

2.2 *Endosomal Uptake*

Virus particles are internalized following association with the surface attachment factors. In some cases, virions may remain in a cell surface-associated state for an extended period of time prior to internalization (Reynard and Volchkov 2015). Although both clathrin-dependent (Aleksandrowicz et al. 2011; Bhattacharyya et al. 2011; Bhattacharyya et al. 2010) and, to a much lesser extent, caveolar (Empig and Goldsmith 2002) endocytic pathways can mediate EBOV entry, virions are primarily taken up by macropinocytosis (Aleksandrowicz et al. 2011; Nanbo et al. 2010; Saeed et al. 2010). In some cell types, this may be mediated by an unconventional dynamin-driven macropinocytosis pathway (Carette et al. 2011). Although it is not clear how CLECs induce endocytosis, PS receptors can induce cellular uptake of PS-containing molecules either by directly binding the envelope phospholipid or through an intermediate connecting the two molecules. The TAM family member and TIM proteins can enhance macropinocytosis through direct or indirect interactions with PS residues in the virion envelope (Brindley et al. 2011; Hunt et al. 2011; Jemielity et al. 2013; Shimojima et al. 2007; Shimojima et al. 2006). In the case of Axl, this is mediated by growth arrest-specific protein 6 precursor (Gas6), which acts as an intermolecular bridge between virion envelope PS with Axl (Brindley et al. 2011). Because this process is thought to be a normal homeostatic mechanism for clearing apoptotic cells with exposed PS on the surface, this entry strategy has been termed "apoptotic mimicry" (Moller-Tank et al. 2013; Morizono and Chen 2014). Another membrane phospholipid that is also exposed by apoptosis in certain cell types, phosphatidylethanolamine (PE), can also bind a subset of PS receptors and facilitate virus internalization (Richard et al. 2015). Cellular regulators of actin-dependent endocytosis and vesicular transport such as Rho and Rac GTPases are primarily responsible for trafficking of virus-containing endosomes (Quinn et al. 2009; Saeed et al. 2010; Saeed et al. 2008).

2.3 *Glycoprotein Processing and Fusion*

Endosomal acidification activates GP cleavage by host proteases, primarily cathepsins L and B (Chandran et al. 2005; Kaletsky et al. 2007; Schornberg et al. 2006). Proteolytic processing by host cathepsins is important for generating fusion-ready forms of GP, although specific protease dependency can be altered by some GP mutations and host cell type (Misasi et al. 2012; Wong et al. 2010). Cathepsin cleavage also induces conformational changes in GP that promote membrane binding and fusion (Brecher et al. 2012). Other endosomal proteases can functionally substitute for cathepsins for cleaving GP (Marzi et al. 2012; Schornberg et al. 2006), although cathepsins are required for GP-mediated entry enhanced by cell-to-cell transfer (Miao et al. 2016).

GP molecules are disulfide-linked heterodimers consisting of two subunits, GP_1 (receptor-binding domain) and GP_2 (fusion domain). GP proteolysis removes the mucin-like domain and the glycan cap from GP_1. This exposes the receptor-binding domain and binding to the endosomal cholesterol transporter Niemann–Pick C1 (NPC1) (Bornholdt et al. 2016a). Two independently performed screens demonstrated that NPC1 is indispensable for EBOV infection (Carette et al. 2011; Cote et al. 2011). Although NPC1 is often characterized as a cellular receptor, it is not involved in extracellular virus uptake, but rather fusion with the endosomal membrane and release of the viral nucleocapsid (NC) into the cytoplasm. The ability of NPC1 to mediate EBOV fusion does not depend on cholesterol transport function (Carette et al. 2011; Cote et al. 2011). The interaction between NPC1 and GP is absolutely required for productive infection (Mingo et al. 2015; Spence et al. 2016), and NPC1 expression determines cell type-specific susceptibility to infection (Martinez et al. 2013). Furthermore, this interaction may determine host species restriction, as distinct amino acid changes in NPC1 determine susceptibility and entry efficiency in reptiles and bats (Hoffmann et al. 2016; Ndungo et al. 2016; Ng et al. 2015). Computational analysis has demonstrated that NPC1 modification by phosphorylation or glycosylation at specific sites may also be critically important for facilitating EBOV entry (Basharat and Yasmin 2015).

The precise mechanism for triggering fusion is not known, although it almost certainly requires contributions from host endosomal proteins or processes. After endosomal acidification, an unknown fusion trigger occurs, causing a fist-like structure at the tip of the hydrophobic fusion loop of GP_2 to insert into the membrane (Gregory et al. 2011; Gregory et al. 2014). GP cleavage is insufficient to induce fusion at physiological temperatures (Bale et al. 2011), although processed GP can undergo fusion at higher temperatures, low pH, and/or reducing conditions (Brecher et al. 2012). However, low pH alone cannot trigger fusion in the presence of uncleaved GP (Markosyan et al. 2016). Combined NPC1 and TIM1 binding cleaved GP in the late endosome may contribute to triggering fusion (Kuroda et al. 2015), while NC release into the cytoplasm also requires interaction between cleaved GP and two-pore channel 2 (TPC2), an intracellular calcium channel (Sakurai et al. 2015). Cysteine proteases also appear to be required for formation of fusion pores needed to release NC into the cytoplasm, but not for fusion triggering (Spence et al. 2016). However, many chronological and mechanistic details about these events and interactions remain uncharacterized and warrant further study.

3 Host Factors Required for Viral Transcription and Translation

Although viral replication has been studied in some detail, very little is known regarding the specific host factors that are required for viral gene or protein expression, or genome synthesis. Most studies have investigated the function of viral proteins and interactions with *cis*-acting regulatory elements, such as promoter sequences and RNA secondary structures, in the EBOV genome itself. Because EBOV is highly pathogenic, all work with infectious virus must be carried out in maximum biocontainment; thus many studies directed at mechanisms of replication have been performed with individual viral proteins, minigenomes, or replicons expressed in cells rather than in the context of a productive infection. These studies have been effective for determining functions of certain viral proteins, but many of the host factors involved in virus replication remain obscure.

3.1 Viral Gene Expression and Genome Replication

Following fusion, the viral NC is released into the cytoplasm for viral protein synthesis and genome replication. The NC is an RNA-protein complex composed of the viral proteins VP30, VP35, the viral RNA-dependent RNA polymerase (L), and the nucleoprotein (NP)-encapsidated $\sim$19 kb single-stranded, negative sense EBOV genome (Muhlberger et al. 1999). The first event following fusion is viral gene mRNA transcription using the encapsidated virus genome as a template. Host factors required for EBOV viral gene expression are poorly elucidated. However, transcriptional complex assembly and function critically depends on interactions with host proteins. The transcriptional activator VP30 is phosphorylated at several N-terminal serine and threonine residues. This phosphorylation inhibits viral mRNA transcription (though not viral genome synthesis), and must be removed by host cellular phosphatases for viral gene expression to proceed (Martinez et al. 2011; Modrof et al. 2002). VP30 dephosphorylation is essential for initiating transcription, as this enables VP30 to bind RNA, bind VP35, and to stabilize the VP35/L transcriptase complex on the template genome (Biedenkopf et al. 2016b). The pleiotropic regulator protein phosphatase 1 (PP1) controls VP30 phosphorylation state (Ilinykh et al. 2014). VP30 phosphorylation—particularly at the S29 serine residue—acts as a dynamic molecular switch that determines whether the polymerase complex acts as a mRNA transcriptase or a genome replicase (Biedenkopf et al. 2013; Biedenkopf et al. 2016a).

A defining trait of EBOV replication is the formation of inclusion bodies in the cytoplasm. These structures are associated with genome replication, but not viral mRNA transcription (Hoenen et al. 2012). EBOV inclusion bodies are enriched with viral proteins and NC components, and likely also contain host proteins, though these are not well characterized. Importin-α7 is involved in interferon (IFN) antagonism and also facilitates inclusion body formation. However, importin-α7 deletion does not effect on survival in a mouse model of lethal EBOV infection and does not appear to be a determinant of pathogenesis (Gabriel et al. 2015). Host nuclear proteins have been identified as regulators of EBOV transcriptional activity, and this may be linked to the observation that large inclusion bodies form in close proximity to the nucleus (Groseth et al. 2009; Nanbo et al. 2013). DNA topoisomerase I (TOP1) interacts directly with L and is required for viral polymerase activity (Takahashi et al. 2013). Another protein that binds nucleic acid, interleukin enhancing factor 3 (ILF3, more commonly called double-stranded RNA binding protein, DRBP76), binds both L and VP35, and can inhibit RdRp activity and block viral transcription (Shabman et al. 2011b; Takahashi et al. 2013).

Other proteins have been implicated in EBOV transcription, but the mechanisms by which these impact transcription are largely unknown. Interactions between VP35 and the 8-kDa dynein light chain (LC8) stabilize the N-terminal oligomerization domain of VP35 and improving its function as a polymerase co-factor (Kubota et al. 2009; Luthra et al. 2015). The endoplasmic reticulum (ER) membrane protein translocator subunit Sec61α co-localizes with VP24 and regulates genome replication by altering viral polymerase activity (Iwasa et al. 2011). Additionally, heat shock protein family A (Hsp70) member 8 (HSPA8) interacts with a stem-loop structure in the EBOV non-coding trailer region. The HSPA8 interaction appears to be a co-factor for the stem loop to function as a *cis*-acting replication enhancer (Sztuba-Solinska et al. 2016). HSPA8 also interacts with VP40 (Yamayoshi et al. 2008), suggesting that this protein may have an additional role in assembly and egress. A recent proteomic screen intended to characterize the host cellular interactome with NP identified a number of Hsp70 family members (including HSPA8) and heat shock protein 90 (Hsp90) family members (Garcia-Dorival et al. 2016), indicating that host chaperone proteins may be needed to stabilize transcriptional complexes. Treatment with an Hsp90 inhibitor (Smith et al. 2010) or a broad-spectrum Hsp70/Hsp90 inhibitor (Booth et al. 2016) impaired virus growth, confirming the importance of chaperone proteins in the EBOV replication.

3.2 Viral Protein Synthesis

Translational control of EBOV protein synthesis has not been studied extensively, as the monocistronic, polyadenylated viral transcripts resemble cellular mRNAs. However, they do contain large, complex untranslated regions at the 5′ and 3′ ends of EBOV mRNAs that appear to regulate cap-dependent translation (Neumann et al. 2009; Shabman et al. 2013). Specific host factors that regulate viral protein synthesis are largely unknown, although they presumably include the cellular translational machinery. Recently, hypusination of eukaryotic initiation factor 5A (eIF5A) is shown to regulate viral transcription by modulating VP30 protein accumulation in infected cells (Olsen et al. 2016). Viral protein accumulation in the cytoplasm is thought to instigate a switch from viral gene expression and mRNA production to RNA genome replication. Hijacking cellular protein synthesis machinery to control viral protein expression represents a means of regulating viral transcription and replication. EBOV also can evade innate antiviral defenses that induce stress granule formation and subsequently shut down translation by sequestering stress granules in cytoplasmic inclusions (Nelson et al. 2016).

Transcriptional regulation of viral protein expression is also employed by EBOV to regulate levels of GP. Most of the GP transcripts in the cell encode a small, non-structural form of GP (sGP) (Volchkova et al. 1998), which undergoes furin cleavage prior to secretion (Volchkova et al. 1999). RNA editing at a site consisting of 7 adenosine nucleotides in the genomic template generates the membrane-bound, structural form of GP. At this site, "stuttering" of the L polymerase inserts an extra adenosine residue in approximately 20% of GP mRNAs (Mehedi et al. 2013; Sanchez et al. 1996; Shabman et al. 2014; Volchkov et al. 1995). The editing site also acts as a cryptic transcriptional terminator and polyadenylation signal (Volchkova et al. 2015b). This mechanism for regulating different forms of GP is an important determinant of pathogenesis (Volchkova et al. 2015a), and likely requires contributions from host proteins involved in RNA binding and modification. RNA editing has been shown to occur differentially during serial passage in a host-dependent fashion (Volchkova et al. 2011), suggesting that this process occurs with cell type- and species-specific efficiencies. Further study of these processes will define the specific host proteins that regulate GP mRNA transcription and accumulation of different forms of GP.

4 Host Factors Required for Assembly and Egress

In addition to acting as sites of transcription, cytoplasmic inclusion bodies are also enriched sources of NC components in close proximity to newly transcribed full-length RNA genomes, thus enabling NC assembly (Hoenen et al. 2012; Nanbo et al. 2013; Noda et al. 2011). Assembled NCs are then transported to the cell surface, where progeny virions complete assembly and bud from the plasma membrane. The host cell machinery is essential to virion assembly and egress, and numerous cellular factors assist with all aspects of this process.

The primary viral regulator of assembly and secretion is the viral matrix protein, VP40. Expressing VP40 alone is sufficient to induce virus-like particles (VLPs) that bud from cells in a process that imitates secretion of infectious virions (Jasenosky et al. 2001). VP40 is a major regulator of intracellular transport required to bring virion components to sites of assembly at the inner surface of the plasma membrane. VP40 facilitates transport of NC from cytoplasmic or perinuclear viral inclusions to the cell surface via interactions with the actin cytoskeleton (Adu-Gyamfi et al. 2012; Han and Harty 2005; Lu et al. 2013; Schudt et al. 2015). VP40 interacts directly with the ubiquitous scaffold protein IQ motif containing GTPase activating protein 1 (IQGAP1). IQGAP1 binds and stabilizes activated GTP-bound Rac and Cdc42 GTPases, and coordinates signaling related to actin cytoskeletal organization. VP40 and IQGAP1 interact directly (Lu et al. 2013), and this association likely enables transport of NC across the long distance from viral inclusion to assembly sites (Schudt et al. 2015). VP40 associates with Sec24C, a component of the COPII vesicular transport system (Yamayoshi et al. 2008) through a proline-rich region near the C-terminus of VP40 (Reynard et al. 2011). Anterograde ER-to-Golgi transport is responsible for transporting VP40 to the cell surface.

The VP40 late assembly (L) domain then mediates membrane fission or "pinching off" of budding virions through a variety of interactions. The VP40 L domain does this by recruiting a variety of host proteins, primarily endosomal sorting complexes required for transport (ESCRT) machinery such as cell death 6 interacting protein (PDCD6IP, also called ALIX) (Han et al. 2015b), tumor susceptibility gene 101 (Tsg101) (Martin-Serrano et al. 2001), and vacuolar sorting protein 4 (VSP-4) (Licata et al. 2003; Silvestri et al. 2007). Tsg101 is the best characterized of the ESCRT proteins mediating viral budding through VP40. Interactions with Tsg101 are controlled by VP40 ubiquitination by the E3 protein-ubiquitin ligase neural precursor cell expressed, developmentally down-regulated 4 (Nedd4) (Timmins et al. 2003; Yasuda et al. 2003). Recently, the related protein itchy E3 ubiquitin protein ligase (ITCH) has been shown to serve an analogous function to Nedd4 (Han et al. 2016), although it is not known whether ITCH replaces Nedd4 or serves as an additional VP40 regulator. Intracellular calcium signaling and mitogen-activated protein kinase (MAPK) activity have also been implicated in VP40 VLP budding (Han and Harty 2007), as has the calcium channel ORAI calcium release-activated calcium modulator 1 (Orai1) (Han et al. 2015a), but the significance to Tsg101-dependent egress in the context of infection is unclear. VP40 ubiquitination can be blocked by interferon-stimulated gene 15 (ISG15), a small, ubiquitin-like modifier that can undergo ISGylation, or attachment of ISG15 to lysine residues in a similar manner as ubiquitination. ISGylation of the VP40 L domain inhibits budding (Okumura et al. 2008).

Somewhat surprisingly, the budding process also involves components of the inflammatory signaling machinery. Suppressor of cytokine signaling 3 (SOCS3), a negative regulator of cytokine signaling, binds VP40 to facilitate egress (Okumura et al. 2015). Additionally, GP binds the pattern recognition receptor (PRR) Toll-like receptor 4 (TLR4). TLR4 signaling ultimately activates nuclear factor κ B (NFKB) signaling and inflammatory gene expression upon entry. TLR4 interactions with GP also enhance VP40 VLP egress (Okumura et al. 2010), suggesting that viral hijacking of RNA virus sensing and inflammatory signaling machinery can enhance virus production and restricts cellular antiviral responses.

The final stages of assembly and budding are regulated by direct interactions between VP40 and phospholipids in the plasma membrane. VP40 exists as a dimer until it reorganizes at the plasma membrane into a hexameric form. In this form, VP40 associates with lipid rafts (Adu-Gyamfi et al. 2013; Panchal et al. 2003; Soni et al. 2013), typically at locations enriched for particular host membrane phospholipids. VP oligomerization occurs in association with PS in the inner leaflet of the cell membrane (Adu-Gyamfi et al. 2015) and selectively induces vesiculation of PS-enriched membranes (Soni and Stahelin 2014). VP40 hexamers are then stabilized by associating with phosphatidylinositol 4,5-bisphosphate (PI(4,5)P2) in the membrane, enhancing membrane curvature and facilitating budding (Gc et al. 2016; Johnson et al. 2016).

While VP40 is a major regulator of budding, virus egress is also determined by interactions with or modifications by host proteins. GP plays an important role in virion secretion as well as in entry. After GP is synthesized in the ER, it is cleaved to its mature form by the proprotein convertase furin (Volchkov et al. 1998), and transported to the cell surface. The host surface protein bone marrow stromal cell antigen 2 (BST-2, also called tetherin) restricts budding by enveloped viruses. GP blocks BST-2 interactions with VP40 by steric shielding through the mucin domain, and enhances VLP budding (Gustin et al. 2015; Kaletsky et al. 2009). The membrane-spanning domain and glycan cap domains of GP are also important for tetherin antagonism and efficient egress (Gnirss et al. 2014; Kuhl et al. 2011; Vande Burgt et al. 2015). However, high cellular levels of $GP_{1,2}$ impair assembly, and reduce virus production (Mohan et al. 2015), resulting in shedding of GP from the cell surface. Shed GP is a truncated form that is cleaved from the surface by tumor necrosis factor α converting enzyme, and can reduce cytotoxicity caused by high levels of cellular GP (Dolnik et al. 2015). Shedding may be a regulatory mechanism to balance GP cytotoxicity with virus production.

The minor matrix protein VP24 is also integrally involved in assembly and secretion. Together with NP, VP40, and VP35, VP24 mediates NC formation (Bharat et al. 2012; Huang et al. 2002; Mateo et al. 2011a), and appears to be important for packaging RNA genomes of the proper length (Watt et al. 2014). Linkages between VP24 and VP35 also stabilize NC structures inside the virion, dependent on post-translational NP modification by host glycosyltransferases (Beniac et al. 2012; Huang et al. 2002) or kinases (Peyrol et al. 2013). These processes likely require either direct or indirect interactions with host proteins, to modify, scaffold, or transport virion components. However, the specific host factors have not been characterized. There remains much to be determined about the host factors that contribute to the budding process. A proteomic screen identified 8 proteins incorporated into EBOV virions that were significantly associated with EBOV replication in subsequent siRNA knockdown experiments (Spurgers et al. 2010). One of these proteins, heat shock protein family A (Hsp70) member 5 (HSPA5), is an ER chaperone protein that is important to VP40-mediated budding, although its specific function is not known (Reid et al. 2014).

5 Antagonism of Antiviral Responses

Virus pathogenicity depends on the ability to subvert the host's innate immune responses. Ebolaviruses have evolved multiple strategies to antagonize or evade antiviral immunity and the IFN system, which both improve the efficiency of viral replication and enhance virulence and disease pathology. EBOV effectively eludes host defenses by targeting multiple cellular and molecular key immune mediators.

5.1 Innate Immune Antagonism

The innate antiviral response is a critical determinant of pathogenesis. The host interferon (IFN) response in particular is a potent tool for blocking or attenuating virus infection, and cells have numerous ways of sensing virus, transducing signals, and producing interferon-stimulated genes (ISGs). There are hundreds to thousands of ISGs depending on cell type, and these genes encode diverse effector molecules that prevent entry, disrupt replication, impair assembly, or mitigate pathology. EBOV has thus developed mechanisms to counteract IFN responses at various critical points in the pathway. Gene expression studies of hepatic cells or macrophages infected with ebolaviruses of varying pathogenicity indicate extensive disruption of Toll-like receptor (TLR), IFN, and NFκB signaling, suggesting that the intensity of the host antiviral response is a substantial virulence determinant (Kash et al. 2006; Melanson et al. 2015; Wahl-Jensen et al. 2011).

The primary viral immune antagonists are VP35 and VP24. VP35 antagonizes innate immunity in multiple places in the pathway, both by blocking sensing by pattern recognition receptors (PRRs), inactivating downstream transcriptional activators required for ISG expression, and directly inhibiting ISG activity. PRRs are a fundamental component of initiating innate immunity, as they recognize pathogen-associated molecular patterns (PAMPs), or molecular motifs common to certain classes of pathogens. Double-stranded RNA (dsRNA) is a potent PAMP associated with RNA virus infections, as this is generally associated with viral genomes or replicative intermediates. Numerous classes of PRRs recognize dsRNA, including the DExD/H-box helicase 58 (DDX58), more commonly known as retinoic acid-inducible gene I (RIG-I). RIG-I is a helicase that recognizes dsRNA in the cytoplasm, where EBOV RNA replication occurs and dsRNA replication intermediates are likely to be present. The VP35 C-terminal domain coats the phosphodiester backbone and caps the ends of double-stranded RNA (dsRNA), sequestering it from RIG-I, and preventing sensing and IFN induction (Bale et al. 2013; Cardenas et al. 2006; Edwards et al. 2016; Kimberlin et al. 2010; Leung et al. 2009; Leung et al. 2010).

VP35 has a general affinity for interacting with dsRNA or dsRNA-binding proteins, and this appears to be important for its function as an IFN antagonist. VP35 mutations in the dsRNA-binding domain attenuate EBOV lethality in a guinea pig model, demonstrating that dsRNA binding is critical for virulence (Prins et al. 2010). VP35 also inhibits activation of eukaryotic translation initiation factor 2 α kinase 2 (EIF2AK2), more commonly known as interferon-induced double-stranded RNA protein kinase (PKR). During a viral infection, PKR phosphorylates eukaryotic initiation factor 2 α (eIF2α), which shuts down translation. VP35 prevents PKR activation and thus prevents shutdown of viral protein translation (Schumann et al. 2009). Furthermore, VP35 also interacts with the cellular dsRNA-binding protein activator of interferon-induced protein kinase EIF2AK2 (PRKRA, also known as PACT). PACT can also activate RIG-I, and VP35 can bind PACT and disrupt RIG-I ATPase activity. PACT, however, can also reduce EBOV replication (Luthra et al. 2013), similar to other dsRNA-binding proteins. DRBP76, which is phosphorylated by PKR (Shabman et al. 2011b), interacts with VP35 and may disrupt EBOV replication and IFN antagonism.

Interference with RNA sensing is not the sole VP35-specific means of innate antiviral antagonism. VP35 can also directly inactivate transcriptional activators required to induce IFN gene expression, specifically interferon regulatory factors 3 and 7 (IRF3; IRF7). Typically, cytoplasmic IRF-3 and IRF-7 are phosphorylated by TANK binding kinase 1 (TBK1) and inhibitor of κ light chain polypeptide gene enhancer in B cells kinase ε (IKBKE, commonly known as IκKε). Activated IRF3 then either homodimerizes or heterodimerizes with activated IRF7. These complexes translocate to the nucleus and activate IFN-β transcription. VP35 prevents TBK1 and IκKε from phosphorylating IRF3 (Prins et al. 2010), hindering type I IFN transcriptional activation (Basler et al. 2003; Hartman et al. 2008). This inhibition is a critical virulence determinant in vivo (Hartman et al. 2008). VP35 inhibits IRF7-dependent IFN transcription by a different mechanism: it modulates the activity of the E3 SUMO ligase protein inhibitor of activated STAT1 (PIAS1). This increases IRF7 SUMOylation and targets the protein for degradation (Chang et al. 2009).

The antiviral functions of VP35 result in immune cell inhibition. Blocking RIG-I sensing and the subsequent failure to induce type I IFNs prevents both inflammatory gene expression and dendritic cell (DC) maturation (Yen et al. 2014; Yen and Basler 2016). EBOV infection induces weak expression of IFNs and DC maturation marker genes (Ilinykh et al. 2015). Mutations that impair or disable the C-terminal domain of VP35 completely restore DC maturation, as well as enable normal migration from peripheral sites to secondary lymphoid organs (Lubaki et al. 2013). Additionally, VP35 can block DC maturation induced by other viruses, exogenous IFN, or other PAMPs such as lipopolysaccharide (LPS) (Jin et al. 2010). These effects are specific to conventional DCs (Leung et al. 2011b), and are a major cause of T cell suppression during EBOV infection.

While VP35 generally acts to inhibit virus sensing and IFN induction, VP24 desensitizes host cells to IFNs to inhibit ISG induction. Several ISGs have been implicated in EBOV restriction, including cholesterol-25-hydroxylase and various interferon-inducible transmembrane protein (IFITM) family members inhibit EBOV entry (Huang et al. 2011; Liu et al. 2013; Wrensch et al. 2015) and tetherin inhibits viral egress (Gnirss et al. 2014; Gustin et al. 2015; Kaletsky et al. 2009; Kuhl et al. 2011; Vande Burgt et al. 2015). VP24 reduces ISG effector expression by multiple mechanisms. First, VP24 blocks signal transduction through IFN receptors to inhibit ISG expression. IFN receptor-binding results in the activation of Janus kinase (JAK) family adaptor proteins Jak1 or tyrosine kinase 2 (Tyk2), which are associated with the IFN receptor intracellular domain. Jak1/Tyk2 then phosphorylate signal transducer and activator of transcription 1 and 2 (STAT1, STAT2), which subsequently form homo- or heterodimers and associate with IRF9. This complex then translocates to the nucleus and activates ISG transcription. VP24 interferes with JAK/STAT signaling in two ways: by binding STAT1 directly (Zhang et al. 2012) and by preventing nuclear import of activated STAT dimers by acting on karyopherin α (KPNA) family members, other nuclear importins, and associated host factors (Gabriel et al. 2015; Garcia-Dorival et al. 2014; Mateo et al. 2010; Reid et al. 2006; Reid et al. 2007; Shabman et al. 2011a; Xu et al. 2014). As with VP35, innate immune antagonism by VP24 is a major molecular virulence determinant (Mateo et al. 2011b).

VP24 can also block induction of inflammatory responses. VP24 impairs MAPK activation in a cell type-specific manner, and can block IFN-β-dependent p38-α phosphorylation (Halfmann et al. 2011). MAPK signaling is critical for regulating inflammatory responses, and the systemic loss of inflammatory regulation is thought to be a major mechanism of pathology in EVD. EBOV production can be induced in persistently infected cells by modulating Ras/MAPK signaling (Strong et al. 2008); thus, modulation of MAPK signaling may serve the dual function of deregulating cellular inflammation and stimulating virus replication.

5.2 Adaptive Immune Antagonism

Viral proteins can directly interfere with lymphocyte-mediated antigen-specific immune responses, as well. VP35 interferes with DC maturation and blocks surface expression of many T cell costimulatory molecules (Ilinykh et al. 2015; Jin et al. 2010; Lubaki et al. 2013; Yen et al. 2014; Yen and Basler 2016), and likely plays a significant role in quelling effective T cell immunity to EBOV. GP in the membrane of DC and other antigen-presenting cells (APC) shields major histocompatibility complex (MHC) molecules (Reynard et al. 2009) and effectively impairs antigen presentation to T cells, resulting in widespread suppression of T cell function.

Another mechanism for evading cellular immunity is the widespread lymphocyte death that occurs during infection. Lymphocytes are not susceptible to EBOV infection, but undergo bystander apoptosis in cultured PBMC and in vivo (Baize et al. 1999; Bradfute et al. 2007; Bradfute et al. 2010; Cross et al. 2015; Geisbert et al. 2000; Gupta et al. 2007; Reed et al. 2004; Wauquier et al. 2010). Both the Fas death receptor (extrinsic) and mitochondrial (intrinsic) pathways trigger lymphocyte apoptosis (Bradfute et al. 2010; Gupta et al. 2007). However, this lymphocyte death is not absolute. Populations of CD8 + T cells emerge late in infection, indicating that some lymphocyte subsets survive and function normally as immune effector cells (Bradfute et al. 2008). This likely has an effect on disease severity, as sustained T cell function is associated with EVD survival (Baize et al. 1999; Dahlke et al. 2016). The means by which certain T cell subsets survive is not known. However, reduced expression of CD45 resulted in reduced lymphocyte apoptosis and improved survival in a mouse model of lethal EBOV infection (Panchal et al. 2009). CD45 is a tyrosine phosphatase that acts as a pleiotropic regulator of lymphocyte receptor signaling, suggesting that T and B cell fate may be linked to regulation of signaling through antigen receptors. Interference with antigen presentation and expression of T cell costimulatory molecules can further limit effector function in surviving T cells.

Due to both B cell depletion and impaired CD4 + T cell help, EBOV infection also damages humoral immune responses. Antibody responses can be reduced by elimination of B cells by apoptosis or interfering with B cell signaling and differentiation processes. Immunity to EBOV is dependent on CD4 + -dependent neutralizing antibody responses in hamsters and non-human primates (Fisher-Hoch et al. 1992b; Prescott et al. 2015). However, as with T cell responses, diverse and sustained neutralizing antibody responses are observed in survivors of EBOV or Sudan virus (SUDV) (Bornholdt et al. 2016b; Flyak et al. 2016; Natesan et al. 2016; Schibler et al. 2015; Sobarzo et al. 2013; Wauquier et al. 2009), suggesting that antibody responses are important for controlling EBOV. Sequence analysis of EBOV strains from the 2014 outbreak demonstrated that major antigenic epitopes were under greater selection pressure to accumulate mutations leading to escape variants (Ramaiah and Arumugaswami 2016), although this varies and does not include all epitopes (Ni et al. 2016). Thus, a combination of virus evolution and impaired antibody production provide a means for EBOV to escape host humoral immunity.

Another means of tempering host antibody responses is antigenic subversion by EBOV sGP. Because the majority of GP transcripts encode sGP, high levels of sGP are secreted from infected cells into circulation and dominate the pool of potential antigens. This skews antibody responses to immunodominant epitopes shared by sGP and membrane-bound GP on the virion surface (Mohan et al. 2012). Membrane-bound GP also can be concealed from antibody recognition by steric shielding mediated by the mucin domain (Francica et al. 2010; Noyori et al. 2013; Reynard et al. 2009).

EBOV can also affect other types of immune cells, in some cases, eliminating them through apoptosis. Monocytes and natural killer cells are both depleted in human cases and animal models (Bradfute et al. 2007; Geisbert et al. 2003b; Ludtke et al. 2016; Reed et al. 2004). EBOV infection also reportedly activates triggering receptor expressed on myeloid cells 1 (TREM1) on neutrophils, resulting in degranulation and potent induction of inflammatory responses (Mohamadzadeh et al. 2006).

6 Cell and Tissue-Specific Host Factors Involved in Pathogenesis

Although EVD has been studied in non-human primates for over two decades, many aspects of EVD pathogenesis remain mysterious. EVD is a complex, systemic disease that targets many different organs and cell types. Severe EVD results from sequential infection of multiple distinct cell types and tissue compartments, and clinical outcome is likely determined by combined mechanisms (Fig. 2). Transmission usually occurs from exposure of infectious body fluids to wounds or breaks in the skin, mucosal surfaces, or parenteral injury (Sanchez et al. 2007). The

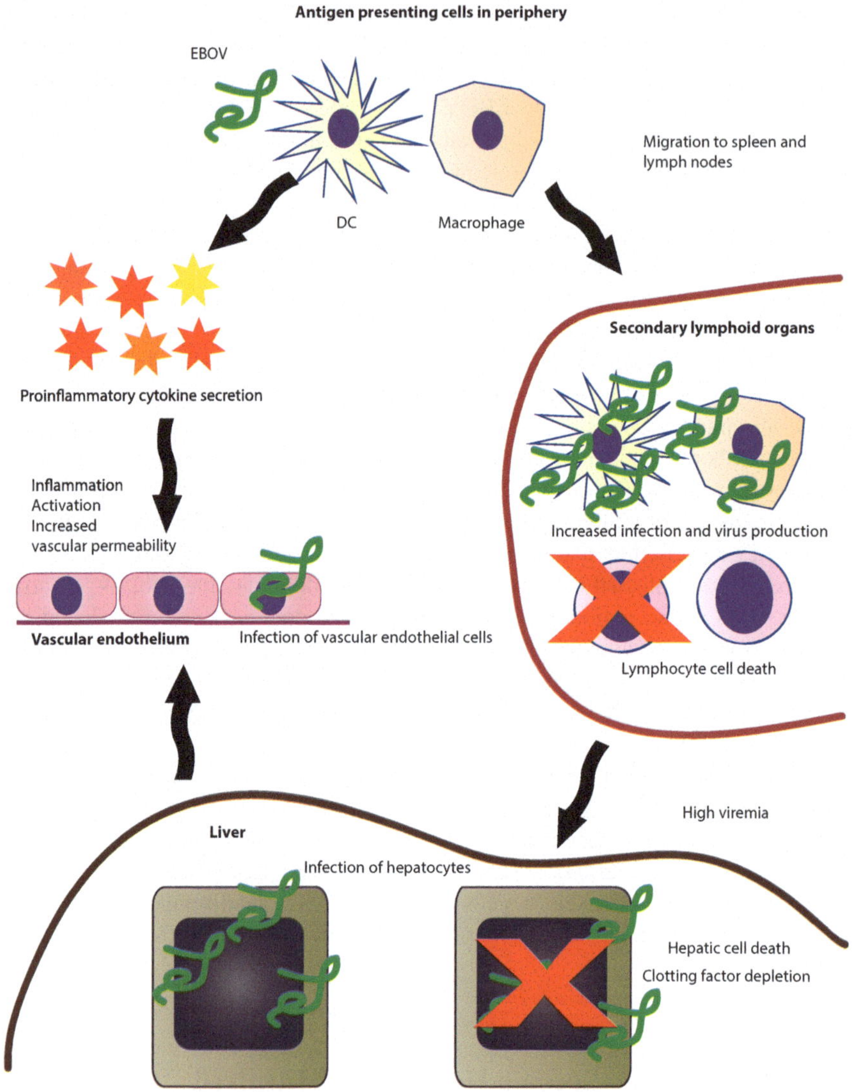

Fig. 2 Antigen presenting cells in periphery

first targets of infection are myeloid-derived APCs, usually macrophages and dendritic cells (DCs) (Connolly et al. 1999; Ebihara et al. 2013; Geisbert et al. 2003b; Gibb et al. 2001; Ryabchikova et al. 1999). Plasmacytoid DC (pDCs) do not support efficient virus entry, possibly as a mechanism to subvert pDC IFN secretion (Leung et al. 2011a; Leung et al. 2011b). Monocytes are not susceptible to EBOV infection, but become susceptible upon differentiation into macrophages (Martinez et al. 2013).

Infected macrophages and DCs exert major influences on pathogenesis and disease outcome. Infection triggers strong proinflammatory responses as they simultaneously repress induction of antiviral cellular functions. Chemokine secretion recruits other susceptible APCs, which then subsequently become infected. Inflammatory gene expression occurs early in infection, stimulated by GP during cell entry (Wahl-Jensen et al. 2011; Wahl-Jensen et al. 2005a). The GP mucin domain stimulates NFκB and MAPK activation, resulting in enhanced cytokine production in DCs (Martinez et al. 2007). Additionally, EBOV GP induces Syk-dependent inflammatory signaling via the CLEC LSECtin in DCs, triggering the release of proinflammatory cytokines such as tumor necrosis factor α (TNFα) and interleukin 6 (IL-6). The massive unregulated inflammatory response is thought to be the primary cause of the severe immunopathology associated with EVD. EBOV infection stimulates potent inflammatory gene expression in a variety of myeloid lineage cells, including human macrophages (Wahl-Jensen et al. 2011), monocyte-derived DCs (Ilinykh et al. 2015), mouse splenocytes (Cilloniz et al. 2011), and peripheral blood mononuclear cells (PBMC) from non-human primates (Caballero et al. 2016; Rubins et al. 2007). Inflammation is further exacerbated through other pathways, including MAPK deregulation (Halfmann et al. 2011), TREM1-mediated neutrophil degranulation (Mohamadzadeh et al. 2006), and inflammatory gene expression enhanced by TOP1, which also is involved in viral genome replication (Rialdi et al. 2016).

Infected APCs will migrate to the lymph nodes, where additional naive cells are infected, amplifying virus production, producing viremia, and inducing bystander apoptosis in lymphocytes. Although lymphocyte apoptosis disrupts adaptive immune responses to EBOV infection, it is not required for disease progression or severity (Bradfute et al. 2010). Massive lymphocyte death resulting in extremely low peripheral levels of CD4 + T cells was associated with fatal outcome in EVD patients during an outbreak of EBOV in Gabon (Wauquier et al. 2010). Sustained immune activation was observed in EVD patients, even after recovery from acute disease (McElroy et al. 2015), consistent with prior experiments in mice (Bradfute et al. 2008). EBOV spread throughout the lymphatic system intricately balances host immunosuppression with virus-induced immune activation.

Liver cells are infected after the host becomes viremic. Hepatocytes are the main target for infection, although resident liver macrophages (Kupffer cells) are also susceptible. Infection has numerous negative consequences for hepatocytes, resulting in highly dysregulated intracellular signaling and ultimately widespread cell death (Bradfute et al. 2010). Transforming growth factor β (TGFβ) signaling is selectively activated in hepatocytes (Ebihara et al. 2013; Kindrachuk et al. 2014), which induces cellular differentiation and an epithelial-to-mesenchymal transition in infected hepatocytes (Kindrachuk et al. 2014). This enables virus spread throughout the liver, further accounting for aberrant hepatic function.

Late in infection, endothelial cells lining blood vessels become infected and simultaneously activated. This results in increased cardiovascular barrier permeability, and ultimately vascular leakage, bleeding, and severely impaired coagulation. Blood vessel barrier collapse does not result from directly infected, dying vascular endothelial cells (Geisbert et al. 2003d). Vascular leakage and hemorrhagic disease more likely result from systemic inflammation and interactions between both viral proteins and the endothelium. Different forms of EBOV GP can exert effects directly on barrier function (Wahl-Jensen et al. 2005b), possibly by stimulating signaling through inflammatory receptors such as TLR4 (Escudero-Perez et al. 2014; Okumura et al. 2010).

Most EBOV infections do not produce severe hemorrhagic disease. Some ebolavirus species, such as Taï Forest virus (TAFV) and Reston virus (RESTV) do not apparently cause severe disease in humans, despite their high pathogenicity in non-human primates (Fisher-Hoch et al. 1992a; Le Guenno et al. 1995). For this reason, EVD replaced EBOV hemorrhagic fever (EHF) as the preferred term for describing the diverse spectrum of disease presentations caused by EBOV infection. During the 2014 West African outbreak, gastrointestinal symptoms (particularly diarrhea) were more prevalent than bleeding characteristic of classic EHF (Bah et al. 2015; Chertow et al. 2014; Dallatomasina et al. 2015; Lado et al. 2015; Schieffelin et al. 2014). Many EVD patients suffer from fatal shock or organ failure with no evidence of hemorrhagic disease or bleeding, and some patients are less genetically predisposed to coagulopathy and vascular leakage. Animal models produce variant EVD phenotypes in different non-human primate species infected with different ebolavirus species (Ebihara et al. 2011; Fisher-Hoch et al. 1992a; Martins et al. 2015; Perry et al.2012; Ryabchikova et al. 1999; Zumbrun et al. 2012), as well as in laboratory mice (Ebihara et al. 2006; Gibb et al. 2001), genetically diverse mice (Rasmussen et al. 2014; Zumbrun et al. 2012), humanized mice (Bird et al. 2016; Spengler et al. 2016), Hamsters (2013), guinea pigs (Cross et al. 2015), and ferrets (Cross et al. 2016; Kozak et al. 2016). Observable disease phenotypes observed in these models range from weight loss and lethality in conventional laboratory mice to complete recapitulation of severe EVD with hemorrhagic syndrome in non-human primates. Thus, host and virus genetics cooperatively determine EVD pathological phenotype and disease outcome.

Hemorrhagic EVD results from two convergent pathologic processes: coagulopathy and vascular leakage. Coagulopathy is characterized by depletion of serum clotting factors, thrombocytopenia, and serum hypofibrinogenemia, all resulting in deficient blood coagulation. This happens in part due to disseminated intravascular coagulation (DIC) caused by excess systemic proinflammatory cytokines. This widespread inflammatory milieu stimulates simultaneous activation of the blood coagulation cascade, fibrinolysis, and fibrin D-dimer accumulation in tissues. DIC in the bloodstream deposits microscopic blood clots called microthrombi throughout the vasculature, obstructing small blood vessels and capillaries. Eventually, this process exhausts the available supply of both platelets and coagulation factors, which are predominantly synthesized in the liver. At this stage of infection, most hepatocytes are infected and dying, and thus cannot restore a fresh supply of clotting factors. DIC is thought to occur from a loss of the coagulants tissue factor (TF) and activated protein C (Cross et al. 2015; Ebihara et al. 2011; Geisbert et al. 2003c). Consequently, coagulation is defective and prolonged. Clotting factor therapy significantly improved outcome and survival in experimentally infected macaques (Geisbert et al. 2003a; Hensley et al. 2007), while recombinant protein C treatment induced coagulation-related gene expression correlated with survival (Yen et al. 2011). Transcripts associated with fibrin D-dimer clearance have also been observed in EBOV-infected non-human primates (Rubins et al. 2007), indicating that hemorrhagic disease progression is associated with compensatory increases in coagulation cascade regulators.

Vascular leakage is the second mechanism underlying EVD hemorrhagic syndrome. Although EBOV infects and destroys endothelial cells late in infection, leakage appears to be the consequence of uncontrolled inflammation rather than erosion of the vascular endothelium (Geisbert et al. 2003d). EBOV does directly cause vascular pathology, although the role in hemorrhagic disease is not clear. EBOV GP alters vascular barrier function (Wahl-Jensen et al. 2005b), and causes endothelial detachment by a cholesterol-dependent mechanism (Hacke et al. 2015). GP is clipped by TNFα converting enzyme and shed profusely from the cell surface during infection (Dolnik et al. 2004; Dolnik et al. 2015), and this leads to increased vascular permeability (Escudero-Perez et al. 2014). Interactions between TLR4 and shed GP trigger potent proinflammatory responses and reduce barrier function (Escudero-Perez et al. 2014; Okumura et al. 2010), possibly by upregulating inducible nitric oxide synthase (iNOS) downstream of TLR4 signaling. iNOS produces nitric oxide (NO), a powerful inducer of endothelial permeability. Increased circulating NO concentrations were observed in EBOV-infected macaques (Geisbert et al. 2003b; Hensley et al. 2002) and EVD patients (Sanchez et al. 2004). A number of signaling pathways relevant have been implicated in causing hemorrhagic disease, including vascular endothelial growth factor (VEGF) and angiopoietin signaling (Kindrachuk et al. 2014; Rasmussen et al. 2014). Mechanisms of vascular leakage in EVD are not understood in great detail, and identifying means of reversing barrier collapse represent a novel therapeutic paradigm for patients with EVD as well as other coagulation disorders.

Another disease presentation that is being increasingly recognized as a manifestation of severe EVD is gastrointestinal disease. During the 2014 EBOV outbreak, vomiting and diarrhea were more common than hemorrhagic disease (Arranz et al. 2016; Baize et al. 2014; Dietz et al. 2015; Lado et al. 2015; Lu et al. 2015; Ohuabunwo et al. 2016; WHO 2014). These symptoms were associated with severe disease and mortality both in West Africa (Bah et al. 2015; Chertow et al. 2014; Dallatomasina et al. 2015; Fitzgerald et al. 2016; Fitzpatrick et al. 2015; Haaskjold et al. 2016; Moole et al. 2015; Qin et al. 2015; Schieffelin et al. 2014) and in a contemporary outbreak in the Democratic Republic of Congo (Nanclares et al. 2016). Gastrointestinal symptoms have been reported in a majority of patients in historical outbreaks of ebolaviruses (Bwaka et al. 1999; Georges et al. 1999; Khan et al. 1999; Kratz et al. 2015; MacNeil et al. 2010; Roddy et al. 2012; WHO 1978a, b). Hypovolemia induced by excessive vomiting or diarrhea is thought to have caused severe electrolyte imbalances in EVD patients (Bah et al. 2015). Supportive care, including intravenous fluids and renal replacement therapy, is associated with positive clinical outcomes (Liddell et al. 2015; O'Shea et al. 2016; Uyeki et al. 2016; Wolf et al. 2015; Wong et al. 2015), suggesting that this may be a useful therapeutic strategy in future EBOV outbreaks.

Going forward, EVD pathogenesis studies must further explore the host responses that produce gastrointestinal pathology, especially considering that treatment with readily available antidiarrheal and antiemetic medications and supportive care may substantially improve patient outcomes.

7 Conclusion

Despite the celebrity status of EBOV compared to other emerging viruses, many of the mechanistic details regarding host cell-dependent EBOV replication and pathogenesis remain relatively obscure. This is in part due to the challenges inherent in conducting research on highly pathogenic viruses. There are limited maximum containment laboratories available to safely perform studies on infectious EBOV, and such studies are expensive and challenging. Surrogate systems such as minigenomes and replicons have enabled detailed investigations into the function of viral proteins and interactions with host molecules conscripted to augment virus replication. Considerable progress has been made, particularly regarding mechanisms of virus entry and egress, and viral antagonism of host immune responses. However, much remains to be studied. Very little is known about host contributions to virus transcription and translation, fundamental processes in the viral replication cycle that could be targeted for host-directed treatments. Targeting host proteins rather than viral proteins is a promising therapeutic strategy, as these have the potential to reverse disease pathology and are unlikely to evolve resistance.

The vast majority of studies regarding filovirus pathogenesis have focused on Zaire ebolaviruses, which is one of five known species within the genus *Ebolavirus*. Two other viruses, SUDV and Bundibugyo virus (BDBV), also cause EVD in humans, while RESTV and TAFV cause severe EVD in non-human primates. The basis of this species-specificity is not clear, but it is probably due in part to contributions from the host. The *Filoviridae* family also includes two other genera, *Marburgvirus* and *Cuevavirus*. Marburgviruses are also human pathogens that cause a severe disease sharing many clinical features with EVD, called Marburg virus (MARV) disease (MVD). Although they are also filoviruses with similarly organized genomes, marburgviruses have distinct mechanisms for carrying out these functions. For example, MARV VP40 serves an analogous function to EBOV VP24 in IFN antagonism by JAK-STAT signaling (Valmas et al. 2010), while MARV VP24 modulates NFκB signaling (Edwards and Basler 2015). It is not known whether Lloviu virus (LLOV), the only cuevavirus isolated to date, is pathogenic in any species; however, LLOV proteins carry out analogous functions to other filoviruses, such as NPC1-dependent cell entry (Ng et al. 2014) and IFN antagonism (Feagins and Basler 2015). Host factors most likely determine pathogenicity, tropism, and host range for all the filoviruses. Future efforts to identify critical host contributors to filoviral disease will lead to improved strategies for preventing and treating these important emerging pathogens.

References

Adu-Gyamfi E, Digman MA, Gratton E, Stahelin RV (2012) Single-particle tracking demonstrates that actin coordinates the movement of the Ebola virus matrix protein. Biophys J 103(9):L41–L43. doi:10.1016/j.bpj.2012.09.026

Adu-Gyamfi E, Soni SP, Xue Y, Digman MA, Gratton E, Stahelin RV (2013) The Ebola virus matrix protein penetrates into the plasma membrane: a key step in viral protein 40 (VP40) oligomerization and viral egress. J Biol Chem 288(8):5779–5789. doi:10.1074/jbc.M112.443960

Adu-Gyamfi E, Johnson KA, Fraser ME, Scott JL, Soni SP, Jones KR, Digman MA, Gratton E, Tessier CR, Stahelin RV (2015) Host Cell Plasma Membrane Phosphatidylserine Regulates the Assembly and Budding of Ebola Virus. J Virol 89(18):9440–9453. doi:10.1128/JVI.01087-15

Aleksandrowicz P, Marzi A, Biedenkopf N, Beimforde N, Becker S, Hoenen T, Feldmann H, Schnittler HJ (2011) Ebola virus enters host cells by macropinocytosis and clathrin-mediated endocytosis. J Infect Dis 204(Suppl 3):S957–S967. doi:10.1093/infdis/jir326

Alvarez CP, Lasala F, Carrillo J, Muniz O, Corbi AL, Delgado R (2002) C-type lectins DC-SIGN and L-SIGN mediate cellular entry by Ebola virus in cis and in trans. J Virol 76(13):6841–6844

Arranz J, Lundeby KM, Hassan S, Zabala Fuentes LM, San Jose Garces P, Haaskjold YL, Bolkan HA, Krogh KO, Jongopi J, Mellesmo S, Josendal O, Opstad A, Svensen E, Kamara AS, Roberts DP, Stamper PD, Austin P, Moosa AJ, Marke D, Berg A, Blomberg B, Riera M (2016) Clinical features of suspected Ebola cases referred to the Moyamba ETC, Sierra Leone: challenges in the later stages of the 2014 outbreak. BMC Infect Dis 16:308. doi:10.1186/s12879-016-1609-9

Bah EI, Lamah MC, Fletcher T, Jacob ST, Brett-Major DM, Sall AA, Shindo N, Fischer WA 2nd, Lamontagne F, Saliou SM, Bausch DG, Moumie B, Jagatic T, Sprecher A, Lawler JV, Mayet T, Jacquerioz FA, Mendez Baggi MF, Vallenas C, Clement C, Mardel S, Faye O, Faye O, Soropogui B, Magassouba N, Koivogui L, Pinto R, Fowler RA (2015) Clinical presentation of patients with Ebola virus disease in Conakry, Guinea. The N Engl J Med 372 (1):40–47. doi:10.1056/NEJMoa1411249

Baize S, Leroy EM, Georges-Courbot MC, Capron M, Lansoud-Soukate J, Debre P, Fisher-Hoch SP, McCormick JB, Georges AJ (1999) Defective humoral responses and extensive intravascular apoptosis are associated with fatal outcome in Ebola virus-infected patients. Nat Med 5(4):423–426. doi:10.1038/7422

Baize S, Pannetier D, Oestereich L, Rieger T, Koivogui L, Magassouba N, Soropogui B, Sow MS, Keita S, De Clerck H, Tiffany A, Dominguez G, Loua M, Traore A, Kolie M, Malano ER, Heleze E, Bocquin A, Mely S, Raoul H, Caro V, Cadar D, Gabriel M, Pahlmann M, Tappe D, Schmidt-Chanasit J, Impouma B, Diallo AK, Formenty P, Van Herp M, Gunther S (2014) Emergence of Zaire Ebola virus disease in Guinea. The N Engl J Med 371(15):1418–1425. doi:10.1056/NEJMoa1404505

Bale S, Liu T, Li S, Wang Y, Abelson D, Fusco M, Woods VL Jr, Saphire EO (2011) Ebola virus glycoprotein needs an additional trigger, beyond proteolytic priming for membrane fusion. PLoS neglected tropical diseases 5(11):e1395. doi:10.1371/journal.pntd.0001395

Bale S, Julien JP, Bornholdt ZA, Krois AS, Wilson IA, Saphire EO (2013) Ebolavirus VP35 coats the backbone of double-stranded RNA for interferon antagonism. J Virol 87(18):10385–10388. doi:10.1128/JVI.01452-13

Basharat Z, Yasmin A (2015) In silico assessment of phosphorylation and O-beta-GlcNAcylation sites in human NPC1 protein critical for Ebola virus entry. Infect Genet Evol 34:326–338. doi:10.1016/j.meegid.2015.06.001

Basler CF, Mikulasova A, Martinez-Sobrido L, Paragas J, Muhlberger E, Bray M, Klenk HD, Palese P, Garcia-Sastre A (2003) The Ebola virus VP35 protein inhibits activation of interferon regulatory factor 3. J Virol 77(14):7945–7956

Beniac DR, Melito PL, Devarennes SL, Hiebert SL, Rabb MJ, Lamboo LL, Jones SM, Booth TF (2012) The organisation of Ebola virus reveals a capacity for extensive, modular polyploidy. PLoS ONE 7(1):e29608. doi:10.1371/journal.pone.0029608

Bharat TA, Noda T, Riches JD, Kraehling V, Kolesnikova L, Becker S, Kawaoka Y, Briggs JA (2012) Structural dissection of Ebola virus and its assembly determinants using cryo-electron tomography. Proc Natl Acad Sci USA 109(11):4275–4280. doi:10.1073/pnas.1120453109

Bhattacharyya S, Warfield KL, Ruthel G, Bavari S, Aman MJ, Hope TJ (2010) Ebola virus uses clathrin-mediated endocytosis as an entry pathway. Virology 401(1):18–28. doi:10.1016/j.virol.2010.02.015

Bhattacharyya S, Hope TJ, Young JA (2011) Differential requirements for clathrin endocytic pathway components in cellular entry by Ebola and Marburg glycoprotein pseudovirions. Virology 419(1):1–9. doi:10.1016/j.virol.2011.07.018

Bhattacharyya S, Zagorska A, Lew ED, Shrestha B, Rothlin CV, Naughton J, Diamond MS, Lemke G, Young JA (2013) Enveloped viruses disable innate immune responses in dendritic cells by direct activation of TAM receptors. Cell Host Microbe 14(2):136–147. doi:10.1016/j.chom.2013.07.005

Biedenkopf N, Hartlieb B, Hoenen T, Becker S (2013) Phosphorylation of Ebola virus VP30 influences the composition of the viral nucleocapsid complex: impact on viral transcription and replication. J Biol Chem 288(16):11165–11174. doi:10.1074/jbc.M113.461285

Biedenkopf N, Lier C, Becker S (2016a) Dynamic Phosphorylation of VP30 Is Essential for Ebola Virus Life Cycle. J Virol 90(10):4914–4925. doi:10.1128/JVI.03257-15

Biedenkopf N, Schlereth J, Grunweller A, Becker S, Hartmann RK (2016b) RNA binding of Ebola virus VP30 is essential for activating viral transcription. J Virol 90(16):7481–7496. doi:10.1128/JVI.00271-16

Bird BH, Spengler JR, Chakrabarti AK, Khristova ML, Sealy TK, Coleman-McCray JD, Martin BE, Dodd KA, Goldsmith CS, Sanders J, Zaki SR, Nichol ST, Spiropoulou CF (2016) Humanized mouse model of ebola virus disease mimics the immune responses in human disease. J Infect Dis 213(5):703–711. doi:10.1093/infdis/jiv538

Booth L, Roberts JL, Ecroyd H, Tritsch SR, Bavari S, Reid SP, Proniuk S, Zukiwski A, Jacob A, Sepulveda CS, Giovannoni F, Garcia CC, Damonte E, Gonzalez-Gallego J, Tunon MJ, Dent P (2016) AR-12 inhibits multiple chaperones concomitant with stimulating autophagosome formation collectively preventing virus replication. J Cell Physiol 231(10):2286–2302. doi:10.1002/jcp.25431

Bornholdt ZA, Ndungo E, Fusco ML, Bale S, Flyak AI, Crowe JE Jr, Chandran K, Saphire EO (2016a) Host-primed ebola virus GP exposes a hydrophobic NPC1 receptor-binding pocket, revealing a target for broadly neutralizing antibodies. mBio 7(1):e02154–e02115. doi:10.1128/mBio.02154-15

Bornholdt ZA, Turner HL, Murin CD, Li W, Sok D, Souders CA, Piper AE, Goff A, Shamblin JD, Wollen SE, Sprague TR, Fusco ML, Pommert KB, Cavacini LA, Smith HL, Klempner M, Reimann KA, Krauland E, Gerngross TU, Wittrup KD, Saphire EO, Burton DR, Glass PJ, Ward AB, Walker LM (2016b) Isolation of potent neutralizing antibodies from a survivor of the 2014 Ebola virus outbreak. Science 351(6277):1078–1083. doi:10.1126/science.aad5788

Bradfute SB, Braun DR, Shamblin JD, Geisbert JB, Paragas J, Garrison A, Hensley LE, Geisbert TW (2007) Lymphocyte death in a mouse model of Ebola virus infection. J Infect Dis 196(Suppl 2):S296–S304. doi:10.1086/520602

Bradfute SB, Warfield KL, Bavari S (2008) Functional CD8 + T cell responses in lethal Ebola virus infection. J Immunol 180(6):4058–4066

Bradfute SB, Swanson PE, Smith MA, Watanabe E, McDunn JE, Hotchkiss RS, Bavari S (2010) Mechanisms and consequences of ebolavirus-induced lymphocyte apoptosis. J Immunol 184(1):327–335. doi:10.4049/jimmunol.0901231

Brecher M, Schornberg KL, Delos SE, Fusco ML, Saphire EO, White JM (2012) Cathepsin cleavage potentiates the Ebola virus glycoprotein to undergo a subsequent fusion-relevant conformational change. J Virol 86(1):364–372. doi:10.1128/JVI.05708-11

Brindley MA, Hunt CL, Kondratowicz AS, Bowman J, Sinn PL, McCray PB Jr, Quinn K, Weller ML, Chiorini JA, Maury W (2011) Tyrosine kinase receptor Axl enhances entry of Zaire ebolavirus without direct interactions with the viral glycoprotein. Virology 415(2):83–94. doi:10.1016/j.virol.2011.04.002

Brudner M, Karpel M, Lear C, Chen L, Yantosca LM, Scully C, Sarraju A, Sokolovska A, Zariffard MR, Eisen DP, Mungall BA, Kotton DN, Omari A, Huang IC, Farzan M, Takahashi K, Stuart L, Stahl GL, Ezekowitz AB, Spear GT, Olinger GG, Schmidt EV, Michelow IC (2013) Lectin-dependent enhancement of Ebola virus infection via soluble and transmembrane C-type lectin receptors. PLoS One 8(4):e60838. doi:10.1371/journal.pone.0060838

Bwaka MA, Bonnet MJ, Calain P, Colebunders R, De Roo A, Guimard Y, Katwiki KR, Kibadi K, Kipasa MA, Kuvula KJ, Mapanda BB, Massamba M, Mupapa KD, Muyembe-Tamfum JJ, Ndaberey E, Peters CJ, Rollin PE, Van den Enden E, Van den Enden E (1999) Ebola hemorrhagic fever in Kikwit, Democratic Republic of the Congo: clinical observations in 103 patients. J Infect Dis 179(Suppl 1):S1–S7. doi:10.1086/514308

Caballero IS, Honko AN, Gire SK, Winnicki SM, Mele M, Gerhardinger C, Lin AE, Rinn JL, Sabeti PC, Hensley LE, Connor JH (2016) In vivo Ebola virus infection leads to a strong innate response in circulating immune cells. BMC Genom 17:707. doi:10.1186/s12864-016-3060-0

Cardenas WB, Loo YM, Gale M Jr, Hartman AL, Kimberlin CR, Martinez-Sobrido L, Saphire EO, Basler CF (2006) Ebola virus VP35 protein binds double-stranded RNA and inhibits alpha/beta interferon production induced by RIG-I signaling. J Virol 80(11):5168–5178. doi:10.1128/JVI.02199-05

Carette JE, Raaben M, Wong AC, Herbert AS, Obernosterer G, Mulherkar N, Kuehne AI, Kranzusch PJ, Griffin AM, Ruthel G, Dal Cin P, Dye JM, Whelan SP, Chandran K, Brummelkamp TR (2011) Ebola virus entry requires the cholesterol transporter Niemann-Pick C1. Nature 477(7364):340–343. doi:10.1038/nature10348

Chandran K, Sullivan NJ, Felbor U, Whelan SP, Cunningham JM (2005) Endosomal proteolysis of the Ebola virus glycoprotein is necessary for infection. Science 308(5728):1643–1645. doi:10.1126/science.1110656

Chang TH, Kubota T, Matsuoka M, Jones S, Bradfute SB, Bray M, Ozato K (2009) Ebola Zaire virus blocks type I interferon production by exploiting the host SUMO modification machinery. PLoS Pathog 5(6):e1000493. doi:10.1371/journal.ppat.1000493

Chertow DS, Kleine C, Edwards JK, Scaini R, Giuliani R, Sprecher A (2014) Ebola virus disease in West Africa–clinical manifestations and management. The N Engl J Med 371(22):2054–2057. doi:10.1056/NEJMp1413084

Cilloniz C, Ebihara H, Ni C, Neumann G, Korth MJ, Kelly SM, Kawaoka Y, Feldmann H, Katze MG (2011) Functional genomics reveals the induction of inflammatory response and metalloproteinase gene expression during lethal Ebola virus infection. J Virol 85(17):9060–9068. doi:10.1128/JVI.00659-11

Connolly BM, Steele KE, Davis KJ, Geisbert TW, Kell WM, Jaax NK, Jahrling PB (1999) Pathogenesis of experimental Ebola virus infection in guinea pigs. J Infect Dis 179(Suppl 1): S203–S217. doi:10.1086/514305

Cote M, Misasi J, Ren T, Bruchez A, Lee K, Filone CM, Hensley L, Li Q, Ory D, Chandran K, Cunningham J (2011) Small molecule inhibitors reveal Niemann-Pick C1 is essential for Ebola virus infection. Nature 477(7364):344–348. doi:10.1038/nature10380

Cross RW, Fenton KA, Geisbert JB, Mire CE, Geisbert TW (2015) Modeling the disease course of Zaire ebolavirus infection in the outbred guinea pig. J Infect Dis 212(Suppl 2):S305–S315. doi:10.1093/infdis/jiv237

Cross RW, Mire CE, Borisevich V, Geisbert JB, Fenton KA, Geisbert TW (2016) The domestic ferret (Mustela putorius furo) as a Lethal Infection Model for 3 Species of Ebolavirus. J Infect Dis 214(4):565–569. doi:10.1093/infdis/jiw209

Dahlke C, Lunemann S, Kasonta R, Kreuels B, Schmiedel S, Ly ML, Fehling SK, Strecker T, Becker S, Altfeld M, Sow A, Lohse AW, Munoz-Fontela C, Addo MM (2016) Comprehensive

characterization of cellular immune responses following Ebola virus infection. J Infect Dis. doi:10.1093/infdis/jiw508
Dahlmann F, Biedenkopf N, Babler A, Jahnen-Dechent W, Karsten CB, Gnirss K, Schneider H, Wrensch F, O'Callaghan CA, Bertram S, Herrler G, Becker S, Pohlmann S, Hofmann-Winkler H (2015) Analysis of Ebola Virus Entry Into Macrophages. J Infect Dis 212(Suppl 2):S247–S257. doi:10.1093/infdis/jiv140
Dallatomasina S, Crestani R, Sylvester Squire J, Declerk H, Caleo GM, Wolz A, Stinson K, Patten G, Brechard R, Gbabai OB, Spreicher A, Van Herp M, Zachariah R (2015) Ebola outbreak in rural West Africa: epidemiology, clinical features and outcomes. Tropical medicine & international health: TM & IH 20(4):448–454. doi:10.1111/tmi.12454
Dietz PM, Jambai A, Paweska JT, Yoti Z, Ksiazek TG (2015) Epidemiology and risk factors for Ebola virus disease in Sierra Leone-23 May 2014–31 January 2015. Clinical infectious diseases: an official publication of the Infectious Diseases Society of America 61(11):1648–1654. doi:10.1093/cid/civ568
Dolnik O, Volchkova V, Garten W, Carbonnelle C, Becker S, Kahnt J, Stroher U, Klenk HD, Volchkov V (2004) Ectodomain shedding of the glycoprotein GP of Ebola virus. EMBO J 23 (10):2175–2184. doi:10.1038/sj.emboj.7600219
Dolnik O, Volchkova VA, Escudero-Perez B, Lawrence P, Klenk HD, Volchkov VE (2015) Shedding of ebola virus surface glycoprotein is a mechanism of self-regulation of cellular cytotoxicity and has a direct effect on virus infectivity. J Infect Dis 212(Suppl 2):S322–S328. doi:10.1093/infdis/jiv268
Dube D, Schornberg KL, Shoemaker CJ, Delos SE, Stantchev TS, Clouse KA, Broder CC, White JM (2010) Cell adhesion-dependent membrane trafficking of a binding partner for the ebolavirus glycoprotein is a determinant of viral entry. Proc Natl Acad Sci USA 107 (38):16637–16642. doi:10.1073/pnas.1008509107
Ebihara H, Takada A, Kobasa D, Jones S, Neumann G, Theriault S, Bray M, Feldmann H, Kawaoka Y (2006) Molecular determinants of Ebola virus virulence in mice. PLoS Pathog 2 (7):e73. doi:10.1371/journal.ppat.0020073
Ebihara H, Rockx B, Marzi A, Feldmann F, Haddock E, Brining D, LaCasse RA, Gardner D, Feldmann H (2011) Host response dynamics following lethal infection of rhesus macaques with Zaire ebolavirus. J Infect Dis 204(Suppl 3):S991–S999. doi:10.1093/infdis/jir336
Ebihara H, Zivcec M, Gardner D, Falzarano D, LaCasse R, Rosenke R, Long D, Haddock E, Fischer E, Kawaoka Y, Feldmann H (2013) A Syrian golden hamster model recapitulating ebola hemorrhagic fever. J Infect Dis 207(2):306–318. doi:10.1093/infdis/jis626
Edwards MR, Basler CF (2015) Marburg virus VP24 protein relieves suppression of the NF-kappaB Pathway Through Interaction With Kelch-like ECH-Associated Protein 1. J Infect Dis 212(Suppl 2):S154–S159. doi:10.1093/infdis/jiv050
Edwards MR, Liu G, Mire CE, Sureshchandra S, Luthra P, Yen B, Shabman RS, Leung DW, Messaoudi I, Geisbert TW, Amarasinghe GK, Basler CF (2016) Differential regulation of interferon responses by Ebola and Marburg virus VP35 proteins. Cell Rep 14(7):1632–1640. doi:10.1016/j.celrep.2016.01.049
Empig CJ, Goldsmith MA (2002) Association of the caveola vesicular system with cellular entry by filoviruses. J Virol 76(10):5266–5270
Escudero-Perez B, Volchkova VA, Dolnik O, Lawrence P, Volchkov VE (2014) Shed GP of Ebola virus triggers immune activation and increased vascular permeability. PLoS Pathog 10(11): e1004509. doi:10.1371/journal.ppat.1004509
Favier AL, Gout E, Reynard O, Ferraris O, Kleman JP, Volchkov V, Peyrefitte C, Thielens NM (2016) Enhancement of ebola virus infection via Ficolin-1 Interaction with the mucin domain of GP glycoprotein. J Virol 90(11):5256–5269. doi:10.1128/JVI.00232-16
Feagins AR, Basler CF (2015) Lloviu virus VP24 and VP35 proteins function as innate immune antagonists in human and bat cells. Virology 485:145–152. doi:10.1016/j.virol.2015.07.010
Fisher-Hoch SP, Brammer TL, Trappier SG, Hutwagner LC, Farrar BB, Ruo SL, Brown BG, Hermann LM, Perez-Oronoz GI, Goldsmith CS et al (1992a) Pathogenic potential of filoviruses: role of geographic origin of primate host and virus strain. J Infect Dis 166(4):753–763

Fisher-Hoch SP, Perez-Oronoz GI, Jackson EL, Hermann LM, Brown BG (1992b) Filovirus clearance in non-human primates. Lancet 340(8817):451–453

Fitzgerald F, Naveed A, Wing K, Gbessay M, Ross JC, Checchi F, Youkee D, Jalloh MB, Baion D, Mustapha A, Jah H, Lako S, Oza S, Boufkhed S, Feury R, Bielicki JA, Gibb DM, Klein N, Sahr F, Yeung S (2016) Ebola virus disease in children, sierra leone, 2014–2015. Emerg Infect Dis 22(10):1769–1777. doi:10.3201/eid2210.160579

Fitzpatrick G, Vogt F, Moi Gbabai OB, Decroo T, Keane M, De Clerck H, Grolla A, Brechard R, Stinson K, Van Herp M (2015) The contribution of ebola viral load at admission and other patient characteristics to mortality in a Medecins Sans Frontieres Ebola Case Management Centre, Kailahun, Sierra Leone, June–October 2014. J Infect Dis 212(11):1752–1758. doi:10.1093/infdis/jiv304

Flyak AI, Shen X, Murin CD, Turner HL, David JA, Fusco ML, Lampley R, Kose N, Ilinykh PA, Kuzmina N, Branchizio A, King H, Brown L, Bryan C, Davidson E, Doranz BJ, Slaughter JC, Sapparapu G, Klages C, Ksiazek TG, Saphire EO, Ward AB, Bukreyev A, Crowe JE Jr (2016) Cross-reactive and potent neutralizing antibody responses in human survivors of natural ebolavirus infection. Cell 164(3):392–405. doi:10.1016/j.cell.2015.12.022

Francica JR, Varela-Rohena A, Medvec A, Plesa G, Riley JL, Bates P (2010) Steric shielding of surface epitopes and impaired immune recognition induced by the ebola virus glycoprotein. PLoS Pathog 6(9):e1001098. doi:10.1371/journal.ppat.1001098

Gabriel G, Feldmann F, Reimer R, Thiele S, Fischer M, Hartmann E, Bader M, Ebihara H, Hoenen T, Feldmann H (2015) Importin-alpha7 is involved in the formation of ebola virus inclusion bodies but is not essential for pathogenicity in mice. J Infect Dis 212(Suppl 2):S316–S321. doi:10.1093/infdis/jiv240

Garcia-Dorival I, Wu W, Dowall S, Armstrong S, Touzelet O, Wastling J, Barr JN, Matthews D, Carroll M, Hewson R, Hiscox JA (2014) Elucidation of the Ebola virus VP24 cellular interactome and disruption of virus biology through targeted inhibition of host-cell protein function. J Proteome Res 13(11):5120–5135. doi:10.1021/pr500556d

Garcia-Dorival I, Wu W, Armstrong SD, Barr JN, Carroll MW, Hewson R, Hiscox JA (2016) Elucidation of the cellular interactome of the Ebolavirus nucleoprotein and identification of therapeutic targets. J Proteome Res. doi:10.1021/acs.jproteome.6b00337

Gc JB, Gerstman BS, Stahelin RV, Chapagain PP (2016) The Ebola virus protein VP40 hexamer enhances the clustering of PI(4,5)P2 lipids in the plasma membrane. Phys Chem Chem Phys 18 (41):28409–28417. doi:10.1039/c6cp03776c

Geisbert TW, Hensley LE, Gibb TR, Steele KE, Jaax NK, Jahrling PB (2000) Apoptosis induced in vitro and in vivo during infection by Ebola and Marburg viruses. Lab Invest 80(2):171–186

Geisbert TW, Hensley LE, Jahrling PB, Larsen T, Geisbert JB, Paragas J, Young HA, Fredeking TM, Rote WE, Vlasuk GP (2003a) Treatment of Ebola virus infection with a recombinant inhibitor of factor VIIa/tissue factor: a study in rhesus monkeys. Lancet 362 (9400):1953–1958. doi:10.1016/S0140-6736(03)15012-X

Geisbert TW, Hensley LE, Larsen T, Young HA, Reed DS, Geisbert JB, Scott DP, Kagan E, Jahrling PB, Davis KJ (2003b) Pathogenesis of Ebola hemorrhagic fever in cynomolgus macaques: evidence that dendritic cells are early and sustained targets of infection. Am J Pathol 163(6):2347–2370. doi:10.1016/S0002-9440(10)63591-2

Geisbert TW, Young HA, Jahrling PB, Davis KJ, Kagan E, Hensley LE (2003c) Mechanisms underlying coagulation abnormalities in ebola hemorrhagic fever: overexpression of tissue factor in primate monocytes/macrophages is a key event. J Infect Dis 188(11):1618–1629. doi:10.1086/379724

Geisbert TW, Young HA, Jahrling PB, Davis KJ, Larsen T, Kagan E, Hensley LE (2003d) Pathogenesis of Ebola hemorrhagic fever in primate models: evidence that hemorrhage is not a direct effect of virus-induced cytolysis of endothelial cells. Am J Pathol 163(6):2371–2382. doi:10.1016/S0002-9440(10)63592-4

Georges AJ, Leroy EM, Renaut AA, Benissan CT, Nabias RJ, Ngoc MT, Obiang PI, Lepage JP, Bertherat EJ, Benoni DD, Wickings EJ, Amblard JP, Lansoud-Soukate JM, Milleliri JM, Baize S, Georges-Courbot MC (1999) Ebola hemorrhagic fever outbreaks in Gabon, 1994–

1997: epidemiologic and health control issues. J Infect Dis 179(Suppl 1):S65–S75. doi:10.1086/514290

Gibb TR, Bray M, Geisbert TW, Steele KE, Kell WM, Davis KJ, Jaax NK (2001) Pathogenesis of experimental Ebola Zaire virus infection in BALB/c mice. J Comp Pathol 125(4):233–242. doi:10.1053/jcpa.2001.0502

Gnirss K, Fiedler M, Kramer-Kuhl A, Bolduan S, Mittler E, Becker S, Schindler M, Pohlmann S (2014) Analysis of determinants in filovirus glycoproteins required for tetherin antagonism. Viruses 6(4):1654–1671. doi:10.3390/v6041654

Gregory SM, Harada E, Liang B, Delos SE, White JM, Tamm LK (2011) Structure and function of the complete internal fusion loop from Ebolavirus glycoprotein 2. Proc Natl Acad Sci USA 108 (27):11211–11216. doi:10.1073/pnas.1104760108

Gregory SM, Larsson P, Nelson EA, Kasson PM, White JM, Tamm LK (2014) Ebolavirus entry requires a compact hydrophobic fist at the tip of the fusion loop. J Virol 88(12):6636–6649. doi:10.1128/JVI.00396-14

Groseth A, Charton JE, Sauerborn M, Feldmann F, Jones SM, Hoenen T, Feldmann H (2009) The Ebola virus ribonucleoprotein complex: a novel VP30-L interaction identified. Virus Res 140 (1–2):8–14. doi:10.1016/j.virusres.2008.10.017

Gupta M, Spiropoulou C, Rollin PE (2007) Ebola virus infection of human PBMCs causes massive death of macrophages, CD4 and CD8 T cell sub-populations in vitro. Virology 364 (1):45–54. doi:10.1016/j.virol.2007.02.017

Gustin JK, Bai Y, Moses AV, Douglas JL (2015) Ebola virus glycoprotein promotes enhanced viral egress by preventing ebola vp40 from associating with the host restriction factor BST2/tetherin. J Infect Dis 212(Suppl 2):S181–S190. doi:10.1093/infdis/jiv125

Haaskjold YL, Bolkan HA, Krogh KO, Jongopi J, Lundeby KM, Mellesmo S, Garces PS, Josendal O, Opstad A, Svensen E, Fuentes LM, Kamara AS, Riera M, Arranz J, Roberts DP, Stamper PD, Austin P, Moosa AJ, Marke D, Hassan S, Eide GE, Berg A, Blomberg B (2016) Clinical features of and risk factors for fatal ebola virus disease, Moyamba District, Sierra Leone, December 2014–February 2015. Emerg Infect Dis 22(9):1537–1544. doi:10.3201/eid2209.151621

Hacke M, Bjorkholm P, Hellwig A, Himmels P, de Almodovar CR, Brugger B, Wieland F, Ernst AM (2015) Inhibition of Ebola virus glycoprotein-mediated cytotoxicity by targeting its transmembrane domain and cholesterol. Nat Commun 6:7688. doi:10.1038/ncomms8688

Halfmann P, Neumann G, Kawaoka Y (2011) The Ebolavirus VP24 protein blocks phosphorylation of p38 mitogen-activated protein kinase. J Infect Dis 204(Suppl 3):S953–S956. doi:10.1093/infdis/jir325

Han Z, Harty RN (2005) Packaging of actin into Ebola virus VLPs. Virol J 2:92. doi:10.1186/1743-422X-2-92

Han Z, Harty RN (2007) Influence of calcium/calmodulin on budding of Ebola VLPs: implications for the involvement of the Ras/Raf/MEK/ERK pathway. Virus Genes 35(3):511–520. doi:10.1007/s11262-007-0125-9

Han Z, Madara JJ, Herbert A, Prugar LI, Ruthel G, Lu J, Liu Y, Liu W, Liu X, Wrobel JE, Reitz AB, Dye JM, Harty RN, Freedman BD (2015a) Calcium regulation of hemorrhagic fever virus budding: Mechanistic implications for host-oriented therapeutic intervention. PLoS Pathog 11(10):e1005220. doi:10.1371/journal.ppat.1005220

Han Z, Madara JJ, Liu Y, Liu W, Ruthel G, Freedman BD, Harty RN (2015b) ALIX rescues budding of a double PTAP/PPEY L-domain deletion mutant of Ebola VP40: a role for ALIX in Ebola virus egress. J Infect Dis 212(Suppl 2):S138–S145. doi:10.1093/infdis/jiu838

Han Z, Sagum CA, Bedford MT, Sidhu SS, Sudol M, Harty RN (2016) ITCH E3 ubiquitin ligase interacts with ebola virus VP40 to regulate budding. J Virol 90(20):9163–9171. doi:10.1128/JVI.01078-16

Hartman AL, Bird BH, Towner JS, Antoniadou ZA, Zaki SR, Nichol ST (2008) Inhibition of IRF-3 activation by VP35 is critical for the high level of virulence of ebola virus. J Virol 82 (6):2699–2704. doi:10.1128/JVI.02344-07

Hensley LE, Young HA, Jahrling PB, Geisbert TW (2002) Proinflammatory response during Ebola virus infection of primate models: possible involvement of the tumor necrosis factor receptor superfamily. Immunol Lett 80(3):169–179

Hensley LE, Stevens EL, Yan SB, Geisbert JB, Macias WL, Larsen T, Daddario-DiCaprio KM, Cassell GH, Jahrling PB, Geisbert TW (2007) Recombinant human activated protein C for the postexposure treatment of Ebola hemorrhagic fever. J Infect Dis 196(Suppl 2):S390–S399. doi:10.1086/520598

Hoenen T, Shabman RS, Groseth A, Herwig A, Weber M, Schudt G, Dolnik O, Basler CF, Becker S, Feldmann H (2012) Inclusion bodies are a site of ebolavirus replication. J Virol 86 (21):11779–11788. doi:10.1128/JVI.01525-12

Hoffmann M, Gonzalez Hernandez M, Berger E, Marzi A, Pohlmann S (2016) The glycoproteins of all filovirus species use the same host factors for entry into bat and human cells but entry efficiency is species dependent. PLoS ONE 11(2):e0149651. doi:10.1371/journal.pone.0149651

Huang Y, Xu L, Sun Y, Nabel GJ (2002) The assembly of Ebola virus nucleocapsid requires virion-associated proteins 35 and 24 and posttranslational modification of nucleoprotein. Mol Cell 10(2):307–316

Huang IC, Bailey CC, Weyer JL, Radoshitzky SR, Becker MM, Chiang JJ, Brass AL, Ahmed AA, Chi X, Dong L, Longobardi LE, Boltz D, Kuhn JH, Elledge SJ, Bavari S, Denison MR, Choe H, Farzan M (2011) Distinct patterns of IFITM-mediated restriction of filoviruses, SARS coronavirus, and influenza A virus. PLoS Pathog 7(1):e1001258. doi:10.1371/journal.ppat.1001258

Hunt CL, Kolokoltsov AA, Davey RA, Maury W (2011) The Tyro3 receptor kinase Axl enhances macropinocytosis of Zaire ebolavirus. J Virol 85(1):334–347. doi:10.1128/JVI.01278-09

Ilinykh PA, Tigabu B, Ivanov A, Ammosova T, Obukhov Y, Garron T, Kumari N, Kovalskyy D, Platonov MO, Naumchik VS, Freiberg AN, Nekhai S, Bukreyev A (2014) Role of protein phosphatase 1 in dephosphorylation of Ebola virus VP30 protein and its targeting for the inhibition of viral transcription. J Biol Chem 289(33):22723–22738. doi:10.1074/jbc.M114.575050

Ilinykh PA, Lubaki NM, Widen SG, Renn LA, Theisen TC, Rabin RL, Wood TG, Bukreyev A (2015) Different temporal effects of ebola virus VP35 and VP24 proteins on global gene expression in human dendritic cells. J Virol 89(15):7567–7583. doi:10.1128/JVI.00924-15

Iwasa A, Halfmann P, Noda T, Oyama M, Kozuka-Hata H, Watanabe S, Shimojima M, Watanabe T, Kawaoka Y (2011) Contribution of Sec61alpha to the life cycle of Ebola virus. J Infect Dis 204(Suppl 3):S919–S926. doi:10.1093/infdis/jir324

Jasenosky LD, Neumann G, Lukashevich I, Kawaoka Y (2001) Ebola virus VP40-induced particle formation and association with the lipid bilayer. J Virol 75(11):5205–5214. doi:10.1128/JVI.75.11.5205-5214.2001

Jemielity S, Wang JJ, Chan YK, Ahmed AA, Li W, Monahan S, Bu X, Farzan M, Freeman GJ, Umetsu DT, Dekruyff RH, Choe H (2013) TIM-family proteins promote infection of multiple enveloped viruses through virion-associated phosphatidylserine. PLoS Pathog 9(3):e1003232. doi:10.1371/journal.ppat.1003232

Ji X, Olinger GG, Aris S, Chen Y, Gewurz H, Spear GT (2005) Mannose-binding lectin binds to Ebola and Marburg envelope glycoproteins, resulting in blocking of virus interaction with DC-SIGN and complement-mediated virus neutralization. J Gen Virol 86(Pt 9):2535–2542. doi:10.1099/vir.0.81199-0

Jin H, Yan Z, Prabhakar BS, Feng Z, Ma Y, Verpooten D, Ganesh B, He B (2010) The VP35 protein of Ebola virus impairs dendritic cell maturation induced by virus and lipopolysaccharide. J Gen Virol 91(Pt 2):352–361. doi:10.1099/vir.0.017343-0

Johnson KA, Taghon GJ, Scott JL, Stahelin RV (2016) The Ebola Virus matrix protein, VP40, requires phosphatidylinositol 4,5-bisphosphate (PI(4,5)P2) for extensive oligomerization at the plasma membrane and viral egress. Sci Rep 6:19125. doi:10.1038/srep19125

Kaletsky RL, Simmons G, Bates P (2007) Proteolysis of the Ebola virus glycoproteins enhances virus binding and infectivity. J Virol 81(24):13378–13384. doi:10.1128/JVI.01170-07

Kaletsky RL, Francica JR, Agrawal-Gamse C, Bates P (2009) Tetherin-mediated restriction of filovirus budding is antagonized by the Ebola glycoprotein. Proc Natl Acad Sci USA 106 (8):2886–2891. doi:10.1073/pnas.0811014106

Kash JC, Muhlberger E, Carter V, Grosch M, Perwitasari O, Proll SC, Thomas MJ, Weber F, Klenk HD, Katze MG (2006) Global suppression of the host antiviral response by Ebola- and Marburgviruses: increased antagonism of the type I interferon response is associated with enhanced virulence. J Virol 80(6):3009–3020. doi:10.1128/JVI.80.6.3009-3020.2006

Khan AS, Tshioko FK, Heymann DL, Le Guenno B, Nabeth P, Kerstiens B, Fleerackers Y, Kilmarx PH, Rodier GR, Nkuku O, Rollin PE, Sanchez A, Zaki SR, Swanepoel R, Tomori O, Nichol ST, Peters CJ, Muyembe-Tamfum JJ, Ksiazek TG (1999) The reemergence of Ebola hemorrhagic fever, Democratic Republic of the Congo, 1995. Commission de Lutte contre les Epidemies a Kikwit. The Journal of infectious diseases 179 Suppl 1:S76-86. doi:10.1086/514306

Kimberlin CR, Bornholdt ZA, Li S, Woods VL Jr, MacRae IJ, Saphire EO (2010) Ebolavirus VP35 uses a bimodal strategy to bind dsRNA for innate immune suppression. Proc Natl Acad Sci USA 107(1):314–319. doi:10.1073/pnas.0910547107

Kindrachuk J, Wahl-Jensen V, Safronetz D, Trost B, Hoenen T, Arsenault R, Feldmann F, Traynor D, Postnikova E, Kusalik A, Napper S, Blaney JE, Feldmann H, Jahrling PB (2014) Ebola virus modulates transforming growth factor beta signaling and cellular markers of mesenchyme-like transition in hepatocytes. J Virol 88(17):9877–9892. doi:10.1128/JVI.01410-14

Kondratowicz AS, Lennemann NJ, Sinn PL, Davey RA, Hunt CL, Moller-Tank S, Meyerholz DK, Rennert P, Mullins RF, Brindley M, Sandersfeld LM, Quinn K, Weller M, McCray PB Jr, Chiorini J, Maury W (2011) T-cell immunoglobulin and mucin domain 1 (TIM-1) is a receptor for Zaire Ebolavirus and Lake Victoria Marburgvirus. Proc Natl Acad Sci USA 108(20):8426–8431. doi:10.1073/pnas.1019030108

Kozak R, He S, Kroeker A, de La Vega MA, Audet J, Wong G, Urfano C, Antonation K, Embury-Hyatt C, Kobinger GP, Qiu X (2016) Ferrets infected with bundibugyo virus or ebola virus recapitulate important aspects of human filovirus disease. J Virol 90(20):9209–9223. doi:10.1128/JVI.01033-16

Kratz T, Roddy P, Tshomba Oloma A, Jeffs B, Pou Ciruelo D, de la Rosa O, Borchert M (2015) Ebola virus disease outbreak in Isiro, Democratic Republic of the Congo, 2012: Signs and Symptoms. Management and Outcomes. PloS one 10(6):e0129333. doi:10.1371/journal.pone.0129333

Kubota T, Matsuoka M, Chang TH, Bray M, Jones S, Tashiro M, Kato A, Ozato K (2009) Ebolavirus VP35 interacts with the cytoplasmic dynein light chain 8. J Virol 83(13):6952–6956. doi:10.1128/JVI.00480-09

Kuhl A, Banning C, Marzi A, Votteler J, Steffen I, Bertram S, Glowacka I, Konrad A, Sturzl M, Guo JT, Schubert U, Feldmann H, Behrens G, Schindler M, Pohlmann S (2011) The Ebola virus glycoprotein and HIV-1 Vpu employ different strategies to counteract the antiviral factor tetherin. J Infect Dis 204(Suppl 3):S850–S860. doi:10.1093/infdis/jir378

Kuroda M, Fujikura D, Nanbo A, Marzi A, Noyori O, Kajihara M, Maruyama J, Matsuno K, Miyamoto H, Yoshida R, Feldmann H, Takada A (2015) Interaction between TIM-1 and NPC1 is important for cellular entry of ebola virus. J Virol 89(12):6481–6493. doi:10.1128/JVI.03156-14

Lado M, Walker NF, Baker P, Haroon S, Brown CS, Youkee D, Studd N, Kessete Q, Maini R, Boyles T, Hanciles E, Wurie A, Kamara TB, Johnson O, Leather AJ (2015) Clinical features of patients isolated for suspected Ebola virus disease at Connaught Hospital, Freetown, Sierra Leone: a retrospective cohort study. Lancet Infect Dis 15(9):1024–1033. doi:10.1016/S1473-3099(15)00137-1

Le Guenno B, Formenty P, Wyers M, Gounon P, Walker F, Boesch C (1995) Isolation and partial characterisation of a new strain of Ebola virus. Lancet 345(8960):1271–1274

Lennemann NJ, Rhein BA, Ndungo E, Chandran K, Qiu X, Maury W (2014) Comprehensive functional analysis of N-linked glycans on Ebola virus GP1. mBio 5(1):e00862–e00813. doi:10.1128/mBio.00862-13

Leung DW, Ginder ND, Fulton DB, Nix J, Basler CF, Honzatko RB, Amarasinghe GK (2009) Structure of the Ebola VP35 interferon inhibitory domain. Proc Natl Acad Sci USA 106 (2):411–416. doi:10.1073/pnas.0807854106

Leung DW, Prins KC, Borek DM, Farahbakhsh M, Tufariello JM, Ramanan P, Nix JC, Helgeson LA, Otwinowski Z, Honzatko RB, Basler CF, Amarasinghe GK (2010) Structural basis for dsRNA recognition and interferon antagonism by Ebola VP35. Nat Struct Mol Biol 17(2):165–172. doi:10.1038/nsmb.1765

Leung LW, Martinez O, Reynard O, Volchkov VE, Basler CF (2011a) Ebola virus failure to stimulate plasmacytoid dendritic cell interferon responses correlates with impaired cellular entry. J Infect Dis 204(Suppl 3):S973–S977. doi:10.1093/infdis/jir331

Leung LW, Park MS, Martinez O, Valmas C, Lopez CB, Basler CF (2011b) Ebolavirus VP35 suppresses IFN production from conventional but not plasmacytoid dendritic cells. Immunol Cell Biol 89(7):792–802. doi:10.1038/icb.2010.169

Licata JM, Simpson-Holley M, Wright NT, Han Z, Paragas J, Harty RN (2003) Overlapping motifs (PTAP and PPEY) within the Ebola virus VP40 protein function independently as late budding domains: involvement of host proteins TSG101 and VPS-4. J Virol 77(3):1812–1819

Liddell AM, Davey RT Jr, Mehta AK, Varkey JB, Kraft CS, Tseggay GK, Badidi O, Faust AC, Brown KV, Suffredini AF, Barrett K, Wolcott MJ, Marconi VC, Lyon GM 3rd, Weinstein GL, Weinmeister K, Sutton S, Hazbun M, Albarino CG, Reed Z, Cannon D, Stroher U, Feldman M, Ribner BS, Lane HC, Fauci AS, Uyeki TM (2015) Characteristics and clinical management of a cluster of 3 patients with ebola virus disease, including the first domestically acquired cases in the United States. Ann Intern Med 163(2):81–90. doi:10.7326/M15-0530

Lin G, Simmons G, Pohlmann S, Baribaud F, Ni H, Leslie GJ, Haggarty BS, Bates P, Weissman D, Hoxie JA, Doms RW (2003) Differential N-linked glycosylation of human immunodeficiency virus and Ebola virus envelope glycoproteins modulates interactions with DC-SIGN and DC-SIGNR. J Virol 77(2):1337–1346

Liu SY, Aliyari R, Chikere K, Li G, Marsden MD, Smith JK, Pernet O, Guo H, Nusbaum R, Zack JA, Freiberg AN, Su L, Lee B, Cheng G (2013) Interferon-inducible cholesterol-25-hydroxylase broadly inhibits viral entry by production of 25-hydroxycholesterol. Immunity 38(1):92–105. doi:10.1016/j.immuni.2012.11.005

Lu J, Qu Y, Liu Y, Jambusaria R, Han Z, Ruthel G, Freedman BD, Harty RN (2013) Host IQGAP1 and Ebola virus VP40 interactions facilitate virus-like particle egress. J Virol 87 (13):7777–7780. doi:10.1128/JVI.00470-13

Lu HJ, Qian J, Kargbo D, Zhang XG, Yang F, Hu Y, Sun Y, Cao YX, Deng YQ, Su HX, Dafae F, Sun Y, Wang CY, Nie WM, Bai CQ, Xia ZP, Liu K, Kargbo B, Gao GF, Jiang JF (2015) Ebola virus outbreak investigation, Sierra Leone, September 28–November 11 2014. Emerg Infect Dis 21(11):1921–1927. doi:10.3201/eid2111.150582

Lubaki NM, Ilinykh P, Pietzsch C, Tigabu B, Freiberg AN, Koup RA, Bukreyev A (2013) The lack of maturation of Ebola virus-infected dendritic cells results from the cooperative effect of at least two viral domains. J Virol 87(13):7471–7485. doi:10.1128/JVI.03316-12

Ludtke A, Ruibal P, Becker-Ziaja B, Rottstegge M, Wozniak DM, Cabeza-Cabrerizo M, Thorenz A, Weller R, Kerber R, Idoyaga J, Magassouba N, Gabriel M, Gunther S, Oestereich L, Munoz-Fontela C (2016) Ebola virus disease is characterized by poor activation and reduced levels of circulating cd16 + monocytes. J Infect Dis 214(suppl 3):S275–S280. doi:10.1093/infdis/jiw260

Luthra P, Ramanan P, Mire CE, Weisend C, Tsuda Y, Yen B, Liu G, Leung DW, Geisbert TW, Ebihara H, Amarasinghe GK, Basler CF (2013) Mutual antagonism between the Ebola virus VP35 protein and the RIG-I activator PACT determines infection outcome. Cell Host Microbe 14(1):74–84. doi:10.1016/j.chom.2013.06.010

Luthra P, Jordan DS, Leung DW, Amarasinghe GK, Basler CF (2015) Ebola virus VP35 interaction with dynein LC8 regulates viral RNA synthesis. J Virol 89(9):5148–5153. doi:10.1128/JVI.03652-14

MacNeil A, Farnon EC, Wamala J, Okware S, Cannon DL, Reed Z, Towner JS, Tappero JW, Lutwama J, Downing R, Nichol ST, Ksiazek TG, Rollin PE (2010) Proportion of deaths and clinical features in Bundibugyo Ebola virus infection. Uganda Emerg Infect Dis 16(12):1969–1972. doi:10.3201/eid1612.100627

Markosyan RM, Miao C, Zheng YM, Melikyan GB, Liu SL, Cohen FS (2016) Induction of cell-cell fusion by Ebola virus glycoprotein: low pH Is not a trigger. PLoS Pathog 12(1): e1005373. doi:10.1371/journal.ppat.1005373

Martinez O, Valmas C, Basler CF (2007) Ebola virus-like particle-induced activation of NF-kappaB and Erk signaling in human dendritic cells requires the glycoprotein mucin domain. Virology 364(2):342–354. doi:10.1016/j.virol.2007.03.020

Martinez MJ, Volchkova VA, Raoul H, Alazard-Dany N, Reynard O, Volchkov VE (2011) Role of VP30 phosphorylation in the Ebola virus replication cycle. J Infect Dis 204(Suppl 3):S934–S940. doi:10.1093/infdis/jir320

Martinez O, Johnson JC, Honko A, Yen B, Shabman RS, Hensley LE, Olinger GG, Basler CF (2013) Ebola virus exploits a monocyte differentiation program to promote its entry. J Virol 87 (7):3801–3814. doi:10.1128/JVI.02695-12

Martins K, Cooper C, Warren T, Wells J, Bell T, Raymond J, Stuthman K, Benko J, Garza N, van Tongeren S, Donnelly G, Retterer C, Dong L, Bavari S (2015) Characterization of clinical and immunological parameters during Ebola virus infection of rhesus macaques. Viral Immunol 28 (1):32–41. doi:10.1089/vim.2014.0085

Martin-Serrano J, Zang T, Bieniasz PD (2001) HIV-1 and Ebola virus encode small peptide motifs that recruit Tsg101 to sites of particle assembly to facilitate egress. Nat Med 7(12):1313–1319. doi:10.1038/nm1201-1313

Marzi A, Akhavan A, Simmons G, Gramberg T, Hofmann H, Bates P, Lingappa VR, Pohlmann S (2006) The signal peptide of the ebolavirus glycoprotein influences interaction with the cellular lectins DC-SIGN and DC-SIGNR. J Virol 80(13):6305–6317. doi:10.1128/JVI.02545-05

Marzi A, Moller P, Hanna SL, Harrer T, Eisemann J, Steinkasserer A, Becker S, Baribaud F, Pohlmann S (2007) Analysis of the interaction of Ebola virus glycoprotein with DC-SIGN (dendritic cell-specific intercellular adhesion molecule 3-grabbing nonintegrin) and its homologue DC-SIGNR. J Infect Dis 196(Suppl 2):S237–S246. doi:10.1086/520607

Marzi A, Reinheckel T, Feldmann H (2012) Cathepsin B & L are not required for ebola virus replication. PLoS neglected tropical diseases 6(12):e1923. doi:10.1371/journal.pntd.0001923

Mateo M, Reid SP, Leung LW, Basler CF, Volchkov VE (2010) Ebolavirus VP24 binding to karyopherins is required for inhibition of interferon signaling. J Virol 84(2):1169–1175. doi:10.1128/JVI.01372-09

Mateo M, Carbonnelle C, Martinez MJ, Reynard O, Page A, Volchkova VA, Volchkov VE (2011a) Knockdown of Ebola virus VP24 impairs viral nucleocapsid assembly and prevents virus replication. J Infect Dis 204(Suppl 3):S892–S896. doi:10.1093/infdis/jir311

Mateo M, Carbonnelle C, Reynard O, Kolesnikova L, Nemirov K, Page A, Volchkova VA, Volchkov VE (2011b) VP24 is a molecular determinant of Ebola virus virulence in guinea pigs. J Infect Dis 204(Suppl 3):S1011–S1020. doi:10.1093/infdis/jir338

Matsuno K, Nakayama E, Noyori O, Marzi A, Ebihara H, Irimura T, Feldmann H, Takada A (2010) C-type lectins do not act as functional receptors for filovirus entry into cells. Biochem Biophys Res Commun 403(1):144–148. doi:10.1016/j.bbrc.2010.10.136

McElroy AK, Akondy RS, Davis CW, Ellebedy AH, Mehta AK, Kraft CS, Lyon GM, Ribner BS, Varkey J, Sidney J, Sette A, Campbell S, Stroher U, Damon I, Nichol ST, Spiropoulou CF, Ahmed R (2015) Human Ebola virus infection results in substantial immune activation. Proc Natl Acad Sci USA 112(15):4719–4724. doi:10.1073/pnas.1502619112

Mehedi M, Hoenen T, Robertson S, Ricklefs S, Dolan MA, Taylor T, Falzarano D, Ebihara H, Porcella SF, Feldmann H (2013) Ebola virus RNA editing depends on the primary editing site sequence and an upstream secondary structure. PLoS Pathog 9(10):e1003677. doi:10.1371/journal.ppat.1003677

Melanson VR, Kalina WV, Williams P (2015) Ebola virus infection induces irregular dendritic cell gene expression. Viral Immunol 28(1):42–50. doi:10.1089/vim.2014.0091

Miao C, Li M, Zheng YM, Cohen FS, Liu SL (2016) Cell-cell contact promotes Ebola virus GP-mediated infection. Virology 488:202–215. doi:10.1016/j.virol.2015.11.019

Mingo RM, Simmons JA, Shoemaker CJ, Nelson EA, Schornberg KL, D'Souza RS, Casanova JE, White JM (2015) Ebola virus and severe acute respiratory syndrome coronavirus display late cell entry kinetics: evidence that transport to NPC1 + endolysosomes is a rate-defining step. J Virol 89(5):2931–2943. doi:10.1128/JVI.03398-14

Misasi J, Chandran K, Yang JY, Considine B, Filone CM, Cote M, Sullivan N, Fabozzi G, Hensley L, Cunningham J (2012) Filoviruses require endosomal cysteine proteases for entry but exhibit distinct protease preferences. J Virol 86(6):3284–3292. doi:10.1128/JVI.06346-11

Modrof J, Muhlberger E, Klenk HD, Becker S (2002) Phosphorylation of VP30 impairs ebola virus transcription. J Biol Chem 277(36):33099–33104. doi:10.1074/jbc.M203775200

Mohamadzadeh M, Coberley SS, Olinger GG, Kalina WV, Ruthel G, Fuller CL, Swenson DL, Pratt WD, Kuhns DB, Schmaljohn AL (2006) Activation of triggering receptor expressed on myeloid cells-1 on human neutrophils by marburg and ebola viruses. J Virol 80(14):7235–7244. doi:10.1128/JVI.00543-06

Mohan GS, Li W, Ye L, Compans RW, Yang C (2012) Antigenic subversion: a novel mechanism of host immune evasion by Ebola virus. PLoS Pathog 8(12):e1003065. doi:10.1371/journal.ppat.1003065

Mohan GS, Ye L, Li W, Monteiro A, Lin X, Sapkota B, Pollack BP, Compans RW, Yang C (2015) Less is more: ebola virus surface glycoprotein expression levels regulate virus production and infectivity. J Virol 89(2):1205–1217. doi:10.1128/JVI.01810-14

Moller-Tank S, Kondratowicz AS, Davey RA, Rennert PD, Maury W (2013) Role of the phosphatidylserine receptor TIM-1 in enveloped-virus entry. J Virol 87(15):8327–8341. doi:10.1128/JVI.01025-13

Moole H, Chitta S, Victor D, Kandula M, Moole V, Ghadiam H, Akepati A, Yerasi C, Uppu A, Dharmapuri S, Boddireddy R, Fischer J, Lynch T (2015) Association of clinical signs and symptoms of Ebola viral disease with case fatality: a systematic review and meta-analysis. J Community Hosp Intern Med Perspect 5(4):28406. doi:10.3402/jchimp.v5.28406

Morizono K, Chen IS (2014) Role of phosphatidylserine receptors in enveloped virus infection. J Virol 88(8):4275–4290. doi:10.1128/JVI.03287-13

Muhlberger E, Weik M, Volchkov VE, Klenk HD, Becker S (1999) Comparison of the transcription and replication strategies of marburg virus and Ebola virus by using artificial replication systems. J Virol 73(3):2333–2342

Nanbo A, Imai M, Watanabe S, Noda T, Takahashi K, Neumann G, Halfmann P, Kawaoka Y (2010) Ebolavirus is internalized into host cells via macropinocytosis in a viral glycoprotein-dependent manner. PLoS Pathog 6(9):e1001121. doi:10.1371/journal.ppat.1001121

Nanbo A, Watanabe S, Halfmann P, Kawaoka Y (2013) The spatio-temporal distribution dynamics of Ebola virus proteins and RNA in infected cells. Scientific reports 3:1206. doi:10.1038/srep01206

Nanclares C, Kapetshi J, Lionetto F, de la Rosa O, Tamfun JJ, Alia M, Kobinger G, Bernasconi A (2016) Ebola Virus Disease, Democratic Republic of the Congo, 2014. Emerg Infect Dis 22(9):1579–1586. doi:10.3201/eid2209.160354

Natesan M, Jensen SM, Keasey SL, Kamata T, Kuehne AI, Stonier SW, Lutwama JJ, Lobel L, Dye JM, Ulrich RG (2016) Human survivors of disease outbreaks caused by Ebola or marburg virus exhibit cross-reactive and long-lived antibody responses. Clin Vaccine Immunol 23(8):717–724. doi:10.1128/CVI.00107-16

Ndungo E, Herbert AS, Raaben M, Obernosterer G, Biswas R, Miller EH, Wirchnianski AS, Carette JE, Brummelkamp TR, Whelan SP, Dye JM, Chandran K (2016) A single residue in ebola virus receptor npc1 influences cellular host range in reptiles. mSphere 1 (2). doi:10.1128/mSphere.00007-16

Nelson EV, Schmidt KM, Deflube LR, Doganay S, Banadyga L, Olejnik J, Hume AJ, Ryabchikova E, Ebihara H, Kedersha N, Ha T, Muhlberger E (2016) Ebola virus does not induce stress granule formation during infection and sequesters stress granule proteins within viral inclusions. J Virol 90(16):7268–7284. doi:10.1128/JVI.00459-16

Neumann G, Watanabe S, Kawaoka Y (2009) Characterization of Ebolavirus regulatory genomic regions. Virus Res 144(1–2):1–7. doi:10.1016/j.virusres.2009.02.005

Ng M, Ndungo E, Jangra RK, Cai Y, Postnikova E, Radoshitzky SR, Dye JM, Ramirez de Arellano E, Negredo A, Palacios G, Kuhn JH, Chandran K (2014) Cell entry by a novel European filovirus requires host endosomal cysteine proteases and Niemann-Pick C1. Virology 468–470:637–646. doi:10.1016/j.virol.2014.08.019

Ng M, Ndungo E, Kaczmarek ME, Herbert AS, Binger T, Kuehne AI, Jangra RK, Hawkins JA, Gifford RJ, Biswas R, Demogines A, James RM, Yu M, Brummelkamp TR, Drosten C, Wang LF, Kuhn JH, Muller MA, Dye JM, Sawyer SL, Chandran K (2015) Filovirus receptor NPC1 contributes to species-specific patterns of ebolavirus susceptibility in bats. Elife 4. doi:10.7554/eLife.11785

Ni M, Chen C, Qian J, Xiao HX, Shi WF, Luo Y, Wang HY, Li Z, Wu J, Xu PS, Chen SH, Wong G, Bi Y, Xia ZP, Li W, Lu HJ, Ma J, Tong YG, Zeng H, Wang SQ, Gao GF, Bo XC, Liu D (2016) Intra-host dynamics of Ebola virus during 2014. Nat Microbiol 1(11):16151. doi:10.1038/nmicrobiol.2016.151

Noda T, Kolesnikova L, Becker S, Kawaoka Y (2011) The importance of the NP: VP35 ratio in Ebola virus nucleocapsid formation. J Infect Dis 204(Suppl 3):S878–S883. doi:10.1093/infdis/jir310

Noyori O, Matsuno K, Kajihara M, Nakayama E, Igarashi M, Kuroda M, Isoda N, Yoshida R, Takada A (2013) Differential potential for envelope glycoprotein-mediated steric shielding of host cell surface proteins among filoviruses. Virology 446(1–2):152–161. doi:10.1016/j.virol.2013.07.029

O'Hearn A, Wang M, Cheng H, Lear-Rooney CM, Koning K, Rumschlag-Booms E, Varhegyi E, Olinger G, Rong L (2015) Role of EXT1 and glycosaminoglycans in the early stage of filovirus entry. J Virol 89(10):5441–5449. doi:10.1128/JVI.03689-14

Ohuabunwo C, Ameh C, Oduyebo O, Ahumibe A, Mutiu B, Olayinka A, Gbadamosi W, Garcia E, Nanclares C, Famiyesin W, Mohammed A, Nguku P, Koko RI, Obasanya J, Adebayo D, Gbadegesin Y, Idigbe O, Oguntimehin O, Nyanti S, Nzuki C, Abdus-Salam I, Adeyemi J, Onyekwere N, Musa E, Brett-Major D, Shuaib F, Nasidi A (2016) Clinical profile and containment of the Ebola virus disease outbreak in two large West African cities, Nigeria, July-September 2014. Int J Infect Dis. doi:10.1016/j.ijid.2016.08.011

Okumura A, Pitha PM, Harty RN (2008) ISG15 inhibits Ebola VP40 VLP budding in an L-domain-dependent manner by blocking Nedd4 ligase activity. Proc Natl Acad Sci USA 105 (10):3974–3979. doi:10.1073/pnas.0710629105

Okumura A, Pitha PM, Yoshimura A, Harty RN (2010) Interaction between Ebola virus glycoprotein and host toll-like receptor 4 leads to induction of proinflammatory cytokines and SOCS1. J Virol 84(1):27–33. doi:10.1128/JVI.01462-09

Okumura A, Rasmussen AL, Halfmann P, Feldmann F, Yoshimura A, Feldmann H, Kawaoka Y, Harty RN, Katze MG (2015) Suppressor of cytokine signaling 3 is an inducible host factor that regulates virus egress during ebola virus infection. J Virol 89(20):10399–10406. doi:10.1128/JVI.01736-15

Olsen ME, Filone CM, Rozelle D, Mire CE, Agans KN, Hensley L, Connor JH (2016) Polyamines and hypusination are required for ebolavirus gene expression and replication. mBio 7(4). doi:10.1128/mBio.00882-16

O'Shea MK, Clay KA, Craig DG, Moore AJ, Lewis S, Espina M, Praught J, Horne S, Kao R, Johnston AM (2016) A health care worker with ebola virus disease and adverse prognostic factors treated in sierra leone. Am J Trop Med Hyg 94(4):829–832. doi:10.4269/ajtmh.15-0461

Panchal RG, Ruthel G, Kenny TA, Kallstrom GH, Lane D, Badie SS, Li L, Bavari S, Aman MJ (2003) In vivo oligomerization and raft localization of Ebola virus protein VP40 during vesicular budding. Proc Natl Acad Sci USA 100(26):15936–15941. doi:10.1073/pnas.2533915100

Panchal RG, Bradfute SB, Peyser B, Warfield KL, Ruthel G, Lane D, Kenny TA, Anderson AO, Raschke WC, Bavari S (2009) Reduced levels of protein tyrosine phosphatase CD45 protect mice from the lethal effects of Ebola virus infection. Cell Host Microbe 6(2):162–173. doi:10.1016/j.chom.2009.07.003

Perry DL, Bollinger L, White GL (2012) The Baboon (Papio spp.) as a model of human Ebola virus infection. Viruses 4(10):2400–2416. doi:10.3390/v4102400

Peyrol J, Thizon C, Gaillard JC, Marchetti C, Armengaud J, Rollin-Genetet F (2013) Multiple phosphorylable sites in the Zaire Ebolavirus nucleoprotein evidenced by high resolution tandem mass spectrometry. J Virol Methods 187(1):159–165. doi:10.1016/j.jviromet.2012.10.003

Powlesland AS, Fisch T, Taylor ME, Smith DF, Tissot B, Dell A, Pohlmann S, Drickamer K (2008) A novel mechanism for LSECtin binding to Ebola virus surface glycoprotein through truncated glycans. J Biol Chem 283(1):593–602. doi:10.1074/jbc.M706292200

Prescott J, Falzarano D, Feldmann H (2015) Natural immunity to ebola virus in the syrian hamster requires antibody responses. J Infect Dis 212(Suppl 2):S271–S276. doi:10.1093/infdis/jiv203

Prins KC, Delpeut S, Leung DW, Reynard O, Volchkova VA, Reid SP, Ramanan P, Cardenas WB, Amarasinghe GK, Volchkov VE, Basler CF (2010) Mutations abrogating VP35 interaction with double-stranded RNA render Ebola virus avirulent in guinea pigs. J Virol 84(6):3004–3015. doi:10.1128/JVI.02459-09

Qin E, Bi J, Zhao M, Wang Y, Guo T, Yan T, Li Z, Sun J, Zhang J, Chen S, Wu Y, Li J, Zhong Y (2015) Clinical features of patients with ebola virus disease in Sierra Leone. Clin Infect Dis: An Official Publ Infect Dis Soc Am 61(4):491–495. doi:10.1093/cid/civ319

Quinn K, Brindley MA, Weller ML, Kaludov N, Kondratowicz A, Hunt CL, Sinn PL, McCray PB Jr, Stein CS, Davidson BL, Flick R, Mandell R, Staplin W, Maury W, Chiorini JA (2009) Rho GTPases modulate entry of Ebola virus and vesicular stomatitis virus pseudotyped vectors. J Virol 83(19):10176–10186. doi:10.1128/JVI.00422-09

Ramaiah A, Arumugaswami V (2016) Ebolavirus evolves in human to minimize the detection by immune cells by accumulating adaptive mutations. Virusdisease 27(2):136–144. doi:10.1007/s13337-016-0305-0

Rasmussen AL, Okumura A, Ferris MT, Green R, Feldmann F, Kelly SM, Scott DP, Safronetz D, Haddock E, LaCasse R, Thomas MJ, Sova P, Carter VS, Weiss JM, Miller DR, Shaw GD, Korth MJ, Heise MT, Baric RS, de Villena FP, Feldmann H, Katze MG (2014) Host genetic diversity enables Ebola hemorrhagic fever pathogenesis and resistance. Science 346(6212):987–991. doi:10.1126/science.1259595

Reed DS, Hensley LE, Geisbert JB, Jahrling PB, Geisbert TW (2004) Depletion of peripheral blood T lymphocytes and NK cells during the course of ebola hemorrhagic Fever in cynomolgus macaques. Viral Immunol 17(3):390–400. doi:10.1089/vim.2004.17.390

Reid SP, Leung LW, Hartman AL, Martinez O, Shaw ML, Carbonnelle C, Volchkov VE, Nichol ST, Basler CF (2006) Ebola virus VP24 binds karyopherin alpha1 and blocks STAT1 nuclear accumulation. J Virol 80(11):5156–5167. doi:10.1128/JVI.02349-05

Reid SP, Valmas C, Martinez O, Sanchez FM, Basler CF (2007) Ebola virus VP24 proteins inhibit the interaction of NPI-1 subfamily karyopherin alpha proteins with activated STAT1. J Virol 81(24):13469–13477. doi:10.1128/JVI.01097-07

Reid SP, Shurtleff AC, Costantino JA, Tritsch SR, Retterer C, Spurgers KB, Bavari S (2014) HSPA5 is an essential host factor for Ebola virus infection. Antiviral Res 109:171–174. doi:10.1016/j.antiviral.2014.07.004

Reynard O, Volchkov VE (2015) Entry of Ebola virus is an asynchronous process. J Infect Dis 212(Suppl 2):S199–S203. doi:10.1093/infdis/jiv189

Reynard O, Borowiak M, Volchkova VA, Delpeut S, Mateo M, Volchkov VE (2009) Ebolavirus glycoprotein GP masks both its own epitopes and the presence of cellular surface proteins. J Virol 83(18):9596–9601. doi:10.1128/JVI.00784-09

Reynard O, Nemirov K, Page A, Mateo M, Raoul H, Weissenhorn W, Volchkov VE (2011) Conserved proline-rich region of Ebola virus matrix protein VP40 is essential for plasma membrane targeting and virus-like particle release. J Infect Dis 204(Suppl 3):S884–S891. doi:10.1093/infdis/jir359

Rhein BA, Brouillette RB, Schaack GA, Chiorini JA, Maury W (2016) Characterization of human and murine T-cell immunoglobulin mucin domain 4 (TIM-4) IgV domain residues critical for Ebola virus entry. J Virol 90(13):6097–6111. doi:10.1128/JVI.00100-16

Rialdi A, Campisi L, Zhao N, Lagda AC, Pietzsch C, Ho JS, Martinez-Gil L, Fenouil R, Chen X, Edwards M, Metreveli G, Jordan S, Peralta Z, Munoz-Fontela C, Bouvier N, Merad M, Jin J, Weirauch M, Heinz S, Benner C, van Bakel H, Basler C, Garcia-Sastre A, Bukreyev A, Marazzi I (2016) Topoisomerase inhibition suppresses inflammatory genes and protects from death by inflammation. Science 352(6289):aad7993. doi:10.1126/science.aad7993

Richard AS, Zhang A, Park SJ, Farzan M, Zong M, Choe H (2015) Virion-associated phosphatidylethanolamine promotes TIM1-mediated infection by Ebola, dengue, and West Nile viruses. Proc Natl Acad Sci USA 112(47):14682–14687. doi:10.1073/pnas.1508095112

Roddy P, Howard N, Van Kerkhove MD, Lutwama J, Wamala J, Yoti Z, Colebunders R, Palma PP, Sterk E, Jeffs B, Van Herp M, Borchert M (2012) Clinical manifestations and case management of Ebola haemorrhagic fever caused by a newly identified virus strain, Bundibugyo, Uganda, 2007–2008. PLoS ONE 7(12):e52986. doi:10.1371/journal.pone.0052986

Rubins KH, Hensley LE, Wahl-Jensen V, Daddario DiCaprio KM, Young HA, Reed DS, Jahrling PB, Brown PO, Relman DA, Geisbert TW (2007) The temporal program of peripheral blood gene expression in the response of nonhuman primates to Ebola hemorrhagic fever. Genome Biol 8(8):R174. doi:10.1186/gb-2007-8-8-r174

Ryabchikova EI, Kolesnikova LV, Luchko SV (1999) An analysis of features of pathogenesis in two animal models of Ebola virus infection. J Infect Dis 179(Suppl 1):S199–S202. doi:10.1086/514293

Saeed MF, Kolokoltsov AA, Freiberg AN, Holbrook MR, Davey RA (2008) Phosphoinositide-3 kinase-Akt pathway controls cellular entry of Ebola virus. PLoS Pathog 4(8):e1000141. doi:10.1371/journal.ppat.1000141

Saeed MF, Kolokoltsov AA, Albrecht T, Davey RA (2010) Cellular entry of ebola virus involves uptake by a macropinocytosis-like mechanism and subsequent trafficking through early and late endosomes. PLoS Pathog 6(9):e1001110. doi:10.1371/journal.ppat.1001110

Sakurai Y, Kolokoltsov AA, Chen CC, Tidwell MW, Bauta WE, Klugbauer N, Grimm C, Wahl-Schott C, Biel M, Davey RA (2015) Ebola virus. Two-pore channels control Ebola virus host cell entry and are drug targets for disease treatment. Science 347(6225):995–998. doi:10.1126/science.1258758

Salvador B, Sexton NR, Carrion R Jr, Nunneley J, Patterson JL, Steffen I, Lu K, Muench MO, Lembo D, Simmons G (2013) Filoviruses utilize glycosaminoglycans for their attachment to target cells. J Virol 87(6):3295–3304. doi:10.1128/JVI.01621-12

Sanchez A, Trappier SG, Mahy BW, Peters CJ, Nichol ST (1996) The virion glycoproteins of Ebola viruses are encoded in two reading frames and are expressed through transcriptional editing. Proc Natl Acad Sci USA 93(8):3602–3607

Sanchez A, Lukwiya M, Bausch D, Mahanty S, Sanchez AJ, Wagoner KD, Rollin PE (2004) Analysis of human peripheral blood samples from fatal and nonfatal cases of Ebola (Sudan) hemorrhagic fever: cellular responses, virus load, and nitric oxide levels. J Virol 78(19):10370–10377. doi:10.1128/JVI.78.19.10370-10377.2004

Sanchez A, Geisbert TW, Feldmann H (2007) Filoviridae: Marburg and Ebola viruses. In: Knipe DM, Howley PM (eds) Fields Virology, 5th edn. Lippincott Williams & Wilkins, Philadelphia, pp 1409–1448 (Vol. 1)

Schibler M, Vetter P, Cherpillod P, Petty TJ, Cordey S, Vieille G, Yerly S, Siegrist CA, Samii K, Dayer JA, Docquier M, Zdobnov EM, Simpson AJ, Rees PS, Sarria FB, Gasche Y, Chappuis F, Iten A, Pittet D, Pugin J, Kaiser L (2015) Clinical features and viral kinetics in a rapidly cured patient with Ebola virus disease: a case report. Lancet Infect Dis 15(9):1034–1040. doi:10.1016/S1473-3099(15)00229-7

Schieffelin JS, Shaffer JG, Goba A, Gbakie M, Gire SK, Colubri A, Sealfon RS, Kanneh L, Moigboi A, Momoh M, Fullah M, Moses LM, Brown BL, Andersen KG, Winnicki S, Schaffner SF, Park DJ, Yozwiak NL, Jiang PP, Kargbo D, Jalloh S, Fonnie M, Sinnah V,

French I, Kovoma A, Kamara FK, Tucker V, Konuwa E, Sellu J, Mustapha I, Foday M, Yillah M, Kanneh F, Saffa S, Massally JL, Boisen ML, Branco LM, Vandi MA, Grant DS, Happi C, Gevao SM, Fletcher TE, Fowler RA, Bausch DG, Sabeti PC, Khan SH, Garry RF, Program KGHLF, Viral Hemorrhagic Fever C, Team WHOCR (2014) Clinical illness and outcomes in patients with Ebola in Sierra Leone. The New England journal of medicine 371 (22):2092–2100. doi:10.1056/NEJMoa1411680

Schornberg K, Matsuyama S, Kabsch K, Delos S, Bouton A, White J (2006) Role of endosomal cathepsins in entry mediated by the Ebola virus glycoprotein. J Virol 80(8):4174–4178. doi:10.1128/JVI.80.8.4174-4178.2006

Schornberg KL, Shoemaker CJ, Dube D, Abshire MY, Delos SE, Bouton AH, White JM (2009) Alpha5beta1-integrin controls ebolavirus entry by regulating endosomal cathepsins. Proc Natl Acad Sci USA 106(19):8003–8008. doi:10.1073/pnas.0807578106

Schudt G, Dolnik O, Kolesnikova L, Biedenkopf N, Herwig A, Becker S (2015) Transport of ebolavirus nucleocapsids is dependent on actin polymerization: live-cell imaging analysis of ebolavirus-infected cells. J Infect Dis 212(Suppl 2):S160–S166. doi:10.1093/infdis/jiv083

Schumann M, Gantke T, Muhlberger E (2009) Ebola virus VP35 antagonizes PKR activity through its C-terminal interferon inhibitory domain. J Virol 83(17):8993–8997. doi:10.1128/JVI.00523-09

Shabman RS, Gulcicek EE, Stone KL, Basler CF (2011a) The Ebola virus VP24 protein prevents hnRNP C1/C2 binding to karyopherin alpha1 and partially alters its nuclear import. J Infect Dis 204(Suppl 3):S904–S910. doi:10.1093/infdis/jir323

Shabman RS, Leung DW, Johnson J, Glennon N, Gulcicek EE, Stone KL, Leung L, Hensley L, Amarasinghe GK, Basler CF (2011b) DRBP76 associates with Ebola virus VP35 and suppresses viral polymerase function. J Infect Dis 204(Suppl 3):S911–S918. doi:10.1093/infdis/jir343

Shabman RS, Hoenen T, Groseth A, Jabado O, Binning JM, Amarasinghe GK, Feldmann H, Basler CF (2013) An upstream open reading frame modulates ebola virus polymerase translation and virus replication. PLoS Pathog 9(1):e1003147. doi:10.1371/journal.ppat.1003147

Shabman RS, Jabado OJ, Mire CE, Stockwell TB, Edwards M, Mahajan M, Geisbert TW, Basler CF (2014) Deep sequencing identifies noncanonical editing of Ebola and Marburg virus RNAs in infected cells. mBio 5(6):e02011. doi:10.1128/mBio.02011-14

Shimojima M, Takada A, Ebihara H, Neumann G, Fujioka K, Irimura T, Jones S, Feldmann H, Kawaoka Y (2006) Tyro3 family-mediated cell entry of Ebola and Marburg viruses. J Virol 80 (20):10109–10116. doi:10.1128/JVI.01157-06

Shimojima M, Ikeda Y, Kawaoka Y (2007) The mechanism of Axl-mediated Ebola virus infection. J Infect Dis 196(Suppl 2):S259–S263. doi:10.1086/520594

Silvestri LS, Ruthel G, Kallstrom G, Warfield KL, Swenson DL, Nelle T, Iversen PL, Bavari S, Aman MJ (2007) Involvement of vacuolar protein sorting pathway in Ebola virus release independent of TSG101 interaction. J Infect Dis 196(Suppl 2):S264–S270. doi:10.1086/520610

Simmons G, Reeves JD, Grogan CC, Vandenberghe LH, Baribaud F, Whitbeck JC, Burke E, Buchmeier MJ, Soilleux EJ, Riley JL, Doms RW, Bates P, Pohlmann S (2003a) DC-SIGN and DC-SIGNR bind ebola glycoproteins and enhance infection of macrophages and endothelial cells. Virology 305(1):115–123

Simmons G, Rennekamp AJ, Chai N, Vandenberghe LH, Riley JL, Bates P (2003b) Folate receptor alpha and caveolae are not required for Ebola virus glycoprotein-mediated viral infection. J Virol 77(24):13433–13438

Smith DR, McCarthy S, Chrovian A, Olinger G, Stossel A, Geisbert TW, Hensley LE, Connor JH (2010) Inhibition of heat-shock protein 90 reduces Ebola virus replication. Antiviral Res 87 (2):187–194. doi:10.1016/j.antiviral.2010.04.015

Sobarzo A, Groseth A, Dolnik O, Becker S, Lutwama JJ, Perelman E, Yavelsky V, Muhammad M, Kuehne AI, Marks RS, Dye JM, Lobel L (2013) Profile and persistence of the virus-specific neutralizing humoral immune response in human survivors of Sudan ebolavirus (Gulu). J Infect Dis 208(2):299–309. doi:10.1093/infdis/jit162

Soni SP, Stahelin RV (2014) The Ebola virus matrix protein VP40 selectively induces vesiculation from phosphatidylserine-enriched membranes. J Biol Chem 289(48):33590–33597. doi:10.1074/jbc.M114.586396

Soni SP, Adu-Gyamfi E, Yong SS, Jee CS, Stahelin RV (2013) The Ebola virus matrix protein deeply penetrates the plasma membrane: an important step in viral egress. Biophys J 104 (9):1940–1949. doi:10.1016/j.bpj.2013.03.021

Spence JS, Krause TB, Mittler E, Jangra RK, Chandran K (2016) Direct visualization of Ebola virus fusion triggering in the endocytic pathway. mBio 7(1):e01857–e01815. doi:10.1128/mBio.01857-15

Spengler JR, Lavender KJ, Martellaro C, Carmody A, Kurth A, Keck JG, Saturday G, Scott DP, Nichol ST, Hasenkrug KJ, Spiropoulou CF, Feldmann H, Prescott J (2016) Ebola virus replication and disease without immunopathology in mice expressing transgenes to support human Myeloid and Lymphoid cell engraftment. J Infect Dis 214(suppl 3):S308–S318. doi:10.1093/infdis/jiw248

Spurgers KB, Alefantis T, Peyser BD, Ruthel GT, Bergeron AA, Costantino JA, Enterlein S, Kota KP, Boltz RC, Aman MJ, Delvecchio VG, Bavari S (2010) Identification of essential filovirion-associated host factors by serial proteomic analysis and RNAi screen. Mol Cell Proteomics 9(12):2690–2703. doi:10.1074/mcp.M110.003418

Strong JE, Wong G, Jones SE, Grolla A, Theriault S, Kobinger GP, Feldmann H (2008) Stimulation of Ebola virus production from persistent infection through activation of the Ras/MAPK pathway. Proc Natl Acad Sci USA 105(46):17982–17987. doi:10.1073/pnas.0809698105

Sztuba-Solinska J, Diaz L, Kumar MR, Kolb G, Wiley MR, Jozwick L, Kuhn JH, Palacios G, Radoshitzky SR, JLG SF, Johnson RF (2016) A small stem-loop structure of the Ebola virus trailer is essential for replication and interacts with heat-shock protein A8. Nucleic Acids Res. doi:10.1093/nar/gkw825

Takada A, Watanabe S, Ito H, Okazaki K, Kida H, Kawaoka Y (2000) Downregulation of beta1 integrins by Ebola virus glycoprotein: implication for virus entry. Virology 278(1):20–26. doi:10.1006/viro.2000.0601

Takada A, Fujioka K, Tsuiji M, Morikawa A, Higashi N, Ebihara H, Kobasa D, Feldmann H, Irimura T, Kawaoka Y (2004) Human macrophage C-type lectin specific for galactose and N-acetylgalactosamine promotes filovirus entry. J Virol 78(6):2943–2947

Takahashi K, Halfmann P, Oyama M, Kozuka-Hata H, Noda T, Kawaoka Y (2013) DNA topoisomerase 1 facilitates the transcription and replication of the Ebola virus genome. J Virol 87(16):8862–8869. doi:10.1128/JVI.03544-12

Timmins J, Schoehn G, Ricard-Blum S, Scianimanico S, Vernet T, Ruigrok RW, Weissenhorn W (2003) Ebola virus matrix protein VP40 interaction with human cellular factors Tsg101 and Nedd4. J Mol Biol 326(2):493–502

Usami K, Matsuno K, Igarashi M, Denda-Nagai K, Takada A, Irimura T (2011) Involvement of viral envelope GP2 in Ebola virus entry into cells expressing the macrophage galactose-type C-type lectin. Biochem Biophys Res Commun 407(1):74–78. doi:10.1016/j.bbrc.2011.02.110

Uyeki TM, Mehta AK, Davey RT, Jr., Liddell AM, Wolf T, Vetter P, Schmiedel S, Grunewald T, Jacobs M, Arribas JR, Evans L, Hewlett AL, Brantsaeter AB, Ippolito G, Rapp C, Hoepelman AI, Gutman J, Working Group of the USECNoCMoEVDPitUS, Europe (2016) Clinical Management of Ebola Virus Disease in the United States and Europe. The New England journal of medicine 374 (7):636-646. doi:10.1056/NEJMoa1504874

Valmas C, Grosch MN, Schumann M, Olejnik J, Martinez O, Best SM, Krahling V, Basler CF, Muhlberger E (2010) Marburg virus evades interferon responses by a mechanism distinct from ebola virus. PLoS Pathog 6(1):e1000721. doi:10.1371/journal.ppat.1000721

Vande Burgt NH, Kaletsky RL, Bates P (2015) Requirements within the Ebola Viral Glycoprotein for Tetherin Antagonism. Viruses 7(10):5587–5602. doi:10.3390/v7102888

Volchkov VE, Becker S, Volchkova VA, Ternovoj VA, Kotov AN, Netesov SV, Klenk HD (1995) GP mRNA of Ebola virus is edited by the Ebola virus polymerase and by T7 and vaccinia virus polymerases. Virology 214(2):421–430
Volchkov VE, Feldmann H, Volchkova VA, Klenk HD (1998) Processing of the Ebola virus glycoprotein by the proprotein convertase furin. Proc Natl Acad Sci USA 95(10):5762–5767
Volchkova VA, Feldmann H, Klenk HD, Volchkov VE (1998) The nonstructural small glycoprotein sGP of Ebola virus is secreted as an antiparallel-orientated homodimer. Virology 250(2):408–414. doi:10.1006/viro.1998.9389
Volchkova VA, Klenk HD, Volchkov VE (1999) Delta-peptide is the carboxy-terminal cleavage fragment of the nonstructural small glycoprotein sGP of Ebola virus. Virology 265(1):164–171. doi:10.1006/viro.1999.0034
Volchkova VA, Dolnik O, Martinez MJ, Reynard O, Volchkov VE (2011) Genomic RNA editing and its impact on Ebola virus adaptation during serial passages in cell culture and infection of guinea pigs. J Infect Dis 204(Suppl 3):S941–S946. doi:10.1093/infdis/jir321
Volchkova VA, Dolnik O, Martinez MJ, Reynard O, Volchkov VE (2015a) RNA editing of the GP gene of Ebola virus is an important pathogenicity factor. J Infect Dis 212(Suppl 2):S226–S233. doi:10.1093/infdis/jiv309
Volchkova VA, Vorac J, Repiquet-Paire L, Lawrence P, Volchkov VE (2015b) Ebola virus GP gene polyadenylation versus RNA editing. J Infect Dis 212(Suppl 2):S191–S198. doi:10.1093/infdis/jiv150
Wahl-Jensen V, Kurz SK, Hazelton PR, Schnittler HJ, Stroher U, Burton DR, Feldmann H (2005a) Role of Ebola virus secreted glycoproteins and virus-like particles in activation of human macrophages. J Virol 79(4):2413–2419. doi:10.1128/JVI.79.4.2413-2419.2005
Wahl-Jensen VM, Afanasieva TA, Seebach J, Stroher U, Feldmann H, Schnittler HJ (2005b) Effects of Ebola virus glycoproteins on endothelial cell activation and barrier function. J Virol 79(16):10442–10450. doi:10.1128/JVI.79.16.10442-10450.2005
Wahl-Jensen V, Kurz S, Feldmann F, Buehler LK, Kindrachuk J, DeFilippis V, da Silva Correia J, Fruh K, Kuhn JH, Burton DR, Feldmann H (2011) Ebola virion attachment and entry into human macrophages profoundly effects early cellular gene expression. PLoS neglected tropical diseases 5(10):e1359. doi:10.1371/journal.pntd.0001359
Watt A, Moukambi F, Banadyga L, Groseth A, Callison J, Herwig A, Ebihara H, Feldmann H, Hoenen T (2014) A novel life cycle modeling system for Ebola virus shows a genome length-dependent role of VP24 in virus infectivity. J Virol 88(18):10511–10524. doi:10.1128/JVI.01272-14
Wauquier N, Becquart P, Gasquet C, Leroy EM (2009) Immunoglobulin G in Ebola outbreak survivors. Gabon. Emerg Infect Dis 15(7):1136–1137. doi:10.3201/eid1507.090402
Wauquier N, Becquart P, Padilla C, Baize S, Leroy EM (2010) Human fatal zaire ebola virus infection is associated with an aberrant innate immunity and with massive lymphocyte apoptosis. PLoS neglected tropical diseases 4 (10). doi:10.1371/journal.pntd.0000837
WHO (1978a) Ebola haemorrhagic fever in Sudan, 1976. Report of a WHO/International Study Team. Bull World Health Organ 56(2):247–270
WHO (1978b) Ebola haemorrhagic fever in Zaire, 1976. Bull World Health Organ 56(2):271–293
WHO (2014) Ebola virus disease in West Africa–the first 9 months of the epidemic and forward projections. The New England journal of medicine 371(16):1481–1495. doi:10.1056/NEJMoa1411100
WHO (2016) Situation Report-Ebola Virus Disease, 10 June 2016. World Health Organization, World Health Organization Situation Report
Wolf T, Kann G, Becker S, Stephan C, Brodt HR, de Leuw P, Grunewald T, Vogl T, Kempf VA, Keppler OT, Zacharowski K (2015) Severe Ebola virus disease with vascular leakage and multiorgan failure: treatment of a patient in intensive care. Lancet 385(9976):1428–1435. doi:10.1016/S0140-6736(14)62384-9
Wong AC, Sandesara RG, Mulherkar N, Whelan SP, Chandran K (2010) A forward genetic strategy reveals destabilizing mutations in the Ebolavirus glycoprotein that alter its protease dependence during cell entry. J Virol 84(1):163–175. doi:10.1128/JVI.01832-09

Wong KK, Perdue CL, Malia J, Kenney JL, Peng S, Gwathney JK, Raczniak GA, Monrovia Medical U (2015) Supportive Care of the First 2 Ebola Virus Disease Patients at the Monrovia Medical Unit. Clin Infect Dis: an Official Publ Infect Dis Soc Am 61(7):e47–e51. doi:10.1093/cid/civ420

Wrensch F, Karsten CB, Gnirss K, Hoffmann M, Lu K, Takada A, Winkler M, Simmons G, Pohlmann S (2015) Interferon-induced transmembrane protein-mediated inhibition of host cell entry of ebolaviruses. J Infect Dis 212(Suppl 2):S210–S218. doi:10.1093/infdis/jiv255

Xu W, Edwards MR, Borek DM, Feagins AR, Mittal A, Alinger JB, Berry KN, Yen B, Hamilton J, Brett TJ, Pappu RV, Leung DW, Basler CF, Amarasinghe GK (2014) Ebola virus VP24 targets a unique NLS binding site on karyopherin alpha 5 to selectively compete with nuclear import of phosphorylated STAT1. Cell Host Microbe 16(2):187–200. doi:10.1016/j.chom.2014.07.008

Yamayoshi S, Noda T, Ebihara H, Goto H, Morikawa Y, Lukashevich IS, Neumann G, Feldmann H, Kawaoka Y (2008) Ebola virus matrix protein VP40 uses the COPII transport system for its intracellular transport. Cell Host Microbe 3(3):168–177. doi:10.1016/j.chom.2008.02.001

Yasuda J, Nakao M, Kawaoka Y, Shida H (2003) Nedd4 regulates egress of Ebola virus-like particles from host cells. J Virol 77(18):9987–9992

Yen BC, Basler CF (2016) Effects of filovirus interferon antagonists on responses of human monocyte-derived dendritic cells to RNA virus infection. J Virol 90(10):5108–5118. doi:10.1128/JVI.00191-16

Yen JY, Garamszegi S, Geisbert JB, Rubins KH, Geisbert TW, Honko A, Xia Y, Connor JH, Hensley LE (2011) Therapeutics of Ebola hemorrhagic fever: whole-genome transcriptional analysis of successful disease mitigation. J Infect Dis 204(Suppl 3):S1043–S1052. doi:10.1093/infdis/jir345

Yen B, Mulder LC, Martinez O, Basler CF (2014) Molecular basis for ebolavirus VP35 suppression of human dendritic cell maturation. J Virol 88(21):12500–12510. doi:10.1128/JVI.02163-14

Yuan S, Cao L, Ling H, Dang M, Sun Y, Zhang X, Chen Y, Zhang L, Su D, Wang X, Rao Z (2015) TIM-1 acts a dual-attachment receptor for Ebolavirus by interacting directly with viral GP and the PS on the viral envelope. Protein Cell. doi:10.1007/s13238-015-0220-y

Zhang AP, Bornholdt ZA, Liu T, Abelson DM, Lee DE, Li S, Woods VL Jr, Saphire EO (2012) The ebola virus interferon antagonist VP24 directly binds STAT1 and has a novel, pyramidal fold. PLoS Pathog 8(2):e1002550. doi:10.1371/journal.ppat.1002550

Zhao D, Han X, Zheng X, Wang H, Yang Z, Liu D, Han K, Liu J, Wang X, Yang W, Dong Q, Yang S, Xia X, Tang L, He F (2016) The myeloid LSECtin Is a DAP12-coupled receptor that is crucial for inflammatory response induced by Ebola virus glycoprotein. PLoS Pathog 12(3): e1005487. doi:10.1371/journal.ppat.1005487

Zumbrun EE, Bloomfield HA, Dye JM, Hunter TC, Dabisch PA, Garza NL, Bramel NR, Baker RJ, Williams RD, Nichols DK, Nalca A (2012) A characterization of aerosolized Sudan virus infection in African green monkeys, cynomolgus macaques, and rhesus macaques. Viruses 4(10):2115–2136. doi:10.3390/v4102115

Role of Host Genes in Influenza Virus Replication

Megan L. Shaw and Silke Stertz

Abstract At every step of their replication cycle influenza viruses depend heavily on their host cells. The multifaceted interactions that occur between the virus and its host cell determine the outcome of the infection, including efficiency of progeny virus production, tropism, and pathogenicity. In order to understand viral disease and develop therapies for influenza it is therefore pertinent to study the intricate interplay between influenza viruses and their required host factors. Here, we review the current knowledge on host cell factors required by influenza virus at the different stages of the viral replication cycle. We also discuss the roles of host factors in zoonotic transmission of influenza viruses and their potential for developing novel antivirals.

Contents

M.L. Shaw
Department of Microbiology, Icahn School of Medicine at Mount Sinai, New York, NY 10029, USA
e-mail: megan.shaw@mssm.edu

S. Stertz (✉)
Institute of Medical Virology, University of Zurich, 8057 Zurich, Switzerland
e-mail: stertz.silke@virology.uzh.ch

Current Topics in Microbiology and Immunology (2018) 419:151–190
DOI 10.1007/82_2017_30

Published Online: 23 June 2017

1 Introduction

Influenza A and B viruses are the causative agents of influenza in humans leading to an estimated 250,000–500,000 deaths per year worldwide (WHO). Influenza A viruses are classified according to subtype which is based on their surface glycoproteins, hemagglutinin (HA), and neuraminidase (NA) (Shaw and Palese 2013). All subtypes are maintained in the avian reservoir but only three, H1N1, H2N2, and H3N2 have ever circulated in humans (Wright et al. 2013). As members of the Orthomyxovirus family, influenza viruses have a lipid envelope and are characterized by a segmented, negative sense, single-stranded RNA genome (Shaw and Palese 2013). Specifically, the genomes of influenza A and B viruses consist of eight segments, each of which encodes one or more viral proteins. For influenza A virus, which is the main focus of this chapter, 11 major viral proteins have been described and their role in the main stages of the viral replication cycle will be reviewed in each section. Briefly, influenza viruses enter the cell via receptor-mediated endocytosis and are trafficked to the late endosome, where they fuse and release their genome into the cell. The genome segments are then imported into the nucleus, where the viral polymerase catalyzes both transcription and replication reactions, giving rise to new viral proteins and newly synthesized genome copies. The new genome segments are exported from the nucleus and assemble at the cell membrane together with the essential viral proteins that make up the virion. In the final steps, the budding virus particle is pinched off from the plasma membrane and is released into the extracellular environment.

Host factors are undoubtedly involved at every step of this life cycle and include cellular proteins and RNAs which can be broadly divided into those that support virus replication and those that play an antiviral role. In the past decade or so great strides have been made to uncover both categories of factors for influenza virus, but for the purposes of this review we have chosen to focus specifically on those that play a supportive role. Of note, very little is known about the host factors required for influenza B virus, as the vast majority of studies have been performed with influenza A viruses, which are therefore the topic of this review. As a consequence of its small coding capacity influenza virus likely relies on a large repertoire of host functions and for a virus that can cross the species barrier this presents an interesting challenge. These critical virus–host interactions can therefore serve as determinants of species specificity. In addition, host factors that are considered to have druggable properties present potentially new targets for the development of antiviral drugs (Watanabe and Kawaoka 2015; Shaw 2011).

2 Approaches to Identify Host Factors Required by Viruses

Over the past decade, knowledge of cellular proteins that are required for viral infections has increased tremendously due to the development of novel technologies, such as genome-wide RNA interference (RNAi) screening or quantitative proteomic applications. These novel methodologies have enabled us to reveal exciting new insights into the biology of virus infections but also to start assembling a global picture of the pathogen-host interplay.

2.1 Screening Approaches for Host Dependency Factors

In the case of influenza viruses, the genome-wide RNAi screens for host dependency factors represent one prominent example of such a technology-driven development. RNAi describes the ability of RNA molecules to downregulate the expression of a gene, either by degradation or translational inhibition of the targeted RNA (Tijsterman and Plasterk 2004). Specificity and recognition is mediated by sequence complementarity and the RNA is delivered either as small interfering RNA (siRNA) or as small hairpin RNA (shRNA). While shRNA constructs are usually delivered via lentiviral vectors, transcribed, and processed in the host cell, siRNAs are short double-stranded RNA molecules of 20–25 base pairs that are directly transfected into host cells. Large libraries consisting of thousands of different siRNAs or shRNAs designed to target most of the known human genes are commercially available and can be used to assess the impact of individual genes on virus infection. This requires a robust, high-throughput compatible assay to measure virus infection, such as automated fluorescence microscopy readouts for viral protein expression or reporter activity measurements if a recombinant reporter-encoding virus has been used (Brass et al. 2009; Konig et al. 2010).

Several genome-wide or large-scale screens using either siRNAs or shRNAs have been performed over the past years and have generated a vast amount of data on host dependency factors for influenza A virus in mammalian cells (Konig et al. 2010; Brass et al. 2009; Shapira et al. 2009; Karlas et al. 2010; Ward et al. 2012; Tran et al. 2013; Watanabe et al. 2014b; Su et al. 2013). These screens differ substantially in their experimental set-up. For example, different cell lines, virus strains, RNAi libraries, and virus infection assays were chosen but also the bioinformatic analysis of the screening data was performed in many different ways. On the one hand, it is therefore not surprising that each screen yielded a unique set of hits found to be required for influenza A virus in the respective experimental system. On the other hand, it can be assumed that the virus uses a core set of host factors for its replication cycle and one would therefore expect a large number of overlapping hits between screens. However, only a few gene hits were common to multiple screens (Stertz and Shaw 2011). So far, the reason for this is only partially understood but a recent meta-analysis of the screens revealed that uniform bioinformatic analysis of the different sets of primary screening data can uncover hidden overlap and thereby reveal at least part of the core set of host factors required by influenza A virus (Tripathi et al. 2015). In addition, analysis of the pathways required by the virus rather than individual genes also discovered substantial overlap between the screening datasets (de Chassey et al. 2012). In summary, the RNAi screens have already revealed new insights into the interplay between influenza virus and its host cell but the full potential has not been exploited yet as follow-up studies on many of the identified factors are lacking and the differences in screening results are not fully understood. At the same time, the next set of new technologies to screen for host dependency factors is already being implemented and used: haploid screens and CRISPR/Cas9 screening approaches (Perreira et al. 2016). In contrast to RNAi the new methods do not rely on knockdown of gene expression but knockout of individual genes.

Haploid human cell lines, like KBM7 or HAP-1, form the basis of the haploid screens (Carette et al. 2009). These cell lines can be targeted with a retroviral gene trap vector to knock out individual genes. When screening for host dependency factors a lytic virus is added so that most of the cells will be killed. If, however, an essential host factor has been knocked out, the virus cannot infect the cell and the cell will survive and grow out. Using deep sequencing on the surviving pool of cells one can identify such factors and therefore current progress in deep-sequencing technology has greatly helped this methodology (Carette et al. 2011). Already the first description of this screening platform included a screen for host factors of influenza A virus and two components of the glycosylation machinery, CMAS and SLC35A, were identified (Carette et al. 2009). It was hypothesized that knockout of these genes would yield reduced levels of sialic acid on the host cell leading to a block in infection.

The latest developments in screening are CRISPR/Cas9 screens that are based on the endonuclease activity of Cas9 which is targeted to a certain gene by a guide RNA (Wang et al. 2014; Shalem et al. 2014; Zhou et al. 2014). Genome-wide libraries of such guide RNAs have been developed and it is therefore possible to generate huge libraries of cells with different knockouts. Selection and identification of host factors is similar to the haploid screens, as a lytic virus can be used to select for infection-resistant cell clones that can be identified via deep-sequencing. This approach holds great promise and will certainly be used soon to identify host factors for influenza virus. For both types of knockout screens, it will be interesting to see how well hit lists correspond within one technology but also how much overlap with the results from the RNAi screens will be obtained. Having data on host dependency factors available from three different technologies will hopefully enable us to define core factors and help us identify promising drug targets for the next generation of antivirals for influenza.

2.2 *Methods to Identify Interaction Partners of Viral Proteins*

Many different techniques exist to identify interaction partners of proteins but most often yeast-two-hybrid (Y2H) or pulldown approaches followed by mass spectrometry are utilized. The basis for Y2H is the binding of a transcription factor, usually GAL4 from yeast, to a DNA element called upstream activating sequence that induces the transcription of a reporter gene. The transcription factor is split into two parts, one part that mediates DNA binding and one that is responsible for transactivation. These two parts can be fused to two different proteins. If constructs for these fusions proteins are expressed in the same cell, and binding occurs between them, reporter gene activity can be measured. For screening approaches, the protein of interest can carry one part of the transcription factor and this can be combined with a library of constructs carrying the other part. Methods based on Y2H have been utilized to find binding partners of influenza virus proteins and have revealed important interactions between the virus and its host cell (Shapira et al. 2009; Zhu et al. 2016; Gao et al. 2015; de Chassey et al. 2013a; Hsu et al. 2013; Tafforeau et al. 2011; Momose et al. 2001; Ma et al. 2012).

For pulldown approaches the protein of interest is expressed in cells, either at the endogenous level or, more commonly, via overexpression, so that the protein of interest can be fused to a short tag sequence. Biochemical isolation of the protein complexes of interest can either be done by incubating the cell lysate with antibodies to the protein of interest (or the tag) and subsequent precipitation by protein A beads (immuno-precipitation). Alternatively, combinations of high affinity binding partners, such as streptavidin (bound to beads and used to precipitate) and biotin (coupled to the protein of interest) are often used. The so-called tandem-affinity purification method proved to be particularly successful (Puig et al. 2001; Rigaut et al. 1999): In this case the protein of interest is coupled to a tag containing a calmodulin-binding site, a tobacco etch virus (TEV) protease cleavage site and a protein A tag. The protein of interest, together with any associated proteins, can be precipitated and purified in a first round by incubation with IgG beads, then cleaved off the beads via TEV protease and subsequently subjected to a second round of purification using calmodulin-coupled beads.

Such methods, and variations of them, have been extensively used to uncover cellular proteins that interact with proteins of influenza virus and might therefore play a role in the viral life cycle. These studies have identified crucial interactions and thereby contributed substantially to our current understanding of the virus-host interplay. Major efforts have been applied to finding interaction partners of NS1 as this nonstructural viral protein has been shown to antagonize the host defense system by multiple mechanisms (Hale 2014). Its currently known interaction partners include components of the innate immune response, such as RIG-I or PKR, members of the splicing and mRNA processing machinery, such as CPSF30, as well as phosphoinositide-3-kinase PI(3)K and many others (Hale et al. 2006; Nemeroff et al. 1998; Mibayashi et al. 2007; Li et al. 2006; Tawaratsumida et al. 2014; Kuo et al. 2016). In addition, a computational approach to predict high-confidence interactions based on the available structural and interactome data has also been implemented to identify interaction partners of NS1. Subsequent experimental verification for predicted interaction partners validated this new method (de Chassey et al. 2013b).

Recent work has aimed to identify interaction partners of all viral proteins in a parallel approach using the same experimental system but separate pulldowns for each viral protein (Tripathi et al. 2015; Watanabe et al. 2014b). This has resulted in global viral interaction networks that can be combined with data from RNAi screens and thereby help to elucidate the core factors required for influenza virus propagation (Fig. 1). Heaton and colleagues went one step further and developed an elegant method to uncover the interaction partners of different viral proteins in the context of infection (Heaton et al. 2016). By performing insertional mutagenesis for each viral segment combined with virus rescue by reverse genetics they first determined sites that tolerate the insertion of a Flag-tag that can be utilized for immuno-precipitation. This was successful for all structural proteins with the exception of NP and M1, and for the nonstructural protein NS1. Different viruses, each encoding one tagged viral protein, were then used to infect cells and purify the complexes of the tagged viral protein with its interaction partners. Subsequent mass spectrometry revealed at least a snapshot of the interactome of influenza A virus during infection and can therefore help to further elucidate virus-host interactions.

Similar to the small overlap between the hit lists of the RNAi screens, the interaction data also vary substantially between studies. While some of the discrepancies can be explained by the differences in experimental systems, particularly differences in viral strains, it is also expected that each dataset contains false-positive and false-negative hits. In future work, it will be important to integrate various sets of data on interaction partners, but also include available data on host dependency factors.

2.3 Additional Methods to Identify Host Factors Related to Virus Infection

In addition to the described screening methods for host dependency factors and the studies on interaction partners of viral proteins, many more techniques have contributed to our current understanding of the intricate relationship between influenza

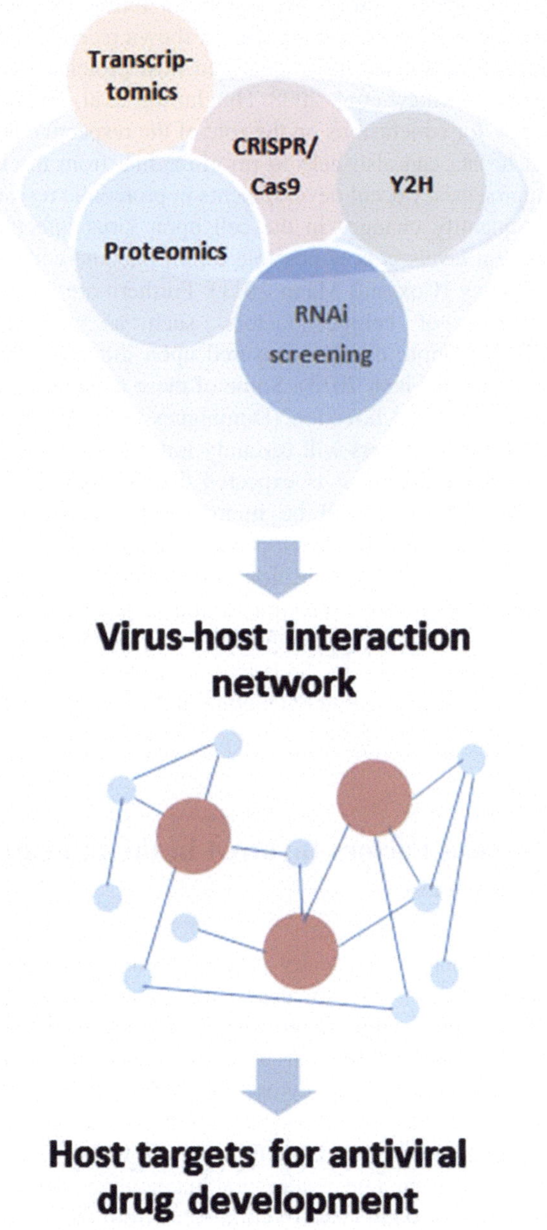

Fig. 1 A virus-host interaction network for influenza virus. Integration of available datasets from RNAi and CRISPR/Cas9 screens, transcriptome, proteomic and yeast-two-hybrid (Y2H) studies results in the assembly of a virus-host interaction network. The connections in this network may reveal candidate host proteins or pathways to target for new influenza antiviral strategies

A virus and its host cell. Transcriptomic approaches, such as DNA micro-arrays or deep sequencing applications, have revealed changes on the transcriptional level induced by infection. While transcriptomic data on their own cannot inform us whether certain genes have a proviral, an antiviral, or no role for the virus, they can help to identify promising factors for follow-up studies when integrated with proteomic or RNAi screening data as shown recently (Chasman et al. 2016). Similarly, proteomic studies have identified host proteins incorporated into influenza virus particles (Shaw et al. 2008; Hutchinson et al. 2014). While incorporation does not allow for conclusions on the role of the respective host factor in the viral life cycle such data can also help to prioritize hits from the interaction or RNAi screening approaches. Recent developments in proteomic research further expand our toolbox to quantify changes in the cell upon virus infection: Quantification of absolute protein levels is now possible on a proteome-wide level with high sensitivity and accuracy (Cox and Mann 2011). Furthermore, changes in posttranslational modifications of cellular factors, such as phosphorylation, ubiquitination, or SUMOylation, can be measured upon different stimuli, including virus infection (Olsen and Mann 2013). Some of these technologies have already been applied to influenza virus infection (Domingues et al. 2015; Soderholm et al. 2016; Dapat et al. 2014); others will certainly follow.

For the future, it is expected that not only the cell culture-based toolbox as discussed above will be improved and expanded but also in vivo screening approaches will be developed and applied. This will likely include RNAi-based methods for the identification of important host factors (Varble et al. 2013; Benitez et al. 2015) but also screening of mouse strains for genetic traits that impact on viral replication capacity and disease outcome (Ferris et al. 2013; Boon et al. 2009). Together with the methods that have already been applied these new approaches will help reveal the global picture of the interplay between influenza A virus and its host cell.

3 Host Factors Involved in the Influenza Virus Entry Process

The influenza virus entry process encompasses several steps that the virus must accomplish in order to successfully enter its host cell [reviewed in (Edinger et al. 2014)]. First, influenza virus binds to its receptor sialic acid via the envelope protein hemagglutinin (HA). In a second step, the virus is internalized mostly via clathrin-mediated endocytosis but alternative, less well-described uptake pathways also exist. This is followed by the third step, endosomal trafficking from early to late endosomes, where the fourth event takes place, fusion of viral and endosomal membranes. In a next step, viral uncoating, which describes the release of the viral ribonucleoprotein (vRNP) complexes from the layer of matrix protein M1, occurs. Lastly, the released vRNPs then get imported into the host cell's nucleus, which marks the completion of successful entry. For all of these steps influenza virus depends critically on its host cell and several of the virus–host interactions during this process have been revealed in recent years (Table 1).

Table 1 Host factors that support influenza virus replication

Stage of virus life cycle	Host factor	Interaction with virus	Reference[a]
Entry/binding	Sialic acid	HA	Shaw and Palese (2013)
Entry/internalization	Clathrin		Chen and Zhuang (2008)
	Epsin-1		Chen and Zhuang (2008)
	Dynamin		Roy et al. (2000)
	EGFR		Eierhoff et al. (2010)
	PLCγ1		Zhu et al. (2014)
	MAPK1		Tripathi et al. (2015)
	MAPK8		Tripathi et al. (2015)
	PAK1		de Vries et al. (2011)
	PI(3)K		Elbahesh et al. (2014)
	Src-family kinases		de Vries et al. (2011)
	FAK/PTK2		Elbahesh et al. (2014)
Entry/endosome trafficking	Actin		Lakadamyali et al. (2003)
	Dynein		Lakadamyali et al. (2003)
	RAB5		Sieczkarski and Whittaker (2003)
	RAB7		Sieczkarski and Whittaker (2003)
	Prolidase		Pohl et al. (2014)
	HDAC8		Yamauchi et al. (2011)
	Cullin-3/SPOPL		Gschweitl et al. (2016)
	PKC		Sieczkarski et al. (2003)
Entry/fusion	v-ATPase complex		Guinea and Carrasco (1995)
	RNASEK		Perreira et al. (2015)
	CD81		He et al. (2013)
	Cathepsin W		Edinger et al. (2015)
	TMPRSS2	HA	Bottcher et al. (2006)
	HAT	HA	Bottcher et al. (2006)
Entry/uncoating	ITCH	M1	Su et al. (2015)
	HDAC6		Banerjee et al. (2014)
Entry/RNP nuclear import	Karyopherin/importin	NP	O'Neill et al. (1995)
	HSP40/DNAJB1		Batra et al. (2016)
Transcription	RNA POL II	polymerase	Engelhardt et al. (2005)
	CHD1	polymerase	Marcos-Villar et al. (2016)
	NXP2/MORC3	polymerase	Ver et al. (2015)
	RRP1B	polymerase	Su et al. (2015)
	SFPQ	polymerase	Landeras-Bueno et al. (2011)
	DDX19	viral transcripts	Diot et al. (2016)

(continued)

Table 1 (continued)

Stage of virus life cycle	Host factor	Interaction with virus	Reference[a]
RNA synthesis	Cyclin T1/CDK9	polymerase	Zhang et al. (2010)
	hCLE/C14ORF166	PA	Rodriguez et al. (2011)
	DNAJA1	PB2, PA	Cao et al. (2014)
Genome replication	ANP32A/PP32	polymerase	Sugiyama et al. (2015)
	ANP32B/APRIL	polymerase	Sugiyama et al. (2015)
Splicing of NS segment	RED	PB1, PB2	Fournier et al. (2014)
	SMU1	polymerase (via RED)	Fournier et al. (2014)
Splicing of M segment	SF2/ASF	M1 transcript	Shih and Krug (1996)
	CLK1		Karlas et al. (2010)
	NS1-BP	NS1	Tsai et al. (2013)
	hnRNPK	M1 transcript	Tsai et al. (2013)
RNP nuclear export	CRM1/XPO1	NEP	Neumann et al. (2000)
	AIMP2	NEP	Gao et al. (2015)
	HSC70	M1	Watanabe et al. (2014a)
	NXT1	NP	Chutiwitoonchai and Aida (2016)
	CHD3	NEP	Hu et al. (2015)
	PRC2	M1	Asaka et al. (2016)
	Nucleolin	NP	Terrier et al. (2016)
	CLUH	PB2, M1	Ando et al. (2016)
	RAF/MEK/ERK pathway		Pleschka et al. (2001)
	SK1		Seo et al. (2013)
	RANBP3		Predicala and Zhou (2013)
	SGK1		Alamares-Sapuay et al. (2013)
	CASPASE 3		Wurzer et al. (2003)
RNP transport	RAB11A/RAB11B	PB2	Kawaguchi et al. (Amorim et al. 2011)
	YB-1	RNP	Kawaguchi et al. (2012)
	HRB	NEP	Eisfeld et al. (2011b)
Membrane protein transport	SEC61	HA, NA	Heaton et al. (2016)
	CDC42		Wang et al. (2012)
	UBR4	M2	(Tripathi et al. 2015)
	TRAPPC6A	M2	Zhu et al. (2016)
	COPI		Sun et al. (2013)
Assembly/budding	Actin	virion	Nayak et al. (2009)
	Cofilin	virion	Liu et al. (2014)
	CD81	virion	He et al. (2013)
	RACK1	M1	Demirov et al. (2012)
	F1Fo-ATPase	NEP	Gorai et al. (2012)

[a]A representative reference is listed. Please see text for more details

3.1 Binding and Internalization of Influenza A Virus

Influenza A virus carries two types of glycoproteins in its lipid envelope, tetramers of neuraminidase (NA) and trimers of hemagglutinin (HA) (Shaw and Palese 2013). HA is the virus' receptor binding protein and recognizes N-acteylneuraminic acid (also called sialic acid), which is the terminal sugar on oligosaccharide chains of glycoproteins- and lipids and is usually linked to galactose as the penultimate sugar. NA can cleave sialic acid from oligosaccharide chains and therefore possesses receptor-destroying activity (RDA). This RDA is thought to be important for entry in vivo when the virus has to get through layers of sialic acid-containing mucus in the respiratory tract in order to reach the respiratory epithelium (Cohen et al. 2013; Yang et al. 2014; Zanin et al. 2015). Binding of HA to sialic acid is of low affinity and therefore multiple molecules of HA need to bind sialic acid at the same time to allow for virus attachment (Sauter et al. 1989). Of note, different types of influenza A virus bind different types of sialic acid. While the HA proteins of avian strains display a strong preference for α-2′3′-linked sialic acid mammalian viruses recognize α-2′6′-linked sialic acid (Weis et al. 1988; Gamblin et al. 2004; Stevens et al. 2006a, b). This difference correlates with the distribution of sialic acid in the respective host: Avian strains replicate mainly in the gastrointestinal tract of birds, where α-2′3′-linked sialic acid is abundant, whereas mammalian strains replicate in the upper respiratory tract where mostly α-2′6′-linked sialic acid is detected (van Riel et al. 2010; Webster et al. 1978).

While sialic acid is present on many different glycoproteins and lipids across the plasma membrane it is thought that successful internalization can only occur at specific sites or patches in the membrane (Carroll and Paulson 1985). However, it is unclear if influenza virus has the ability to specifically target such sites or even induce the formation of them (Rust et al. 2004). Alternatively, rolling of virus particles along the surface of the cell with continuous binding and release of HA-sialic acid interactions until internalization-competent sites are reached could also be a strategy for the virus to accomplish internalization. In either case, the virus fully relies on host machinery to be taken up into the target cell. It has been shown that clathrin-mediated endocytosis is the main internalization route for influenza A virus (Chen and Zhuang 2008). Besides clathrin also the adaptor protein epsin-1 has been shown to be a crucial host factor for this step. It was observed that the formation of clathrin-coated pits at the site of virus binding occurs in combination with the recruitment of epsin-1 (Chen and Zhuang 2008). Scission of the virus-containing vesicles is then mediated by dynamin (Roy et al. 2000). However, the virus can also use alternative entry routes and upon blocking of the clathrin-mediated pathway the other routes can fully compensate (Chen and Zhuang 2008; Sieczkarski and Whittaker 2002). Virus-containing vesicles without a clathrin coat are readily detectable by microscopy techniques but the mechanism of internalization is not entirely clear (Rust et al. 2004; Matlin et al. 1981). Macropinocytosis, which describes the uptake of large-size cargo through actin-dependent formation of vesicles, has been shown to be one of these alternative internalization routes and this seems to be especially relevant for filamentous virions (de Vries et al. 2011; Rossman et al. 2012). In addition to the cellular transport machinery, the virus also requires the function of cellular signaling cascades at the stage of internalization. It has been shown that virus binding to the cell triggers epidermal growth factor receptor (EGFR) signaling but also activation and requirement of downstream components of the cascade, such as phosphoinositide-specific phospholipase $\gamma 1$ (PLC$\gamma 1$) or mitogen-activated kinases MAPK1 and MAPK8, have been observed (Eierhoff et al. 2010; Fujioka et al. 2013; Zhu et al. 2014; Tripathi et al. 2015). In the case of macropinocytic uptake signaling via receptor-tyrosine kinases, PAK1, phosphoinositide-3-kinase (PI(3)K), as well as src-family kinases and focal adhesion kinase, has been found to be important for successful internalization (de Vries et al. 2011; Elbahesh et al. 2014). This illustrates how a simple step of the viral life cycle, such as internalization, requires a plethora of host factors and functions.

3.2 Endosomal Trafficking and Fusion

Once influenza A virus has been internalized, endosomal trafficking from early endosomes (close to the plasma membrane) to perinuclear late endosomes takes place. Initially, the virus-containing early endosomes are transported away from the plasma membrane via actin-dependent processes as observed by live-cell microscopy (Lakadamyali et al. 2003). Next, a rapid dynein-dependent movement is thought to occur, followed by nucleus-directed movement of the endosomal vesicles via microtubules (Lakadamyali et al. 2003). During trafficking the early endosomes undergo a maturation process, which involves cellular factors such as early endosomal antigen 1 (EEA1), Rab5, and PI(3)K (Bucci et al. 1992; Mu et al. 1995; Simonsen et al. 1998; Christoforidis et al. 1999). The transition from early to late endosomes is mediated by fusion of early endosomes with lysosomes or other late endosomes resulting in the formation of intraluminal vesicles (Luzio et al. 2007; Huotari and Helenius 2011). This occurs during the microtubule-dependent transport toward the perinuclear region. The switch from Rab5 to Rab7 marks a successful transition but Rab5 and Rab7 are not only markers of endosomal trafficking, they have also been shown to be important for successful viral entry (Rink et al. 2005; Sieczkarski and Whittaker 2003). Additional host factors required for the endosomal trafficking of the virus include prolidase, which was shown to play a role in early endosomal routing (Pohl et al. 2014), as well as HDAC8 and the Cullin-3/SPOPL, which are both required for correct endosome maturation and trafficking (Yamauchi et al. 2011; Huotari et al. 2012; Gschweitl et al. 2016). At the stage of late endosomes protein kinase C (PKC) has also been implicated in the entry process of influenza viruses (Sieczkarski et al. 2003).

If influenza A virus successfully travels along the endocytic route and finally localizes to a late endosome in the perinuclear area, fusion of the viral envelope and the endosomal membrane can occur. This requires a pH of 5.1–5.8 depending on the strain of influenza A virus, with human isolates generally requiring lower pH values than avian strains (Galloway et al. 2013). The characteristic low pH of late endosomes can be attributed to the v-ATPase complex, which consists of many different subunits and can pump protons from the cytoplasm into the endosomal lumen in an energy-dependent manner. It is therefore not surprising that several subunits of the v-ATPase complex have been identified as critical host factors for entry (Konig et al. 2010; Guinea and Carrasco 1995). In line with these findings, the host protein RNASEK that associates with and is required for the function of the v-ATPase was also found to be important for viral entry (Perreira et al. 2015).

When the HA protein encounters low pH it undergoes conformational rearrangements that culminate in the exposure of the fusion peptide, which can be inserted into the target membrane (the late endosomal membrane) and initiate fusion of the two membranes [reviewed in (Russell 2014)]. This process can be observed in in vitro assays and it was therefore believed to occur without the help of cellular factors. However, recent studies have indicated that the virus also relies on host factors to accomplish this step of the replication cycle. The first example was the tetraspanin CD81, for which it could be demonstrated that fusion preferentially occurs in CD81-positive late endosomes (He et al. 2013). Upon knockdown of CD81 a reduction in fusion events could be observed. While the molecular mechanism is thus far unknown it is interesting to note that CD81 was also identified as cellular component of influenza virions (Shaw et al. 2008). The second example was Cathepsin W. Depletion of this protease was shown to cause an accumulation of influenza A virus in late endosomes and a block in fusion (Edinger et al. 2015). Reintroduction of the catalytically active protease but not a catalytically inactive mutant released this block indicating that proteolytic cleavage of a thus far unknown target protein by Cathepsin W is required for fusion.

In addition, the cellular proteases that mediate cleavage of HA0, the precursor form of HA, can be considered as host factors required for fusion. Early on, it was observed that influenza virions with uncleaved HA are not infectious and that cleavage of HA depends on the host cell in which influenza virus is grown (Klenk et al. 1975). A few years later it could be demonstrated that the decrease in infectivity for uncleaved HA was due to a block at the stage of fusion (Huang et al. 1981). For highly pathogenic avian influenza viruses of the H5 and H7 subtype it was observed that HA cleavage can occur in almost any cell type. This was found to correlate with the presence of a stretch of basic amino acids at the cleavage site that can be recognized by the ubiquitously expressed proteases of the furin family (Bosch et al. 1981; Steinhauer 1999; Klenk et al. 1984). In contrast, the cleavage site of human influenza viruses, as well as low pathogenic avian strains, contains only one basic amino acid and cleavage is restricted to certain cell types (Garten et al. 1981). In recent years, some of the responsible proteases have been revealed: It was found that the Golgi-associated protease TMPRSS2 and the plasma membrane resident HAT protease can both cleave HA of different strains of human or low pathogenic avian strains, including H1N1 and H3N2 (Bottcher et al. 2006). Moreover, it has been revealed that H1N1 viruses as well as H7N9 viruses fully depend on TMPRSS2 for successful propagation, whereas H3N2 viruses can use thus far unknown alternative proteases (Tarnow et al. 2014; Sakai et al. 2014). TMPRSS2 therefore represents a promising drug target for H1N1 and H7N9 viruses.

3.3 Uncoating and Nuclear Import of the vRNPs

Upon fusion of viral and endosomal membranes a so-called fusion pore is formed through which the vRNPs can be released into the cytoplasm and this process is termed uncoating. Successful uncoating requires the function of the viral protein M2 (Wharton et al. 1994). Tetramers of M2 sit in the viral envelope and form a small ion channel that becomes activated by low pH and possesses specificity for monovalent cations, such as protons (Zebedee and Lamb 1988; Holsinger and Lamb 1991; Pinto et al. 1992; Sugrue and Hay 1991; Chizhmakov et al. 1996). During endosomal trafficking when the endosomal lumen becomes acidified and an influx of potassium is observed M2 allows for the protons and potassium ions to enter the viral particle (Stauffer et al. 2014; Wharton et al. 1994). Both, the acidification and the increased potassium levels are thought to trigger conformational changes in M1 which are crucial for uncoating. M1 has an important role in virus assembly as it binds both the envelope with the glycoproteins and also the vRNPs. The changes in M1 conformation triggered by low pH and increased potassium levels lead to reduced interaction between M1 and the vRNPs and thereby enable the subsequent release of the vRNPs through the fusion pore into the cytoplasm.

While the contributions of the viral proteins to this step of the replication cycle are well understood we are only beginning to uncover the role of host factors in this process. Recently, the E3 ubiquitin ligase Itch was found to play a role in uncoating (Su et al. 2013). It could be shown that upon influenza virus infection Itch becomes phosphorylated and translocates to endosomes where it ubiquitinates M1. This posttranslational modification of M1 seems to be important for efficient release of the vRNPs. A second study revealed a role for HDAC6 in uncoating (Banerjee et al. 2014): It was found that influenza A virus particles contain unanchored ubiquitin chains and evidence was provided that these unanchored ubiquitin chains lead to the recruitment of HDAC6 to the uncoating site. As components of the cellular aggresome machinery were also found to be required for uncoating it was hypothesized that HDAC6 can recruit the aggresome machinery which creates the physical forces needed to release the vRNPs from the fusion site.

Early on, it was observed that upon successful uncoating, M1 stays behind in the cytoplasm, while the vRNPs get imported into the nucleus (Martin and Helenius 1991). Import occurs through the nuclear pores via an active, energy-dependent process accomplished by cellular karyopherins/importins (O'Neill et al. 1995). The members of this protein family can bind to the GTP-binding protein Ran and this interaction is important for the directionality of transport. In the cytoplasm karyopherins bind to their cargo and mediate import into the nucleus, where cargo is released upon Ran-GTP binding of karyopherin. In its Ran-GTP-bound form the karyopherin is shuttled back to the cytoplasm, where GTP hydrolysis occurs and the next cycle of import can take place. Cargo recognition is mediated via so-called nuclear localization signals (NLS), which usually contain a stretch of basic amino acids. All three subunits of the polymerase complex possess an NLS but it was found that the NLS on the viral nucleoprotein NP is sufficient for the initial import of vRNPs during viral entry (O'Neill et al. 1995; Cros and Palese 2003). On the side of the karyopherins, the family members alpha-1, alpha-3 and alpha-5 have been described to be involved in the initial import (O'Neill et al. 1995; Wang et al. 1997; Melen et al. 2003). Furthermore, heat shock protein 40 (Hsp40/DnaJB1) was shown to be required for efficient binding of vRNPs by karyopherin alpha (Batra et al. 2016).

Successful nuclear import of the vRNPs marks the end of the viral entry process and the beginning of the nuclear phase of the influenza virus replication cycle.

4 Host Factors Involved in Viral Replication and Transcription

Unlike most RNA viruses, influenza viruses replicate and transcribe their genome in the nucleus of the cell. Therefore, it is expected that the host factors involved in these processes will be located in the nucleus, or possibly be recruited to the nucleus upon virus infection. After entering the nucleus the viral RNPs containing the negative sense vRNA serve as templates for transcription and replication. The heterotrimeric polymerase complex (PB1, PB2, PA) that is associated with each RNP is responsible for initiating primary transcription, and one of the reasons that influenza virus requires the nucleus is that it has a unique method of priming transcription which is referred to as "cap-snatching" (Shaw and Palese 2013). The PB2 protein binds to the cap structure on the 5′ end of cellular pre-mRNAs (predominantly noncoding RNAs (Gu et al. 2015; Koppstein et al. 2015)), and this is followed by cleavage of the pre-mRNA by the endonuclease function of PA. This produces a 5′-capped, 10–13 nucleotide primer which is then used by the RNA-dependent RNA polymerase (PB1) to initiate transcription. All viral transcripts are also polyadenylated and as such they are translated just like host mRNAs. The mRNAs from two of the segments, M and NS, are alternatively spliced, each producing two protein products (Shaw and Palese 2013). This process is dependent on the host cell splicing machinery, and is another reason that influenza virus replicates in the nucleus. The same viral polymerase is responsible for replicating the vRNA via a positive sense cRNA intermediate, which is a complementary copy of the vRNA and lacks any 5′ or 3′ modifications. Also, in comparison to transcription, the initiation of replication occurs without a primer. It is still not entirely clear what drives the polymerase to switch between transcriptional and replication modes but the latest structural data indicate that the polymerase complex undergoes substantial rearrangement (Thierry et al. 2016), potentially allowing for interactions with different sets of cellular factors. Suffice to say that host functions are intimately involved in the production of both viral transcripts and new copies of the viral genome (Table 1).

4.1 Genome-Wide "OMIC" Studies on the Replication Complex

In attempts to define the repertoire of cellular proteins that are involved in transcription and replication, protein interaction studies have been performed on the isolated viral RNPs (Mayer et al. 2007), the polymerase complex (Bradel-Tretheway et al. 2011; Jorba et al. 2008), and the individual protein components (NP, PB1, PB2, PA) (Shapira et al. 2009; Bradel-Tretheway et al. 2011; Watanabe et al. 2014b; Tripathi et al. 2015). The complexity of these studies has increased with advancing technologies, and most recently includes the use of recombinant viruses bearing affinity tagged proteins (Heaton et al. 2016; York et al. 2014), as well as viruses expressing one half of a luciferase complementing system (Munier et al. 2013). In all cases many interacting proteins were identified, and when combined this dataset presents numerous candidates to explore for potential roles in viral polymerase function. So far only a fraction of these have been further investigated for functional roles and we will touch on these in the following sections.

4.2 Requirements for Transcription

Within the nucleus the influenza viral polymerase machinery needs to be located close to sites of cellular transcription in order to steal the 5′-capped RNA primers from nascent host transcripts. As such, one of the key host interactions is between the viral polymerase and cellular RNA polymerase II (Engelhardt et al. 2005). Specifically, the serine-5-phosphorylated form of the Pol II C-terminal domain which is the form involved in initiation of transcription and addition of the 5′-cap structure (Martinez-Alonso et al. 2016). Other interactions that have been implicated in localizing the viral replication machinery to specific nuclear subdomains are associations with chromatin remodelers. CHD1, which is recruited to transcriptionally active genes via interactions with the H3 K4me3 histone mark, has been shown to associate with influenza virus polymerase (Marcos-Villar et al. 2016). RNAi depletion of CHD1 reduces viral polymerase activity, primary transcription, and viral growth, leading to the conclusion that CHD1 is a positive regulator of influenza virus replication (Marcos-Villar et al. 2016). NXP2/MORC3 may also be involved in ensuring close proximity of the virus to sites of cellular transcription in the nucleus as there is evidence that it too recognizes H3 K4me3 marks (Li et al. 2012, 2016). NXP2/MORC3 was originally identified as a polymerase interacting protein in a proteomic study (Jorba et al. 2008), and subsequent work has verified this interaction and shown that it is required for influenza virus transcription (Ver et al. 2015). An emerging common feature of these virus–host interactions is that changes are observed later in infection after viral transcription has taken place. These include degradation of both RNA pol II (Rodriguez et al. 2007; Vreede et al. 2010) and CHD1 (Marcos-Villar et al. 2016) and relocalization of NXP2/MORC3 to the cytoplasm (Ver et al. 2015), and it is thought that this is partly responsible for the host gene shut off observed in influenza virus infected cells. Through this tight regulation the virus manages to balance its need for host transcription early in the life cycle with the need to also prevent induction of the host antiviral response.

Host factors so far implicated specifically in the production of viral transcripts include RRP1B (ribosomal RNA processing 1 homolog B) and SFPQ (splicing factor proline-glutamine rich). RRP1B was identified in an RNAi screen and upon further characterization was found to interact with the trimeric viral polymerase complex and to show partial relocalization from the nucleolus to the nucleoplasm in infected cells (Su et al. 2015). Depletion assays indicate that it is required for optimal transcription and specifically the ability of the polymerase to bind to 5′-capped host mRNAs (Su et al. 2015). SFPQ was found to associate with the influenza virus polymerase complex (Jorba et al. 2008) and knockdown has been shown to impair virus growth and viral gene expression (Landeras-Bueno et al. 2011). Despite the role of SFPQ in cellular mRNA splicing, it was found not to be required for viral mRNA splicing and instead lack of SFPQ results in defective polyadenylation of viral transcripts (Landeras-Bueno et al. 2011). Most recently, it has been shown that DDX19, a DExD-box RNA helicase, associates with viral transcripts in the nucleus and promotes their export (Diot et al. 2016).

4.3 Requirements for Viral RNA Synthesis

Some factors have been implicated in promoting viral transcription due to their link to RNA Pol II but in the absence of data to discriminate between transcription and replication a fair conclusion is that they are required for all viral RNA synthesis. Cyclin T1/CDK9 is one such factor that has been proposed to be required for efficient virus transcription, although reductions in cRNA and vRNA were also observed (Zhang et al. 2010). Interestingly, the kinase activity of cyclin T1/CDK9 is not required for this effect however, it does interact with the vRNP and individually with the PB1, PB2, and PA subunits of the polymerase complex (Zhang et al. 2010). Based on less efficient interaction of the polymerase with the serine-2-phosphorylated form of the RNA Pol II C-terminal domain in the absence of CDK9 it has been proposed that CDK9 serves as an adapter between the viral polymerase and RNA Pol II (Zhang et al. 2010). However, recent data indicating a direct interaction between influenza virus polymerase and RNA Pol II suggests that this may have to be investigated further (Martinez-Alonso et al. 2016). hCLE/C14orf166 is a RNA Pol II transcriptional regulator and was first identified as an interacting partner of the viral PA protein (Huarte et al. 2001). In the absence of hCLE/C14orf166 reduced levels of influenza virus transcription, replication and viral titers are observed (Rodriguez et al. 2011). Most recently, it was reported that hCLE/C14orf166 is incorporated into influenza virions which means that the virus is actively transporting one of its required host factors to the next cell (Rodriguez-Frandsen et al. 2016). DNAJA1, a member of the HSP40 family, interacts with both PB2 and PA subunits of the viral polymerase and has been demonstrated to enhance viral RNA polymerase activity in vitro (Cao et al. 2014). In a cellular context it was observed that DnaJA1 is recruited into the nucleus together with the PB1-PA dimer but that it is not involved in assembly of the newly synthesized polymerase complex (Cao et al. 2014). Therefore, this is an example of a normally cytoplasmic protein that is specifically transported into the nucleus to facilitate viral RNA synthesis.

4.4 Requirements for Replication

The only factors to date that have been described to specifically promote viral genome replication are pp32 and APRIL (also known as ANP32A and ANP32B, respectively) (Sugiyama et al. 2015). These two factors were identified from fractionated nuclear extract and were shown to support the synthesis of vRNA from a cRNA template. Both of these proteins had previously been identified as binding to the polymerase complex (Bradel-Tretheway et al. 2011) and in this later study the interaction was isolated to free polymerase complex not associated with the RNP (Sugiyama et al. 2015). Interestingly, ANP32A has been identified as a species specificity host factor that determines the ability of avian influenza viruses to replicate in mammalian cells (Long et al. 2016), and this will be discussed further in Sect. 6.2.

4.5 Requirements for Viral mRNA Splicing

Two of the influenza virus genomic segments produce multiple protein products by means of alternative splicing. The M segment encodes the M1 protein from an unspliced transcript and the M2 protein from a spliced transcript. Likewise, the NS segment produces NS1 from the unspliced transcript and NS2/NEP from the spliced transcript. Splicing of cellular genes is normally very efficient, but for the virus this process must be regulated to ensure that proteins from both spliced and unspliced mRNAs are made. Interestingly, cellular factors that have been implicated in this regulation appear to have specific effects on either the M or NS segments. For example, the cellular splicing factors RED and SMU1 are preferentially required for splicing of the NS segment (Fournier et al. 2014). RED was identified as interacting with both PB1 and PB2 proteins, and SMU1, which is known to bind to RED, co-purified with the RED-polymerase complex (Fournier et al. 2014). Depletion of either RED or SMU1 was shown to reduce viral titers and a decrease in the NEP to NS1 ratio at both transcript and protein levels were observed (Fournier et al. 2014). Comparatively, more proteins have been implicated in splicing of the M segment. The first host factor described to regulate M splicing, and production of M2 protein, was SF2/ASF (Shih and Krug 1996). Specifically it was shown that SF2/ASF binds to a purine-rich splicing enhancer sequence in the 3′ exon of the M1 transcript, and this controls activation of the relatively weak M2 5′ splice site (Shih and Krug 1996). SF2/ASF is phosphorylated by the kinase, CLK1, thereby regulating its activity (Bullock et al. 2009). So when CLK1 was identified in an RNAi screen as being essential for influenza virus replication, it was hypothesized that CLK1 kinase activity was required for production of M2 (Karlas et al. 2010). Indeed, it was shown that inhibition of CLK1 by siRNA or a small molecule inhibitor, lead to a reduction in M2 mRNA, but had no effect on M1, NS1, or NEP transcripts (Karlas et al. 2010). Another study identified the cellular proteins NS1-BP (so named because it binds to influenza virus NS1) and hnRNPK as regulators of M splicing (Tsai et al. 2013). NS1-BP was shown to associate with several hnRNPs which interact directly with M1 mRNA. Of these, hnRNPK was found to facilitate M splicing as depletion of hnRNPK, as well as NS1-BP, resulted in decreased levels of M2 message and protein (Tsai et al. 2013). Finally, there is evidence that splicing of M and NS may take place in different subcellular locations, which potentially explains their reliance on distinct sets of host factors. This finding comes from a study reporting that splicing of M transcripts is dependent on their localization in nuclear speckles, which are enriched with splicing factors, including NS1-BP and hnRNPK (Mor et al. 2016).

5 Host Factors Involved in Nuclear Export, Trafficking, and Budding

For the final steps of the viral life cycle the virus has to export the newly synthesized vRNPs out of the nucleus and traffic them to the budding site at the plasma membrane (Shaw and Palese 2013). At the same time the viral membrane proteins, HA, NA, and M2, traffic through the ER-Golgi secretory network and accumulate at the site of virus assembly. The RNPs, along with M1 and NEP, are packaged into the budding virion which then separates from the cell membrane and is released. Our knowledge of how host factors contribute to these processes is still rather limited, but several studies have provided more insight in recent years (Pohl et al. 2016). We also know that cellular proteins can be detected in purified influenza virions (Shaw et al. 2008; Hutchinson et al. 2014), and that for some this reflects their functional role in virus assembly. In the following sections, we will review the current information regarding the involvement of host factors in the later stages of the influenza virus life cycle (Table 1).

5.1 Nuclear Export of RNPs

Viral RNPs are exported from the nucleus through the nuclear pore complex in an energy-dependent process that requires the cellular exportin machinery and Ran-GTP. Crm1, also known as XPO1, is the major exportin responsible for transport of influenza virus RNPs from the nucleus to the cytoplasm and this is achieved via an interaction with the nuclear export sequences (NES) in NEP (O'Neill et al. 1998; Neumann et al. 2000; Huang et al. 2013). M1 associates with both NEP (Yasuda et al. 1993; Shimizu et al. 2011) and the RNP (Ye et al. 1999; Baudin et al. 2001), so therefore serves as an adapter between the RNP, NEP, and the cellular export machinery. However, an alternative model has been proposed where NEP makes contact with the polymerase on the RNP as well as M1, and enhances the M1/RNP association (Brunotte et al. 2014). Regardless, it is clear that both NEP and M1 are critical for facilitating the interaction between the RNP and the nuclear export machinery, and not surprisingly this is regulated by several mechanisms involving host factors. First, the levels of M1 help to ensure that RNPs are not exported prematurely. Early in infection, M1 is subject to ubiquitination on lysine 242 which targets it for degradation (Gao et al. 2015), but later it becomes sumoylated on the same residue and is stabilized, allowing it to bind to newly made RNPs in the nucleus (Wu et al. 2011). This switch from ubiquitination to sumoylation appears to be regulated by AIMP2, which is bound and stabilized by NEP, and depletion of AIMP2 leads to decreased virus titers, confirming that this host factor supports virus replication (Gao et al. 2015). Hsc70 has also been implicated in RNP export as RNAi knockdown results in decreased virus production, and M1 and NP remain in the nucleus. Hsc70 was first reported to bind to the C-terminal domain of M1 (Watanabe et al. 2006), and later it was shown that Hsc70 competes with NEP for binding to M1 (Watanabe et al. 2014a), which indicates that it may be influencing the formation of the RNP-M1-NEP complex. Alternatively, it has been suggested that Hsc70 may mediate nuclear export of RNP-M1 complexes independently of NEP, but dependent on an interaction with Crm1 (Watanabe et al. 2008). Similarly, NXT1 has been implicated in an NEP-independent export pathway by binding to NP (Chutiwitoonchai and Aida 2016), which has previously been reported to interact with the Crm1 machinery (Elton et al. 2001). NXT1 is a nuclear export factor that is involved in the final stages of Crm1-dependent export (Black et al. 2001), and it is shown that NP, NXT1, and Crm1 form a complex that promotes the export of NP (Chutiwitoonchai and Aida 2016).

An emerging theme from the literature is the role of host chromatin in RNP export. This was first highlighted by a study which showed that RNPs are tethered to chromatin when Crm1-dependent export is blocked (Chase et al. 2011). In addition, Crm1 was found to relocalize to areas of dense chromatin in influenza virus infected cells, and an increase in Crm1, Ran, and Rcc1 association was observed (Chase et al. 2011). Formation of the Crm1-Ran-Rcc1 complex occurs in order to regenerate Ran-GTP from Ran-GDP, and thus allows recycled Crm1 to bind to new substrates. It is proposed that by targeting RNPs to the chromatin where these new Crm1-Ran-GTP complexes are formed, the virus is able to monopolize the nuclear export machinery. More recently, it has been shown that NEP interacts with chromodomain-helicase-DNA-binding protein 3 (CHD3) via its first NES, and that CHD3 is required to localize both NEP and Crm1 to the dense chromatin (Hu et al. 2015). The polycomb repressive complex 2 (PRC2), which is required for the formation of facultative heterochromatin, has also been implicated in RNP export as depletion leads to nuclear accumulation of RNPs (Asaka et al. 2016). The study shows that PRC2 interacts with M1 and facilitates complex formation between M1 and RNP. Finally, an interaction between NP and nucleolin has been reported to increase the association of RNPs with chromatin (Terrier et al. 2016). Taken together these data suggest that formation of the RNP-M1-NEP-Crm1 complex is localized to dense/facultative chromatin within the nucleus and that this ensures access to recycled Crm1. The host protein, CLUH, which interacts with PB2 and M1, also appears to regulate RNP export as in the absence of CLUH, RNPs, M1, and NEP are retained in the nucleus (Ando et al. 2016). It is not required for CRM1-dependent export but rather it appears that CLUH is important for the subnuclear transport of RNPs and M1 via nuclear speckles prior to their interaction with the export machinery (Ando et al. 2016).

Host cell signaling pathways play an important role in regulating the timing of RNP export. Early on it was recognized that inhibition of the Raf/MEK/ERK signaling pathway leads to accumulation of RNPs in the nucleus (Pleschka et al. 2001). This MAP kinase cascade is activated by virus infection later in infection (from 4 h onwards) and it is hypothesized that certain cellular factors that play an essential role in RNP export become phosphorylated as a result (Pleschka et al. 2001). Influenza virus infection has also been shown to increase the expression and activation of sphingosine kinase (SK) 1, which is known to regulate several pathways, including NF-κB, ERK MAPK, and PI3K/AKT (Seo et al. 2013). It was found that inhibition of SK1 blocked the activation of RanBP3 by ERK MAPK and PI3K/AKT (Seo et al. 2013), and since RanBP3 is a co-factor of Crm1 (and has been shown to be required for influenza virus RNP export (Predicala and Zhou 2013)), this links ERK activity with the Crm1 export machinery. Another kinase, SGK1, has also be implicated in RNP export but it is unclear whether it is regulating a viral or cellular factor (Alamares-Sapuay et al. 2013).

Lastly, an alternative export pathway that is regulated by caspase activity has been proposed. Influenza virus induces apoptosis late in infection and inhibition of caspase 3 was shown to inhibit influenza virus and cause RNPs to accumulate in the nucleus (Wurzer et al. 2003). Subsequently, it was shown that activation of caspase is associated with degradation of Nup153, a subunit of the nuclear pore complex and the size of the nuclear pore was found to be enlarged (Muhlbauer et al. 2015). Thus, it is thought that this may represent an additional, Crm1-independent, pathway by which influenza virus transports RNPs from the nucleus to the cytoplasm, particularly at late stages of infection.

5.2 *Trafficking to the Plasma Membrane*

5.2.1 Cytoplasmic Transport of RNPs

Once in the cytoplasm, the RNPs use the microtubule network and vesicular transport system to reach the budding site at the plasma membrane, where they meet up with the structural proteins HA, NA, and M2 (Hutchinson and Fodor 2013). During this transit it is believed that the RNPs are sorted based on unique packaging signals, and are bundled together so that eight RNPs, each representing one of the eight segments, are presented for packaging at the budding site. M1 may assist in delivery of RNPs to the correct site by associating with the plasma membrane as well as with the cytoplasmic tails of the viral glycoproteins.

Several studies have highlighted the importance of the Rab11 pathway for transporting RNPs to the site of virus assembly. Rab11 GTPases are involved in the transport of recycling endosomes to the apical surface, and two members of this family, Rab11a and Rab11b, have been linked to RNP trafficking. In the absence of Rab11 fewer virus particles are formed (Bruce et al. 2010) and RNPs fail to accumulate at the plasma membrane late in infection (Amorim et al. 2011; Eisfeld et al. 2011a). Moreover, it has been shown that RNPs co-localize with Rab11-positive recycling endosomes and that there is an interaction between RNPs and the GTP-bound form of Rab11 (Eisfeld et al. 2011a; Momose et al. 2011; Amorim et al. 2011). This association appears to be mediated by the viral polymerase, specifically the PB2 subunit (Amorim et al. 2011; Momose et al. 2011; Avilov et al. 2012). It has also been shown that RNPs accumulate at the microtubule organizing center (MTOC) (Momose et al. 2007) and it is proposed that this facilitates access to the Rab11 recycling endosomes which then carry the RNPs to the cell surface along the microtubule network (Amorim et al. 2011). Initial localization to the MTOC is mediated by Y-box-binding protein 1 (YB-1) which interacts with RNPs in the nucleus and upon export it is said to act as a porter to deliver the RNPs to the MTOC (Kawaguchi et al. 2012). Another host factor implicated in the early cytoplasmic transport of RNPs is the human immunodeficiency virus (HIV) Rev-binding protein (HRB). HRB was identified as an NEP-interacting partner (O'Neill et al. 1998) and depletion prevents accumulation of RNPs at the cell surface (Eisfeld et al. 2011b). Co-localization of HRB with NEP is observed in the perinuclear region, and also with RNPs at the MTOC which has led to the proposal that HRB mediates the transfer of RNPs from the nuclear export machinery to the vesicular transport network at the MTOC (Eisfeld et al. 2011b).

So far there is no evidence that Rab11 stays associated with RNPs as they are packaged into virions, so it remains to be seen how RNPs are transferred, or diverted, from Rab11 recycling endosomes to the site of virus assembly at the plasma membrane. It is very likely that additional host factors will be found to facilitate this final step in RNP transport.

5.2.2 Transport of Viral Membrane Proteins

The three integral membrane proteins, HA, NA, and M2, all traffic through the ER-Golgi secretory network and accumulate at the apical surface of the plasma membrane (Shaw and Palese 2013). HA and NA specifically cluster in areas rich in cholesterol and sphingolipids (the so-called lipid rafts), and drive the formation of the bud. In contrast, M2 is found in regions adjacent to lipid rafts and is involved in pinching off the budding virion from the membrane, which is consistent with the finding that there is very little M2 incorporated into virus particles. In the final step, the neuraminidase activity of NA is required to cleave any sialic acid linkages to release the virion from the cell.

It is well known that cholesterol, which is a key component of lipid rafts, is essential for HA and NA localization at the plasma membrane, and for virus assembly (Scheiffele et al. 1997). Likewise, sphingomyelin, which is another lipid raft component, is also important. It has been shown that cells deficient in sphingomyelin synthesis do not support virus assembly and that HA and NA transport is blocked at a point beyond the trans-Golgi network (Tafesse et al. 2013). Thus, lipid synthesis appears to be critical but much less is known about specific host factors that participate in the transport of these viral membrane proteins through the ER-Golgi network and beyond. Starting in the ER, viral proteins are co-translationally folded, and thus rely on host chaperones such as calnexin and calreticulin (Hebert et al. 1997). The translocon, which translocates polypeptides into the ER, is also critical and a major component, Sec61, has been shown to interact with HA and NA (Heaton et al. 2016). RNAi knockdown or small molecule inhibition of Sec61 causes a decrease in virus production accompanied by a lack of HA and NA surface expression, but no loss of M2 expression (Heaton et al. 2016), so potentially this reflects a greater requirement for translocon activity for glycosylated viral proteins. For transit through the ER-Golgi network to the cell surface vesicular trafficking proteins such as Rho family GTPases could be critical, and so far one member, Cdc42, has been shown to be important for the transport of NA from the Golgi to the plasma membrane (Wang et al. 2012). At the moment it is unclear whether Cdc42 also participates in trafficking of HA or M2. UBR4, which interacts with M2, has been shown to be important for M2 trafficking from the ER to the plasma membrane, and is relocalized from the nucleus to the ER in infected cells (Tripathi et al. 2015). It is potentially important for HA and NA too, as decreased surface levels of all three viral membrane proteins are observed in the absence of UBR4, culminating in a loss of virus particle production (Tripathi et al. 2015). It has also been reported that TRAPPC6A, is required for M2 trafficking to the cell surface and interacts with the M2 cytoplasmic tail (Zhu et al. 2016). Finally, pharmacological inhibition of COPI complexes has been shown to inhibit surface expression of viral glycoproteins and virus assembly (Sun et al. 2013), indicating either a direct role in viral protein transport or a role in transport of essential host membrane components.

5.2.3 Budding and Release of Virus Particles

Either HA or NA alone is sufficient to initiate virus budding and it is thought that in infected cells the accumulation of these two glycoproteins in lipid raft domains at the plasma membrane drives formation of the budding virus particle (Nayak et al. 2009). This must be coordinated with delivery of RNPs to the assembly site for genome packaging. Following RNP packaging, M2 is responsible for scission from the cell membrane and then NA cleaves sialic acid linkages with HA to release the virus particle from the cell (Shaw and Palese 2013). Several cellular proteins have been found to be incorporated into influenza virus particles and for some this may reflect a specific role in formation of the virion (Hutchinson et al. 2014; Shaw et al. 2008). Actin is one such example and it is proposed that actin rearrangement may be required to push the cell membrane outwards during bud formation (Nayak et al. 2009), and to maintain lipid raft integrity (Simpson-Holley et al. 2002). It has also been shown that cofilin (also detected in virus particles) is important for regulating these actin dynamics (Liu et al. 2014). CD81, which is incorporated into the membrane of virus particles and accumulates at the neck of the budding virion, was shown to facilitate membrane scission through either direct or indirect mechanisms (He et al. 2013). This same step may involve RACK1 which is a scaffolding protein for signaling cascades, and interacts with the viral M1 protein (Demirov et al. 2012). Loss of this interaction or depletion of RACK1 results in an arrested budding phenotype where virus particles remain attached to the cell membrane. Finally, through investigation of NEP-interacting proteins, F1Fo-ATPase has been shown to promote efficient virus budding in an ATPase dependent manner and it is thought that local concentration of this activity at the plasma membrane helps to induce membrane curvature (Gorai et al. 2012).

6 The Role of Host Factors in Zoonotic Transmission of Influenza Virus

The reservoir of influenza A viruses are aquatic birds, with the exception of the recently identified bat influenza viruses (Tong et al. 2012, 2013). Therefore, influenza viruses have been co-evolving with birds over a long period of time (Wright et al. 2013). They are well adapted to usurp the avian host machinery, including signaling and transport pathways, protein translation, and glycosylation functions and many more. If influenza A virus is transmitted to other hosts, such as pigs or humans, in the majority of cases the virus will not be able to replicate efficiently and start a transmission chain in the new host as it is not adapted to use the new host's cellular functions (Schrauwen and Fouchier 2014). While some host proteins are highly conserved and will allow the virus to interact with them, others will differ substantially between avian and mammalian cells and thereby pose a barrier to cross-species transmission. Thus far, we know of a few examples of host factors that belong to the latter group but we are still far from having a detailed picture of the conserved versus the barrier-posing host dependency factors. As many more tools are available for studying host factors in human cells and such human host factors also represent potential drug targets, most work has been performed in human cells with human viruses. Very little is known about the interaction between avian strains and their avian host cells and it will be important to develop more tools and assays to fill this gap in our understanding of the virus–host interplay.

In this section, we will describe two examples of host factors known to pose a barrier to cross-species transmission of influenza A viruses.

6.1 Restriction of Zoonotic Transmission by Host-Specific Distribution of Sialic Acids

The best known example for a barrier to zoonotic transmission of influenza A virus is the distribution of its receptor, namely sialic acid modifications on glycoproteins and lipids at the plasma membrane of target cells [reviewed in (Cauldwell et al. 2014; de Graaf and Fouchier 2014)]. Sugar modifications do not represent the typical host dependency factor. However, the distribution of the glycans depends on the expression levels and patterns of specific sialyltransferases and these can therefore be considered an example of host factors that pose a barrier to cross-species transmission of influenza viruses.

In the avian host the virus replicates in the gastrointestinal tract, where the epithelial cells display mostly α-2′3′-linked sialic acid. The numbering refers to the linkage of sialic acid to the penultimate sugar of the glycan, which is usually galactose. The hemagglutinin proteins of avian strains therefore bind preferentially to α-2′3′-linked sialic acid (Rogers and Paulson 1983). In contrast, mammalian strains of influenza A virus replicate in the mammalian upper respiratory tract, where mostly α-2′6′-linked sialic acid is present and their hemagglutinins display higher affinity for α-2′6′-linked sialic acid (Gambaryan et al. 1997; Matrosovich et al. 2000). For a successful zoonotic transmission from an avian to a mammalian host the virus needs to accomplish a change in receptor specificity, which is mediated by amino acid substitutions in hemagglutinin that change the preference from α-2′3′- to α-2′6′-linked sialic acid. This switch has been documented for many viruses that underwent successful zoonotic transmission and is considered the most important event in the evolution of a new mammalian lineage of influenza A virus (Liu et al. 2009; Yamada et al. 2006; Glaser et al. 2005; Stevens et al. 2006a).

6.2 Restriction of Zoonotic Transmission by the Host Protein ANP32A

For many years, it has been known that mutations in the viral polymerase-encoding genes are selected during adaptation of avian influenza viruses to mammalian cells. Early on, it was recognized that an amino acid change at position 627 in PB2 is of particular importance: While most avian strains carry a glutamic acid (E) at this position most mammalian isolates have a lysine (K) instead, and this switch has occurred multiple times independently (Subbarao et al. 1993; Steel et al. 2009; Van Hoeven et al. 2009). It was further demonstrated that while avian polymerases display almost no activity in reconstituted polymerase assays in human cells a single change from E627 to K627 is sufficient to confer human-like activity to the avian polymerase (Mehle and Doudna 2009; Moncorge et al. 2010). However, the underlying molecular mechanism was unknown.

Only recently, a key host factor behind the barrier encountered by avian polymerases in human cells was identified. It could be demonstrated that expression of avian ANP32A (acidic (leucine-rich) nuclear phosphoprotein 32 family, member A) but not human ANP32A can rescue the activity of an avian polymerase in human cells but not increase the activity of polymerases derived from human strains of influenza A virus (Long et al. 2016). This difference could be attributed to a stretch of 33 amino acids present in the avian version but not in mammalian ANP32A that was necessary for the polymerase-enhancing effect. Transfer of these 33 amino acids to the human version enabled avian polymerase activity in human cells. Interestingly, an earlier study had shown that human ANP32A and ANP32B are required by human polymerases for RNA replication, specifically for vRNA synthesis from the cRNA template (Sugiyama et al. 2015). Furthermore, it was found that human ANP32A and ANP32B can both interact with the trimeric complex of PB1, PB2, and PA but not the individual subunits or the vRNP complex. Binding properties of avian polymerases to human and avian ANP32 A or B have not been analyzed yet but one can speculate that avian polymerases can only interact with avian ANP32A and rely on it for RNA replication. Adaptation, by the change at position 627 in PB2, could potentially allow the avian polymerase to recruit human ANP32A or B to accomplish vRNA synthesis. However, further studies are required to verify or modify this current model of action.

Many more cellular factors can impact the potential of avian strains to cross the species barrier and successfully replicate in mammalian cells but more work, in particular, in avian experimental systems, is required to identify the key players and elucidate their mechanism of action.

7 Host Factors for Antiviral Drug Development

Knowledge of critical virus–host interactions has potential translational applications as the host factor could be targeted by small molecule inhibitors to achieve antiviral activity (Shaw 2011; Watanabe and Kawaoka 2015). Two of the most promising examples of this for influenza that we will highlight are inhibition of Crm1 and the Raf/MEK/ERK pathway. Verdinexor is an orally bioavailable inhibitor of nuclear export that targets Crm1 (also known as XPO1), which, as described in Sect. 5.1, is required for export of viral RNPs from the nucleus. Strong antiviral activity of verdinexor against influenza viruses has been demonstrated both in tissue culture and in mouse models, where it reduces virus titers in the lungs and increases survival of infected animals (Perwitasari et al. 2014). Although Crm1-dependent export of cellular proteins is also affected by this compound, it and related compounds appear to be tolerated in both animal studies and in human Phase I trials for cancer (Abdul Razak et al. 2016), providing an opportunity for antiviral applications. Of note, due to the acute nature of influenza infections, antiviral therapies are normally of short duration which perhaps provides greater opportunity to explore host-directed drug strategies without adverse effects associated with long-term inhibition of cellular pathways.

The Raf/MEK/ERK signaling pathway (involved in RNP export) has also been targeted as an antiviral strategy, and the MEK-inhibitor U0126 has been demonstrated to inhibit the growth of influenza viruses and shows efficacy in animal studies (Droebner et al. 2011). In addition, when used in combination with oseltamivir (which targets the viral NA), MEK inhibitors exhibit synergistic activity (Haasbach et al. 2013). This could allow lower, less toxic, doses to be used so may prove to be advantageous. While concerns about toxicity will always plague host-targeted approaches to antiviral development, there are some distinct benefits. These include the fact that these compounds are unlikely to be affected by mutations in the virus, so they would be effective against viruses that are resistant to virus-directed antivirals. Also, because the requirement for the host factor is normally conserved, they will probably be effective against a wide-range and possibly all influenza viruses. In fact, sometimes the host pathway is important for several different virus families which provides potential for the development of a broad-spectrum antiviral. With increasing knowledge of the host cell functions that influenza virus depends on it is hoped that we will be able to expand the number of potential antiviral targets and revitalize antiviral drug development for influenza.

References

Abdul Razak AR, Mau-Soerensen M, Gabrail NY, Gerecitano JF, Shields AF, Unger TJ, Saint-Martin JR, Carlson R, Landesman Y, McCauley D, Rashal T, Lassen U, Kim R, Stayner LA, Mirza MR, Kauffman M, Shacham S, Mahipal A (2016) First-in-class,

First-in-human phase I study of selinexor, a selective inhibitor of nuclear export, in patients with advanced solid tumors. J Clin Oncol 34(34):4142–4150

Alamares-Sapuay JG, Martinez-Gil L, Stertz S, Miller MS, Shaw ML, Palese P (2013) Serum- and glucocorticoid-regulated kinase 1 is required for nuclear export of the ribonucleoprotein of influenza A virus. J Virol 87(10):6020–6026

Amorim MJ, Bruce EA, Read EK, Foeglein A, Mahen R, Stuart AD, Digard P (2011) A Rab11- and microtubule-dependent mechanism for cytoplasmic transport of influenza A virus viral RNA. J Virol 85(9):4143–4156

Ando T, Yamayoshi S, Tomita Y, Watanabe S, Watanabe T, Kawaoka Y (2016) The host protein CLUH participates in the subnuclear transport of influenza virus ribonucleoprotein complexes. Nat Microbiol 1(8):16062

Asaka MN, Kawaguchi A, Sakai Y, Mori K, Nagata K (2016) Polycomb repressive complex 2 facilitates the nuclear export of the influenza viral genome through the interaction with M1. Sci Rep 6:33608

Avilov SV, Moisy D, Naffakh N, Cusack S (2012) Influenza A virus progeny vRNP trafficking in live infected cells studied with the virus-encoded fluorescently tagged PB2 protein. Vaccine 30 (51):7411–7417

Banerjee I, Miyake Y, Nobs SP, Schneider C, Horvath P, Kopf M, Matthias P, Helenius A, Yamauchi Y (2014) Influenza A virus uses the aggresome processing machinery for host cell entry. Science 346(6208):473–477

Batra J, Tripathi S, Kumar A, Katz JM, Cox NJ, Lal RB, Sambhara S, Lal SK (2016) Human heat shock protein 40 (Hsp40/DnaJB1) promotes influenza A virus replication by assisting nuclear import of viral ribonucleoproteins. Sci Rep 6:19063

Baudin F, Petit I, Weissenhorn W, Ruigrok RW (2001) In vitro dissection of the membrane and RNP binding activities of influenza virus M1 protein. Virology 281(1):102–108

Benitez AA, Panis M, Xue J, Varble A, Shim JV, Frick AL, Lopez CB, Sachs D, tenOever BR (2015) In Vivo RNAi screening identifies MDA5 as a significant contributor to the cellular defense against Influenza A Virus. Cell Rep 11(11):1714–1726

Black BE, Holaska JM, Levesque L, Ossareh-Nazari B, Gwizdek C, Dargemont C, Paschal BM (2001) NXT1 is necessary for the terminal step of Crm1-mediated nuclear export. J Cell Biol 152(1):141–155

Boon AC, deBeauchamp J, Hollmann A, Luke J, Kotb M, Rowe S, Finkelstein D, Neale G, Lu L, Williams RW, Webby RJ (2009) Host genetic variation affects resistance to infection with a highly pathogenic H5N1 influenza A virus in mice. J Virol 83(20):10417–10426

Bosch FX, Garten W, Klenk HD, Rott R (1981) Proteolytic cleavage of influenza virus hemagglutinins: primary structure of the connecting peptide between HA1 and HA2 determines proteolytic cleavability and pathogenicity of Avian influenza viruses. Virology 113(2):725–735

Bottcher E, Matrosovich T, Beyerle M, Klenk HD, Garten W, Matrosovich M (2006) Proteolytic activation of influenza viruses by serine proteases TMPRSS2 and HAT from human airway epithelium. J Virol 80(19):9896–9898

Bradel-Tretheway BG, Mattiacio JL, Krasnoselsky A, Stevenson C, Purdy D, Dewhurst S, Katze MG (2011) Comprehensive proteomic analysis of influenza virus polymerase complex reveals a novel association with mitochondrial proteins and RNA polymerase accessory factors. J Virol 85(17):8569–8581

Brass AL, Huang IC, Benita Y, John SP, Krishnan MN, Feeley EM, Ryan BJ, Weyer JL, van der Weyden L, Fikrig E, Adams DJ, Xavier RJ, Farzan M, Elledge SJ (2009) The IFITM proteins mediate cellular resistance to influenza A H1N1 virus, West Nile virus, and dengue virus. Cell 139(7):1243–1254

Bruce EA, Digard P, Stuart AD (2010) The Rab11 pathway is required for influenza A virus budding and filament formation. J Virol 84(12):5848–5859

Brunotte L, Flies J, Bolte H, Reuther P, Vreede F, Schwemmle M (2014) The nuclear export protein of H5N1 influenza A viruses recruits Matrix 1 (M1) protein to the viral ribonucleoprotein to mediate nuclear export. J Biol Chem 289(29):20067–20077

Bucci C, Parton RG, Mather IH, Stunnenberg H, Simons K, Hoflack B, Zerial M (1992) The small GTPase rab5 functions as a regulatory factor in the early endocytic pathway. Cell 70(5): 715–728

Bullock AN, Das S, Debreczeni JE, Rellos P, Fedorov O, Niesen FH, Guo K, Papagrigoriou E, Amos AL, Cho S, Turk BE, Ghosh G, Knapp S (2009) Kinase domain insertions define distinct roles of CLK kinases in SR protein phosphorylation. Structure 17(3):352–362

Cao M, Wei C, Zhao L, Wang J, Jia Q, Wang X, Jin Q, Deng T (2014) DnaJA1/Hsp40 is co-opted by influenza A virus to enhance its viral RNA polymerase activity. J Virol 88(24):14078–14089

Carette JE, Guimaraes CP, Varadarajan M, Park AS, Wuethrich I, Godarova A, Kotecki M, Cochran BH, Spooner E, Ploegh HL, Brummelkamp TR (2009) Haploid genetic screens in human cells identify host factors used by pathogens. Science 326(5957):1231–1235

Carette JE, Guimaraes CP, Wuethrich I, Blomen VA, Varadarajan M, Sun C, Bell G, Yuan B, Muellner MK, Nijman SM, Ploegh HL, Brummelkamp TR (2011) Global gene disruption in human cells to assign genes to phenotypes by deep sequencing. Nat Biotechnol 29(6):542–546

Carroll SM, Paulson JC (1985) Differential infection of receptor-modified host cells by receptor-specific influenza viruses. Virus Res 3(2):165–179

Cauldwell AV, Long JS, Moncorge O, Barclay WS (2014) Viral determinants of influenza A host range. J Gen Virol 95(6):1193–1210

Chase GP, Rameix-Welti MA, Zvirbliene A, Zvirblis G, Gotz V, Wolff T, Naffakh N, Schwemmle M (2011) Influenza virus ribonucleoprotein complexes gain preferential access to cellular export machinery through chromatin targeting. PLoS Pathog 7(9):e1002187

Chasman D, Walters KB, Lopes TJ, Eisfeld AJ, Kawaoka Y, Roy S (2016) Integrating transcriptomic and proteomic data using predictive regulatory network models of host response to pathogens. PLoS Comput Biol 12(7):e1005013

Chen C, Zhuang X (2008) Epsin 1 is a cargo-specific adaptor for the clathrin-mediated endocytosis of the influenza virus. Proc Natl Acad Sci USA 105(33):11790–11795

Chizhmakov IV, Geraghty FM, Ogden DC, Hayhurst A, Antoniou M, Hay AJ (1996) Selective proton permeability and pH regulation of the influenza virus M2 channel expressed in mouse erythroleukaemia cells. J Physiol 494(Pt 2):329–336

Christoforidis S, Miaczynska M, Ashman K, Wilm M, Zhao L, Yip SC, Waterfield MD, Backer JM, Zerial M (1999) Phosphatidylinositol-3-OH kinases are Rab5 effectors. Nat Cell Biol 1(4):249–252

Chutiwitoonchai N, Aida Y (2016) NXT1, a novel influenza A NP binding protein, promotes the nuclear export of NP via a CRM1-dependent pathway. Viruses 8(8) pii: E209

Cohen M, Zhang XQ, Senaati HP, Chen HW, Varki NM, Schooley RT, Gagneux P (2013) Influenza A penetrates host mucus by cleaving sialic acids with neuraminidase. Virol J 10:321

Cox J, Mann M (2011) Quantitative, high-resolution proteomics for data-driven systems biology. Annu Rev Biochem 80:273–299

Cros JF, Palese P (2003) Trafficking of viral genomic RNA into and out of the nucleus: influenza, Thogoto and Borna disease viruses. Virus Res 95(1–2):3–12

Dapat C, Saito R, Suzuki H, Horigome T (2014) Quantitative phosphoproteomic analysis of host responses in human lung epithelial (A549) cells during influenza virus infection. Virus Res 179:53–63

de Chassey B, Meyniel-Schicklin L, Aublin-Gex A, Andre P, Lotteau V (2012) Genetic screens for the control of influenza virus replication: from meta-analysis to drug discovery. Mol BioSyst 8 (4):1297–1303

de Chassey B, Aublin-Gex A, Ruggieri A, Meyniel-Schicklin L, Pradezynski F, Davoust N, Chantier T, Tafforeau L, Mangeot PE, Ciancia C, Perrin-Cocon L, Bartenschlager R, Andre P, Lotteau V (2013a) The interactomes of influenza virus NS1 and NS2 proteins identify new host factors and provide insights for ADAR1 playing a supportive role in virus replication. PLoS Pathog 9(7):e1003440 de Chassey B, Meyniel-Schicklin L, Aublin-Gex A, Navratil V, Chantier T, Andre P, Lotteau V (2013b) Structure homology and interaction redundancy for discovering virus-host protein interactions. EMBO Rep 14(10):938–944

de Graaf M, Fouchier RA (2014) Role of receptor binding specificity in influenza A virus transmission and pathogenesis. EMBO J 33(8):823–841

de Vries E, Tscherne DM, Wienholts MJ, Cobos-Jimenez V, Scholte F, Garcia-Sastre A, Rottier PJ, de Haan CA (2011) Dissection of the influenza A virus endocytic routes reveals macropinocytosis as an alternative entry pathway. PLoS Pathog 7(3):e1001329

Demirov D, Gabriel G, Schneider C, Hohenberg H, Ludwig S (2012) Interaction of influenza A virus matrix protein with RACK1 is required for virus release. Cell Microbiol 14(5):774–789

Diot C, Fournier G, Dos Santos M, Magnus J, Komarova A, van der Werf S, Munier S, Naffakh N (2016) Influenza A Virus polymerase recruits the RNA helicase DDX19 to promote the nuclear export of viral mRNAs. Sci Rep 6:33763

Domingues P, Golebiowski F, Tatham MH, Lopes AM, Taggart A, Hay RT, Hale BG (2015) Global reprogramming of host SUMOylation during Influenza Virus Infection. Cell Rep 13 (7):1467–1480

Droebner K, Pleschka S, Ludwig S, Planz O (2011) Antiviral activity of the MEK-inhibitor U0126 against pandemic H1N1v and highly pathogenic avian influenza virus in vitro and in vivo. Antiviral Res 92(2):195–203

Edinger TO, Pohl MO, Stertz S (2014) Entry of influenza A virus: host factors and antiviral targets. J Gen Virol 95(Pt 2):263–277

Edinger TO, Pohl MO, Yanguez E, Stertz S (2015) Cathepsin W Is required for escape of influenza A virus from late endosomes. MBio 6(3):e00297

Eierhoff T, Hrincius ER, Rescher U, Ludwig S, Ehrhardt C (2010) The epidermal growth factor receptor (EGFR) promotes uptake of influenza A viruses (IAV) into host cells. PLoS Pathog 6(9):e1001099

Eisfeld AJ, Kawakami E, Watanabe T, Neumann G, Kawaoka Y (2011a) RAB11A is essential for transport of the influenza virus genome to the plasma membrane. J Virol 85(13):6117–6126

Eisfeld AJ, Neumann G, Kawaoka Y (2011b) Human immunodeficiency virus rev-binding protein is essential for influenza a virus replication and promotes genome trafficking in late-stage infection. J Virol 85(18):9588–9598

Elbahesh H, Cline T, Baranovich T, Govorkova EA, Schultz-Cherry S, Russell CJ (2014) Novel roles of focal adhesion kinase in cytoplasmic entry and replication of influenza A viruses. J Virol 88(12):6714–6728

Elton D, Simpson-Holley M, Archer K, Medcalf L, Hallam R, McCauley J, Digard P (2001) Interaction of the influenza virus nucleoprotein with the cellular CRM1-mediated nuclear export pathway. J Virol 75(1):408–419

Engelhardt OG, Smith M, Fodor E (2005) Association of the influenza A virus RNA-dependent RNA polymerase with cellular RNA polymerase II. J Virol 79(9):5812–5818

Ferris MT, Aylor DL, Bottomly D, Whitmore AC, Aicher LD, Bell TA, Bradel-Tretheway B, Bryan JT, Buus RJ, Gralinski LE, Haagmans BL, McMillan L, Miller DR, Rosenzweig E, Valdar W, Wang J, Churchill GA, Threadgill DW, McWeeney SK, Katze MG, Pardo-Manuel de Villena F, Baric RS, Heise MT (2013) Modeling host genetic regulation of influenza pathogenesis in the collaborative cross. PLoS Pathog 9(2):e1003196

Fournier G, Chiang C, Munier S, Tomoiu A, Demeret C, Vidalain PO, Jacob Y, Naffakh N (2014) Recruitment of RED-SMU1 complex by influenza A virus RNA polymerase to control viral mRNA splicing. PLoS Pathog 10(6):e1004164

Fujioka Y, Tsuda M, Nanbo A, Hattori T, Sasaki J, Sasaki T, Miyazaki T, Ohba Y (2013) A Ca(2 +)-dependent signalling circuit regulates influenza A virus internalization and infection. Nat Commun 4:2763

Galloway SE, Reed ML, Russell CJ, Steinhauer DA (2013) Influenza HA subtypes demonstrate divergent phenotypes for cleavage activation and pH of fusion: implications for host range and adaptation. PLoS Pathog 9(2):e1003151

Gambaryan AS, Tuzikov AB, Piskarev VE, Yamnikova SS, Lvov DK, Robertson JS, Bovin NV, Matrosovich MN (1997) Specification of receptor-binding phenotypes of influenza virus isolates from different hosts using synthetic sialylglycopolymers: non-egg-adapted human H1 and H3 influenza A and influenza B viruses share a common high binding affinity for 6'-sialyl (N-acetyllactosamine). Virology 232(2):345–350

Gamblin SJ, Haire LF, Russell RJ, Stevens DJ, Xiao B, Ha Y, Vasisht N, Steinhauer DA, Daniels RS, Elliot A, Wiley DC, Skehel JJ (2004) The structure and receptor binding properties of the 1918 influenza hemagglutinin. Science 303(5665):1838–1842

Gao S, Wu J, Liu RY, Li J, Song L, Teng Y, Sheng C, Liu D, Yao C, Chen H, Jiang W, Chen S, Huang W (2015) Interaction of NS2 with AIMP2 facilitates the switch from ubiquitination to SUMOylation of M1 in influenza A virus-infected cells. J Virol 89(1):300–311

Garten W, Bosch FX, Linder D, Rott R, Klenk HD (1981) Proteolytic activation of the influenza virus hemagglutinin: the structure of the cleavage site and the enzymes involved in cleavage. Virology 115(2):361–374

Glaser L, Stevens J, Zamarin D, Wilson IA, Garcia-Sastre A, Tumpey TM, Basler CF, Taubenberger JK, Palese P (2005) A single amino acid substitution in 1918 influenza virus hemagglutinin changes receptor binding specificity. J Virol 79(17):11533–11536

Gorai T, Goto H, Noda T, Watanabe T, Kozuka-Hata H, Oyama M, Takano R, Neumann G, Watanabe S, Kawaoka Y (2012) F1Fo-ATPase, F-type proton-translocating ATPase, at the plasma membrane is critical for efficient influenza virus budding. Proc Natl Acad Sci USA 109 (12):4615–4620

Gschweitl M, Ulbricht A, Barnes CA, Enchev RI, Stoffel-Studer I, Meyer-Schaller N, Huotari J, Yamauchi Y, Greber UF, Helenius A, Peter M (2016) A SPOPL/Cullin-3 ubiquitin ligase complex regulates endocytic trafficking by targeting EPS15 at endosomes. eLife 5:e13841

Gu W, Gallagher GR, Dai W, Liu P, Li R, Trombly MI, Gammon DB, Mello CC, Wang JP, Finberg RW (2015) Influenza A virus preferentially snatches noncoding RNA caps. RNA 21 (12):2067–2075

Guinea R, Carrasco L (1995) Requirement for vacuolar proton-ATPase activity during entry of influenza virus into cells. J Virol 69(4):2306–2312

Haasbach E, Hartmayer C, Planz O (2013) Combination of MEK inhibitors and oseltamivir leads to synergistic antiviral effects after influenza A virus infection in vitro. Antiviral Res 98 (2):319–324

Hale BG (2014) Conformational plasticity of the influenza A virus NS1 protein. J Gen Virol 95(Pt 10):2099–2105

Hale BG, Jackson D, Chen YH, Lamb RA, Randall RE (2006) Influenza A virus NS1 protein binds p85beta and activates phosphatidylinositol-3-kinase signaling. Proc Natl Acad Sci USA 103(38):14194–14199

He J, Sun E, Bujny MV, Kim D, Davidson MW, Zhuang X (2013) Dual function of CD81 in influenza virus uncoating and budding. PLoS Pathog 9(10):e1003701

Heaton NS, Moshkina N, Fenouil R, Gardner TJ, Aguirre S, Shah PS, Zhao N, Manganaro L, Hultquist JF, Noel J, Sachs D, Hamilton J, Leon PE, Chawdury A, Tripathi S, Melegari C, Campisi L, Hai R, Metreveli G, Gamarnik AV, Garcia-Sastre A, Greenbaum B, Simon V, Fernandez-Sesma A, Krogan NJ, Mulder LC, van Bakel H, Tortorella D, Taunton J, Palese P, Marazzi I (2016) Targeting viral proteostasis limits influenza virus, HIV, and dengue virus infection. Immunity 44(1):46–58

Hebert DN, Zhang JX, Chen W, Foellmer B, Helenius A (1997) The number and location of glycans on influenza hemagglutinin determine folding and association with calnexin and calreticulin. J Cell Biol 139(3):613–623

Holsinger LJ, Lamb RA (1991) Influenza virus M2 integral membrane protein is a homotetramer stabilized by formation of disulfide bonds. Virology 183(1):32–43

Hsu WB, Shih JL, Shih JR, Du JL, Teng SC, Huang LM, Wang WB (2013) Cellular protein HAX1 interacts with the influenza A virus PA polymerase subunit and impedes its nuclear translocation. J Virol 87(1):110–123

Hu Y, Liu X, Zhang A, Zhou H, Liu Z, Chen H, Jin M (2015) CHD3 facilitates vRNP nuclear export by interacting with NES1 of influenza A virus NS2. Cell Mol Life Sci CMLS 72 (5):971–982

Huang RT, Rott R, Klenk HD (1981) Influenza viruses cause hemolysis and fusion of cells. Virology 110(1):243–247

Huang S, Chen J, Chen Q, Wang H, Yao Y, Chen J, Chen Z (2013) A second CRM1-dependent nuclear export signal in the influenza A virus NS2 protein contributes to the nuclear export of viral ribonucleoproteins. J Virol 87(2):767–778

Huarte M, Sanz-Ezquerro JJ, Roncal F, Ortin J, Nieto A (2001) PA subunit from influenza virus polymerase complex interacts with a cellular protein with homology to a family of transcriptional activators. J Virol 75(18):8597–8604

Huotari J, Helenius A (2011) Endosome maturation. EMBO J 30(17):3481–3500

Huotari J, Meyer-Schaller N, Hubner M, Stauffer S, Katheder N, Horvath P, Mancini R, Helenius A, Peter M (2012) Cullin-3 regulates late endosome maturation. Proc Natl Acad Sci USA 109(3):823–828

Hutchinson EC, Fodor E (2013) Transport of the influenza virus genome from nucleus to nucleus. Viruses 5(10):2424–2446

Hutchinson EC, Charles PD, Hester SS, Thomas B, Trudgian D, Martinez-Alonso M, Fodor E (2014) Conserved and host-specific features of influenza virion architecture. Nat Commun 5:4816

Jorba N, Juarez S, Torreira E, Gastaminza P, Zamarreno N, Albar JP, Ortin J (2008) Analysis of the interaction of influenza virus polymerase complex with human cell factors. Proteomics 8 (10):2077–2088

Karlas A, Machuy N, Shin Y, Pleissner KP, Artarini A, Heuer D, Becker D, Khalil H, Ogilvie LA, Hess S, Maurer AP, Muller E, Wolff T, Rudel T, Meyer TF (2010) Genome-wide RNAi screen identifies human host factors crucial for influenza virus replication. Nature 463(7282):818–822

Kawaguchi A, Matsumoto K, Nagata K (2012) YB-1 functions as a porter to lead influenza virus ribonucleoprotein complexes to microtubules. J Virol 86(20):11086–11095

Klenk HD, Rott R, Orlich M, Blodorn J (1975) Activation of influenza A viruses by trypsin treatment. Virology 68(2):426–439

Klenk HD, Garten W, Rott R (1984) Inhibition of proteolytic cleavage of the hemagglutinin of influenza virus by the calcium-specific ionophore A23187. EMBO J 3(12):2911–2915

Konig R, Stertz S, Zhou Y, Inoue A, Hoffmann HH, Bhattacharyya S, Alamares JG, Tscherne DM, Ortigoza MB, Liang Y, Gao Q, Andrews SE, Bandyopadhyay S, De Jesus P, Tu BP, Pache L, Shih C, Orth A, Bonamy G, Miraglia L, Ideker T, Garcia-Sastre A, Young JA, Palese P, Shaw ML, Chanda SK (2010) Human host factors required for influenza virus replication. Nature 463(7282):813–817

Koppstein D, Ashour J, Bartel DP (2015) Sequencing the cap-snatching repertoire of H1N1 influenza provides insight into the mechanism of viral transcription initiation. Nucleic Acids Res 43(10):5052–5064

Kuo RL, Li ZH, Li LH, Lee KM, Tam EH, Liu HM, Liu HP, Shih SR, Wu CC (2016) Interactome analysis of the NS1 protein encoded by influenza A H1N1 virus reveals a positive regulatory role of host protein PRP19 in viral replication. J Proteome Res 15(5):1639–1648

Lakadamyali M, Rust MJ, Babcock HP, Zhuang X (2003) Visualizing infection of individual influenza viruses. Proc Natl Acad Sci USA 100(16):9280–9285

Landeras-Bueno S, Jorba N, Perez-Cidoncha M, Ortin J (2011) The splicing factor proline-glutamine rich (SFPQ/PSF) is involved in influenza virus transcription. PLoS Pathog 7(11):e1002397

Li S, Min JY, Krug RM, Sen GC (2006) Binding of the influenza A virus NS1 protein to PKR mediates the inhibition of its activation by either PACT or double-stranded RNA. Virology 349(1):13–21

Li X, Foley EA, Molloy KR, Li Y, Chait BT, Kapoor TM (2012) Quantitative chemical proteomics approach to identify post-translational modification-mediated protein-protein interactions. J Am Chem Soc 134(4):1982–1985

Li S, Yen L, Pastor WA, Johnston JB, Du J, Shew CJ, Liu W, Ho J, Stender B, Clark AT, Burlingame AL, Daxinger L, Patel DJ, Jacobsen SE (2016) Mouse MORC3 is a GHKL ATPase that localizes to H3K4me3 marked chromatin. Proc Natl Acad Sci USA 113(35): E5108–5116

Liu J, Stevens DJ, Haire LF, Walker PA, Coombs PJ, Russell RJ, Gamblin SJ, Skehel JJ (2009) Structures of receptor complexes formed by hemagglutinins from the Asian Influenza pandemic of 1957. Proc Natl Acad Sci USA 106(40):17175–17180

Liu G, Xiang Y, Guo C, Pei Y, Wang Y, Kitazato K (2014) Cofilin-1 is involved in regulation of actin reorganization during influenza A virus assembly and budding. Biochem Biophys Res Commun 453(4):821–825

Long JS, Giotis ES, Moncorge O, Frise R, Mistry B, James J, Morisson M, Iqbal M, Vignal A, Skinner MA, Barclay WS (2016) Species difference in ANP32A underlies influenza A virus polymerase host restriction. Nature 529(7584):101–104

Luzio JP, Bright NA, Pryor PR (2007) The role of calcium and other ions in sorting and delivery in the late endocytic pathway. Biochem Soc Trans 35(Pt 5):1088–1091

Ma H, Kien F, Maniere M, Zhang Y, Lagarde N, Tse KS, Poon LL, Nal B (2012) Human annexin A6 interacts with influenza a virus protein M2 and negatively modulates infection. J Virol 86 (3):1789–1801

Marcos-Villar L, Pazo A, Nieto A (2016) Influenza virus and chromatin: role of the CHD1 chromatin remodeler in the virus life cycle. J Virol 90(7):3694–3707

Martin K, Helenius A (1991) Transport of incoming influenza virus nucleocapsids into the nucleus. J Virol 65:232–244

Martinez-Alonso M, Hengrung N, Fodor E (2016) RNA-free and ribonucleoprotein-associated influenza virus polymerases directly bind the serine-5-phosphorylated carboxyl-terminal domain of host RNA polymerase II. J Virol 90(13):6014–6021

Matlin KS, Reggio H, Helenius A, Simons K (1981) Infectious entry pathway of influenza virus in a canine kidney cell line. J Cell Biol 91(3 Pt 1):601–613

Matrosovich M, Tuzikov A, Bovin N, Gambaryan A, Klimov A, Castrucci MR, Donatelli I, Kawaoka Y (2000) Early alterations of the receptor-binding properties of H1, H2, and H3 avian influenza virus hemagglutinins after their introduction into mammals. J Virol 74 (18):8502–8512

Mayer D, Molawi K, Martinez-Sobrido L, Ghanem A, Thomas S, Baginsky S, Grossmann J, Garcia-Sastre A, Schwemmle M (2007) Identification of cellular interaction partners of the influenza virus ribonucleoprotein complex and polymerase complex using proteomic-based approaches. J Proteome Res 6(2):672–682

Mehle A, Doudna JA (2009) Adaptive strategies of the influenza virus polymerase for replication in humans. Proc Natl Acad Sci USA 106(50):21312–21316

Melen K, Fagerlund R, Franke J, Kohler M, Kinnunen L, Julkunen I (2003) Importin alpha nuclear localization signal binding sites for STAT1, STAT2, and influenza A virus nucleoprotein. J Biol Chem 278(30):28193–28200

Mibayashi M, Martinez-Sobrido L, Loo YM, Cardenas WB, Gale M Jr, Garcia-Sastre A (2007) Inhibition of retinoic acid-inducible gene I-mediated induction of beta interferon by the NS1 protein of influenza A virus. J Virol 81(2):514–524

Momose F, Basler CF, O'Neill RE, Iwamatsu A, Palese P, Nagata K (2001) Cellular splicing factor RAF-2p48/NPI-5/BAT1/UAP56 interacts with the influenza virus nucleoprotein and enhances viral RNA synthesis. J Virol 75(4):1899–1908

Momose F, Kikuchi Y, Komase K, Morikawa Y (2007) Visualization of microtubule-mediated transport of influenza viral progeny ribonucleoprotein. Microbes Infect 9(12–13):1422–1433

Momose F, Sekimoto T, Ohkura T, Jo S, Kawaguchi A, Nagata K, Morikawa Y (2011) Apical transport of influenza A virus ribonucleoprotein requires Rab11-positive recycling endosome. PLoS One 6(6):e21123

Moncorge O, Mura M, Barclay WS (2010) Evidence for avian and human host cell factors that affect the activity of influenza virus polymerase. J Virol 84(19):9978–9986

Mor A, White A, Zhang K, Thompson M, Esparza M, Munoz-Moreno R, Koide K, Lynch KW, Garcia-Sastre A, Fontoura BM (2016) Influenza virus mRNA trafficking through host nuclear speckles. Nat Microbiol 1(7):16069

Mu FT, Callaghan JM, Steele-Mortimer O, Stenmark H, Parton RG, Campbell PL, McCluskey J, Yeo JP, Tock EP, Toh BH (1995) EEA1, an early endosome-associated protein. EEA1 is a conserved alpha-helical peripheral membrane protein flanked by cysteine "fingers" and contains a calmodulin-binding IQ motif. J Biol Chem 270(22):13503–13511

Muhlbauer D, Dzieciolowski J, Hardt M, Hocke A, Schierhorn KL, Mostafa A, Muller C, Wisskirchen C, Herold S, Wolff T, Ziebuhr J, Pleschka S (2015) Influenza virus-induced caspase-dependent enlargement of nuclear pores promotes nuclear export of viral ribonucleoprotein complexes. J Virol 89(11):6009–6021

Munier S, Rolland T, Diot C, Jacob Y, Naffakh N (2013) Exploration of binary virus-host interactions using an infectious protein complementation assay. Mol Cell Proteomics MCP 12 (10):2845–2855

Nayak DP, Balogun RA, Yamada H, Zhou ZH, Barman S (2009) Influenza virus morphogenesis and budding. Virus Res 143(2):147–161

Nemeroff ME, Barabino SM, Li Y, Keller W, Krug RM (1998) Influenza virus NS1 protein interacts with the cellular 30 kDa subunit of CPSF and inhibits 3'end formation of cellular pre-mRNAs. Mol Cell 1(7):991–1000

Neumann G, Hughes MT, Kawaoka Y (2000) Influenza A virus NS2 protein mediates vRNP nuclear export through NES-independent interaction with hCRM1. EMBO J 19(24):6751–6758

Olsen JV, Mann M (2013) Status of large-scale analysis of post-translational modifications by mass spectrometry. Mol Cell Proteomics MCP 12(12):3444–3452

O'Neill R, Jaskunas R, Blobel G, Palese P, Moroianu J (1995) Nuclear import of influenza virus RNA can be mediated by viral nucleoprotein and transport factors requiered for protein import. J Biol Chem 270:22701–22704

O'Neill RE, Talon J, Palese P (1998) The influenza virus NEP (NS2 protein) mediates the nuclear export of viral ribonucleoproteins. EMBO J 17(1):288–296

Perreira JM, Aker AM, Savidis G, Chin CR, McDougall WM, Portmann JM, Meraner P, Smith MC, Rahman M, Baker RE, Gauthier A, Franti M, Brass AL (2015) RNASEK Is a V-ATPase-associated factor required for endocytosis and the replication of rhinovirus, influenza A virus, and dengue virus. Cell Rep 12(5):850–863

Perreira JM, Meraner P, Brass AL (2016) Functional genomic strategies for elucidating human-virus interactions: will CRISPR Knockout RNAi and haploid cells? Adv Virus Res 94:1–51

Perwitasari O, Johnson S, Yan X, Howerth E, Shacham S, Landesman Y, Baloglu E, McCauley D, Tamir S, Tompkins SM, Tripp RA (2014) Verdinexor, a novel selective inhibitor of nuclear export, reduces influenza a virus replication in vitro and in vivo. J Virol 88(17):10228–10243

Pinto LH, Holsinger LJ, Lamb RA (1992) Influenza virus M2 protein has ion channel activity. Cell 69:517–528

Pleschka S, Wolff T, Ehrhardt C, Hobom G, Planz O, Rapp UR, Ludwig S (2001) Influenza virus propagation is impaired by inhibition of the Raf/MEK/ERK signalling cascade. Nat Cell Biol 3 (3):301–305

Pohl MO, Edinger TO, Stertz S (2014) Prolidase is required for early trafficking events during influenza A virus entry. J Virol 88(19):11271–11283

Pohl MO, Lanz C, Stertz S (2016) Late stages of the influenza A virus replication cycle-a tight interplay between virus and host. J Gen Virol 97(9):2058–2072

Predicala R, Zhou Y (2013) The role of Ran-binding protein 3 during influenza A virus replication. J Gen Virol 94(Pt 5):977–984

Puig O, Caspary F, Rigaut G, Rutz B, Bouveret E, Bragado-Nilsson E, Wilm M, Seraphin B (2001) The tandem affinity purification (TAP) method: a general procedure of protein complex purification. Methods 24(3):218–229

Rigaut G, Shevchenko A, Rutz B, Wilm M, Mann M, Seraphin B (1999) A generic protein purification method for protein complex characterization and proteome exploration. Nat Biotechnol 17(10):1030–1032

Rink J, Ghigo E, Kalaidzidis Y, Zerial M (2005) Rab conversion as a mechanism of progression from early to late endosomes. Cell 122(5):735–749

Rodriguez A, Perez-Gonzalez A, Nieto A (2007) Influenza virus infection causes specific degradation of the largest subunit of cellular RNA polymerase II. J Virol 81(10):5315–5324

Rodriguez A, Perez-Gonzalez A, Nieto A (2011) Cellular human CLE/C14orf166 protein interacts with influenza virus polymerase and is required for viral replication. J Virol 85(22):12062–12066

Rodriguez-Frandsen A, de Lucas S, Perez-Gonzalez A, Perez-Cidoncha M, Roldan-Gomendio A, Pazo A, Marcos-Villar L, Landeras-Bueno S, Ortin J, Nieto A (2016) hCLE/C14orf166, a cellular protein required for viral replication, is incorporated into influenza virus particles. Sci Rep 6:20744

Rogers GN, Paulson JC (1983) Receptor determinants of human and animal influenza virus isolates: differences in receptor specificity of the H3 hemagglutinin based on species of origin. Virology 127(2):361–373

Rossman JS, Leser GP, Lamb RA (2012) Filamentous influenza virus enters cells via macropinocytosis. J Virol 86(20):10950–10960

Roy AM, Parker JS, Parrish CR, Whittaker GR (2000) Early stages of influenza virus entry into Mv-1 lung cells: involvement of dynamin. Virology 267(1):17–28

Russell CJ (2014) Acid-induced membrane fusion by the hemagglutinin protein and its role in influenza virus biology. Curr Top Microbiol Immunol 385:93–116

Rust MJ, Lakadamyali M, Zhang F, Zhuang X (2004) Assembly of endocytic machinery around individual influenza viruses during viral entry. Nat Struct Mol Biol 11(6):567–573

Sakai K, Ami Y, Tahara M, Kubota T, Anraku M, Abe M, Nakajima N, Sekizuka T, Shirato K, Suzaki Y, Ainai A, Nakatsu Y, Kanou K, Nakamura K, Suzuki T, Komase K, Nobusawa E, Maenaka K, Kuroda M, Hasegawa H, Kawaoka Y, Tashiro M, Takeda M (2014) The host protease TMPRSS2 plays a major role in in vivo replication of emerging H7N9 and seasonal influenza viruses. J Virol 88(10):5608–5616

Sauter NK, Bednarski MD, Wurzburg BA, Hanson JE, Whitesides GM, Skehel JJ, Wiley DC (1989) Hemagglutinins from two influenza virus variants bind to sialic acid derivatives with millimolar dissociation constants: a 500-MHz proton nuclear magnetic resonance study. Biochemistry 28(21):8388–8396

Scheiffele P, Roth MG, Simons K (1997) Interaction of influenza virus haemagglutinin with sphingolipid-cholesterol membrane domains via its transmembrane domain. EMBO J 16 (18):5501–5508

Schrauwen EJ, Fouchier RA (2014) Host adaptation and transmission of influenza A viruses in mammals. Emerg Microbes Infect 3(2):e9

Seo YJ, Pritzl CJ, Vijayan M, Bomb K, McClain ME, Alexander S, Hahm B (2013) Sphingosine kinase 1 serves as a pro-viral factor by regulating viral RNA synthesis and nuclear export of viral ribonucleoprotein complex upon influenza virus infection. PLoS One 8(8):e75005

Shalem O, Sanjana NE, Hartenian E, Shi X, Scott DA, Mikkelsen TS, Heckl D, Ebert BL, Root DE, Doench JG, Zhang F (2014) Genome-scale CRISPR-Cas9 knockout screening in human cells. Science 343(6166):84–87

Shapira SD, Gat-Viks I, Shum BO, Dricot A, de Grace MM, Wu L, Gupta PB, Hao T, Silver SJ, Root DE, Hill DE, Regev A, Hacohen N (2009) A physical and regulatory map of host-influenza interactions reveals pathways in H1N1 infection. Cell 139(7):1255–1267

Shaw ML (2011) The host interactome of influenza virus presents new potential targets for antiviral drugs. Rev Med Virol 21(6):358–369

Shaw ML, Palese P (2013) Orthomyxoviridae. In: Knipe DM, Howley PM (eds) Fields virology, vol 1, 6th edn. Lippincott Williams & Wilkins, Philadelphia, pp 1151–1185

Shaw ML, Stone KL, Colangelo CM, Gulcicek EE, Palese P (2008) Cellular proteins in influenza virus particles. PLoS Pathog 4(6):e1000085

Shih SR, Krug RM (1996) Novel exploitation of a nuclear function by influenza virus: the cellular SF2/ASF splicing factor controls the amount of the essential viral M2 ion channel protein in infected cells. EMBO J 15(19):5415–5427

Shimizu T, Takizawa N, Watanabe K, Nagata K, Kobayashi N (2011) Crucial role of the influenza virus NS2 (NEP) C-terminal domain in M1 binding and nuclear export of vRNP. FEBS Lett 585(1):41–46

Sieczkarski SB, Whittaker GR (2002) Influenza virus can enter and infect cells in the absence of clathrin-mediated endocytosis. J Virol 76(20):10455–10464

Sieczkarski SB, Whittaker GR (2003) Differential requirements of Rab5 and Rab7 for endocytosis of influenza and other enveloped viruses. Traffic 4(5):333–343

Sieczkarski SB, Brown HA, Whittaker GR (2003) Role of protein kinase C betaII in influenza virus entry via late endosomes. J Virol 77(1):460–469

Simonsen A, Lippe R, Christoforidis S, Gaullier JM, Brech A, Callaghan J, Toh BH, Murphy C, Zerial M, Stenmark H (1998) EEA1 links PI(3)K function to Rab5 regulation of endosome fusion. Nature 394(6692):494–498

Simpson-Holley M, Ellis D, Fisher D, Elton D, McCauley J, Digard P (2002) A functional link between the actin cytoskeleton and lipid rafts during budding of filamentous influenza virions. Virology 301(2):212–225

Soderholm S, Kainov DE, Ohman T, Denisova OV, Schepens B, Kulesskiy E, Imanishi SY, Corthals G, Hintsanen P, Aittokallio T, Saelens X, Matikainen S, Nyman TA (2016) Phosphoproteomics to characterize host response during influenza A virus infection of human macrophages. Mol Cell Proteomics MCP 15(10):3203–3219

Stauffer S, Feng Y, Nebioglu F, Heilig R, Picotti P, Helenius A (2014) Stepwise priming by acidic pH and a high K^+ concentration is required for efficient uncoating of influenza A virus cores after penetration. J Virol 88(22):13029–13046

Steel J, Lowen AC, Mubareka S, Palese P (2009) Transmission of influenza virus in a mammalian host is increased by PB2 amino acids 627 K or 627E/701N. PLoS Pathog 5(1):e1000252

Steinhauer DA (1999) Role of hemagglutinin cleavage for the pathogenicity of influenza virus. Virology 258(1):1–20

Stertz S, Shaw ML (2011) Uncovering the global host cell requirements for influenza virus replication via RNAi screening. Microbes Infect 13(5):516–525

Stevens J, Blixt O, Glaser L, Taubenberger JK, Palese P, Paulson JC, Wilson IA (2006a) Glycan microarray analysis of the hemagglutinins from modern and pandemic influenza viruses reveals different receptor specificities. J Mol Biol 355(5):1143–1155

Stevens J, Blixt O, Tumpey TM, Taubenberger JK, Paulson JC, Wilson IA (2006b) Structure and receptor specificity of the hemagglutinin from an H5N1 influenza virus. Science 312 (5772):404–410

Su WC, Chen YC, Tseng CH, Hsu PW, Tung KF, Jeng KS, Lai MM (2013) Pooled RNAi screen identifies ubiquitin ligase Itch as crucial for influenza A virus release from the endosome during virus entry. Proc Natl Acad Sci USA 110(43):17516–17521

Su WC, Hsu SF, Lee YY, Jeng KS, Lai MM (2015) A nucleolar protein, ribosomal RNA processing 1 Homolog B (RRP1B), enhances the recruitment of cellular mRNA in influenza virus transcription. J Virol 89(22):11245–11255

Subbarao EK, London W, Murphy BR (1993) A single amino acid in the PB2-Gene of Influenza-A virus is a determinant of host range. J Virol 67:1761–1764

Sugiyama K, Kawaguchi A, Okuwaki M, Nagata K (2015) pp 32 and APRIL are host cell-derived regulators of influenza virus RNA synthesis from cRNA. eLife 4

Sugrue RJ, Hay AJ (1991) Structural characteristics of the M2 protein of influenza A viruses: evidence that it forms a tetrameric channel. Virology 180:617–624

Sun E, He J, Zhuang X (2013) Dissecting the role of COPI complexes in influenza virus infection. J Virol 87(5):2673–2685

Tafesse FG, Sanyal S, Ashour J, Guimaraes CP, Hermansson M, Somerharju P, Ploegh HL (2013) Intact sphingomyelin biosynthetic pathway is essential for intracellular transport of influenza virus glycoproteins. Proc Natl Acad Sci USA 110(16):6406–6411

Tafforeau L, Chantier T, Pradezynski F, Pellet J, Mangeot PE, Vidalain PO, Andre P, Rabourdin-Combe C, Lotteau V (2011) Generation and comprehensive analysis of an influenza virus polymerase cellular interaction network. J Virol 85(24):13010–13018

Tarnow C, Engels G, Arendt A, Schwalm F, Sediri H, Preuss A, Nelson PS, Garten W, Klenk HD, Gabriel G, Bottcher-Friebertshauser E (2014) TMPRSS2 is a host factor that is essential for pneumotropism and pathogenicity of H7N9 influenza A virus in mice. J Virol 88(9):4744–4751

Tawaratsumida K, Phan V, Hrincius ER, High AA, Webby R, Redecke V, Hacker H (2014) Quantitative proteomic analysis of the influenza A virus nonstructural proteins NS1 and NS2 during natural cell infection identifies PACT as an NS1 target protein and antiviral host factor. J Virol 88(16):9038–9048

Terrier O, Carron C, De Chassey B, Dubois J, Traversier A, Julien T, Cartet G, Proust A, Hacot S, Ressnikoff D, Lotteau V, Lina B, Diaz JJ, Moules V, Rosa-Calatrava M (2016) Nucleolin interacts with influenza A nucleoprotein and contributes to viral ribonucleoprotein complexes nuclear trafficking and efficient influenza viral replication. Sci Rep 6:29006

Thierry E, Guilligay D, Kosinski J, Bock T, Gaudon S, Round A, Pflug A, Hengrung N, El Omari K, Baudin F, Hart DJ, Beck M, Cusack S (2016) Influenza polymerase can adopt an alternative configuration involving a radical repacking of PB2 domains. Mol Cell 61(1): 125–137

Tijsterman M, Plasterk RH (2004) Dicers at RISC; the mechanism of RNAi. Cell 117(1):1–3

Tong S, Li Y, Rivailler P, Conrardy C, Castillo DA, Chen LM, Recuenco S, Ellison JA, Davis CT, York IA, Turmelle AS, Moran D, Rogers S, Shi M, Tao Y, Weil MR, Tang K, Rowe LA, Sammons S, Xu X, Frace M, Lindblade KA, Cox NJ, Anderson LJ, Rupprecht CE, Donis RO (2012) A distinct lineage of influenza A virus from bats. Proc Natl Acad Sci USA 109 (11):4269–4274

Tong S, Zhu X, Li Y, Shi M, Zhang J, Bourgeois M, Yang H, Chen X, Recuenco S, Gomez J, Chen LM, Johnson A, Tao Y, Dreyfus C, Yu W, McBride R, Carney PJ, Gilbert AT, Chang J, Guo Z, Davis CT, Paulson JC, Stevens J, Rupprecht CE, Holmes EC, Wilson IA, Donis RO (2013) New world bats harbor diverse influenza a viruses. PLoS Pathog 9(10):e1003657

Tran AT, Rahim MN, Ranadheera C, Kroeker A, Cortens JP, Opanubi KJ, Wilkins JA, Coombs KM (2013) Knockdown of specific host factors protects against influenza virus-induced cell death. Cell Death Dis 4:e769

Tripathi S, Pohl MO, Zhou Y, Rodriguez-Frandsen A, Wang G, Stein DA, Moulton HM, DeJesus P, Che J, Mulder LC, Yanguez E, Andenmatten D, Pache L, Manicassamy B, Albrecht RA, Gonzalez MG, Nguyen Q, Brass A, Elledge S, White M, Shapira S, Hacohen N, Karlas A, Meyer TF, Shales M, Gatorano A, Johnson JR, Jang G, Johnson T, Verschueren E, Sanders D, Krogan N, Shaw M, Konig R, Stertz S, Garcia-Sastre A, Chanda SK (2015) Meta- and orthogonal integration of influenza "OMICs" data defines a role for UBR4 in virus budding. Cell Host Microbe 18(6):723–735

Tsai PL, Chiou NT, Kuss S, Garcia-Sastre A, Lynch KW, Fontoura BM (2013) Cellular RNA binding proteins NS1-BP and hnRNP K regulate influenza A virus RNA splicing. PLoS Pathog 9(6):e1003460

Van Hoeven N, Pappas C, Belser JA, Maines TR, Zeng H, Garcia-Sastre A, Sasisekharan R, Katz JM, Tumpey TM (2009) Human HA and polymerase subunit PB2 proteins confer transmission of an avian influenza virus through the air. Proc Natl Acad Sci USA 106(9):3366–3371

van Riel D, den Bakker MA, Leijten LM, Chutinimitkul S, Munster VJ, de Wit E, Rimmelzwaan GF, Fouchier RA, Osterhaus AD, Kuiken T (2010) Seasonal and pandemic human influenza viruses attach better to human upper respiratory tract epithelium than avian influenza viruses. Am J Pathol 176(4):1614–1618

Varble A, Benitez AA, Schmid S, Sachs D, Shim JV, Rodriguez-Barrueco R, Panis M, Crumiller M, Silva JM, Sachidanandam R, tenOever BR (2013) An in vivo RNAi screening approach to identify host determinants of virus replication. Cell Host Microbe 14(3):346–356

Ver LS, Marcos-Villar L, Landeras-Bueno S, Nieto A, Ortin J (2015) The cellular factor NXP2/MORC3 is a positive regulator of influenza virus multiplication. J Virol 89(19):10023–10030
Vreede FT, Chan AY, Sharps J, Fodor E (2010) Mechanisms and functional implications of the degradation of host RNA polymerase II in influenza virus infected cells. Virology 396(1): 125–134
Wang P, Palese P, O'Neill RE (1997) The NPI-1/NPI-3 (Karyopherin alpha) binding site on the influenza A virus nucleoprotein NP is a nonconventional nuclear localization signal. J Virol 71:1850–1856
Wang S, Li H, Chen Y, Wei H, Gao GF, Liu H, Huang S, Chen JL (2012) Transport of influenza virus neuraminidase (NA) to host cell surface is regulated by ARHGAP21 and Cdc42 proteins. J Biol Chem 287(13):9804–9816
Wang T, Wei JJ, Sabatini DM, Lander ES (2014) Genetic screens in human cells using the CRISPR-Cas9 system. Science 343(6166):80–84
Ward SE, Kim HS, Komurov K, Mendiratta S, Tsai PL, Schmolke M, Satterly N, Manicassamy B, Forst CV, Roth MG, Garcia-Sastre A, Blazewska KM, McKenna CE, Fontoura BM, White MA (2012) Host modulators of H1N1 cytopathogenicity. PLoS One 7(8):e39284
Watanabe T, Kawaoka Y (2015) Influenza virus-host interactomes as a basis for antiviral drug development. Curr Opin Virol 14:71–78
Watanabe K, Fuse T, Asano I, Tsukahara F, Maru Y, Nagata K, Kitazato K, Kobayashi N (2006) Identification of Hsc70 as an influenza virus matrix protein (M1) binding factor involved in the virus life cycle. FEBS Lett 580(24):5785–5790
Watanabe K, Takizawa N, Noda S, Tsukahara F, Maru Y, Kobayashi N (2008) Hsc70 regulates the nuclear export but not the import of influenza viral RNP: a possible target for the development of anti-influenza virus drugs. Drug Discoveries Ther 2(2):77–84
Watanabe K, Shimizu T, Noda S, Tsukahara F, Maru Y, Kobayashi N (2014a) Nuclear export of the influenza virus ribonucleoprotein complex: Interaction of Hsc70 with viral proteins M1 and NS2. FEBS open bio 4:683–688
Watanabe T, Kawakami E, Shoemaker JE, Lopes TJ, Matsuoka Y, Tomita Y, Kozuka-Hata H, Gorai T, Kuwahara T, Takeda E, Nagata A, Takano R, Kiso M, Yamashita M, Sakai-Tagawa Y, Katsura H, Nonaka N, Fujii H, Fujii K, Sugita Y, Noda T, Goto H, Fukuyama S, Watanabe S, Neumann G, Oyama M, Kitano H, Kawaoka Y (2014b) Influenza virus-host interactome screen as a platform for antiviral drug development. Cell Host Microbe 16(6): 795–805
Webster RG, Yakhno M, Hinshaw VS, Bean WJ, Murti KG (1978) Intestinal influenza: replication and characterization of influenza viruses in ducks. Virology 84:268–278
Weis W, Brown JH, Cusack S, Paulson JC, Skehel JJ, Wiley DC (1988) Structure of the influenza virus haemagglutinin complexed with its receptor, sialic acid. Nature 333(6172):426–431
Wharton SA, Belshe RB, Skehel JJ, Hay AJ (1994) Role of virion M2 protein in influenza virus uncoating: specific reduction in the rate of membrane fusion between virus and liposomes by amantadine. J Gen Virol 75(Pt 4):945–948
Wright PF, Neumann G, Kawaoka Y (2013) Orthomyxoviruses. In: Knipe DM, Howley PM (eds) Fields virology, vol 1. Lippincott Williams & Wilkins, Philadelphia, pp 1186–1243
Wu CY, Jeng KS, Lai MM (2011) The SUMOylation of matrix protein M1 modulates the assembly and morphogenesis of influenza A virus. J Virol 85(13):6618–6628
Wurzer WJ, Planz O, Ehrhardt C, Giner M, Silberzahn T, Pleschka S, Ludwig S (2003) Caspase 3 activation is essential for efficient influenza virus propagation. EMBO J 22(11):2717–2728
Yamada S, Suzuki Y, Suzuki T, Le MQ, Nidom CA, Sakai-Tagawa Y, Muramoto Y, Ito M, Kiso M, Horimoto T, Shinya K, Sawada T, Usui T, Murata T, Lin Y, Hay A, Haire LF, Stevens DJ, Russell RJ, Gamblin SJ, Skehel JJ, Kawaoka Y (2006) Haemagglutinin mutations responsible for the binding of H5N1 influenza A viruses to human-type receptors. Nature 444 (7117):378–382

Yamauchi Y, Boukari H, Banerjee I, Sbalzarini IF, Horvath P, Helenius A (2011) Histone deacetylase 8 is required for centrosome cohesion and influenza A virus entry. PLoS Pathog 7 (10):e1002316

Yang X, Steukers L, Forier K, Xiong R, Braeckmans K, Van Reeth K, Nauwynck H (2014) A beneficiary role for neuraminidase in influenza virus penetration through the respiratory mucus. PLoS One 9(10):e110026

Yasuda J, Nakada S, Kato A, Toyoda T, Ishihama A (1993) Molecular assembly of influenza virus: association of the NS2 protein with virion matrix. Virology 196(1):249–255

Ye Z, Liu T, Offringa DP, McInnis J, Levandowski RA (1999) Association of influenza virus matrix protein with ribonucleoproteins. J Virol 73(9):7467–7473

York A, Hutchinson EC, Fodor E (2014) Interactome analysis of the influenza A virus transcription/replication machinery identifies protein phosphatase 6 as a cellular factor required for efficient virus replication. J Virol 88(22):13284–13299

Zanin M, Marathe B, Wong SS, Yoon SW, Collin E, Oshansky C, Jones J, Hause B, Webby R (2015) Pandemic swine H1N1 influenza viruses with almost undetectable neuraminidase activity are not transmitted via aerosols in ferrets and are inhibited by human mucus but not swine mucus. J Virol 89(11):5935–5948

Zebedee SL, Lamb RA (1988) Influenza A virus M2 protein: monoclonal antibody restriction of virus growth and detection of M2 in virions. J Virol 62(8):2762–2772

Zhang J, Li G, Ye X (2010) Cyclin T1/CDK9 interacts with influenza A virus polymerase and facilitates its association with cellular RNA polymerase II. J Virol 84(24):12619–12627

Zhou Y, Zhu S, Cai C, Yuan P, Li C, Huang Y, Wei W (2014) High-throughput screening of a CRISPR/Cas9 library for functional genomics in human cells. Nature 509(7501):487–491

Zhu L, Ly H, Liang Y (2014) PLC-gamma1 signaling plays a subtype-specific role in postbinding cell entry of influenza A virus. J Virol 88(1):417–424

Zhu P, Liang L, Shao X, Luo W, Jiang S, Zhao Q, Sun N, Zhao Y, Li J, Wang J, Zhou Y, Zhang J, Wang G, Jiang L, Chen H, Li C (2016) Host cellular protein TRAPPC6ADelta interacts with influenza A virus M2 protein and regulates viral propagation by modulating M2 trafficking. J Virol 91(1). pii: e01757-16

A Functional Genomics Approach to Henipavirus Research: The Role of Nuclear Proteins, MicroRNAs and Immune Regulators in Infection and Disease

Cameron R. Stewart, Celine Deffrasnes, Chwan Hong Foo, Andrew G.D. Bean and Lin-Fa Wang

Abstract Hendra and Nipah viruses (family *Paramyxoviridae*, genus *Henipavirus*) are zoonotic RNA viruses that cause lethal disease in humans and are designated as Biosafety Level 4 (BSL4) agents. Moreover, henipaviruses belong to the same group of viruses that cause disease more commonly in humans such as measles, mumps and respiratory syncytial virus. Due to the relatively recent emergence of the henipaviruses and the practical constraints of performing functional genomics studies at high levels of containment, our understanding of the henipavirus infection cycle is incomplete. In this chapter we describe recent loss-of-function (i.e. RNAi) functional genomics screens that shed light on the henipavirus–host interface at a genome-wide level. Further to this, we cross-reference RNAi results with studies probing host proteins targeted by henipavirus proteins, such as nuclear proteins and immune modulators. These functional genomics studies join a growing body of evidence demonstrating that nuclear and nucleolar host proteins play a crucial role in henipavirus infection. Furthermore these studies will underpin future efforts to define the role of nucleolar host–virus interactions in infection and disease.

C.R. Stewart (✉) · C.H. Foo · A.G.D. Bean
CSIRO Health and Biosecurity, Australian Animal Health Laboratory, Geelong, VIC, Australia
e-mail: cameron.stewart@csiro.au

C. Deffrasnes
Department of Microbiology, Monash University, Clayton, VIC, Australia

L.-F. Wang
Programme in Emerging Infectious Diseases, Duke-NUS Medical School, Singapore 169857, Singapore

Current Topics in Microbiology and Immunology (2018) 419:191–214
DOI 10.1007/82_2017_28

Published Online: 04 July 2017

Contents

1 Introduction of Henipaviruses

1.1 Discovery and Classification

Paramyxoviruses (order *Mononegavirales*) are single-stranded RNA viruses of negative polarity that can cause diseases in humans (rabies, measles virus, mumps virus, respiratory syncytial virus, human parainfluenza virus, Ebola virus) and animals (Newcastle disease virus, canine distemper virus, borna disease virus). The family *Paramyxoviridae* is divided into two subfamilies (*Paramyxovirinae* and *Pneumovirinae*), with Hendra virus (HeV) being the foundation member of the genus *Henipavirus* in the subfamily *Paramyxovirinae*. The discovery of the HeV and Nipah virus (NiV) had a striking impact on our understanding of paramyxovirus biology. Henipaviruses have a much wider host range and a significantly larger genome than other paramyxoviruses, and to date are the only biosafety level (BSL)-4 agents within the family. With mortality rates of human infection between 50 and 100%, HeV and NiV are among the most deadly viruses known to infect humans.

HeV emerged in 1994 in the Brisbane suburb of Hendra, Queensland, Australia, where it caused an outbreak of severe respiratory disease in horses that led to the natural death or euthanasia of 14 out of 21 affected animals. Two people who had close contact with the infected horses were infected and one of these patients died

(Murray et al. 1995). Extensive sampling demonstrated that Australian mainland flying foxes (family *Pteropodidae*, genus *pteropus*) were seropositive for neutralising antibodies against HeV (Young et al. 1996), while the virus was subsequently isolated from flying fox uterine fluid and urine (Halpin et al. 2000), providing strong evidence for Australian mainland flying foxes as the HeV reservoir. Sporadic HeV incidents occurred in horses between 1994 and 2010, with 14 events identified. An alarming number of HeV incidents (34 in total) occurred between 2011 and 2013, with 18 of those occurring in 2011 alone, highlighting the unpredictable nature of HeV outbreaks. Seven human cases of HeV disease have been observed, four of which resulted in fatal disease. All recorded cases of HeV transmission to humans have occurred directly from affected horses. The horses are believed to have acquired HeV infection following direct exposure to secretions from flying foxes. More recently, the decline of reported human cases of HeV infection is potentially due to the development of a vaccine to inhibit HeV disease in horses (Middleton et al. 2014).

NiV was first identified during a disease outbreak on the west coast of Peninsular Malaysia in late 1998. Commercial pig farmers suffered disease characterised by febrile encephalitis that was linked to mild respiratory and neurological disease in pigs (Mohd Nor et al. 2000; From the Centers for Disease Control and Prevention 1999). Nucleotide sequencing demonstrated the virus was closely related to HeV, whilst fruit bats of the *Pteropodidae* family, *Pteropus* genus, were confirmed as the natural reservoir (Yob et al. 2001). Epidemiological evidence suggested that human infections were caused by transmission from pigs which likely had prior contact with fruit bats (Update: outbreak of Nipah virus–Malaysia and Singapore 1999). By mid-1999, cases of human infection were reported in Singapore, where abattoir workers developed NiV infection associated with contact with pigs imported from Malaysia. This initial outbreak of NiV in Malaysia resulted in 265 human cases reported with 105 deaths. Since 2001, NiV outbreaks have been reported almost every year in selected districts of Bangladesh (Hossain et al. 2008; Luby et al. 2009a). Unlike HeV, human-to-human transmission of NiV has been documented (Luby et al. 2009b), including in a hospital setting.

An increasing focus on flying foxes as viral reservoirs has led to the discovery of new henipaviruses. The genus was expanded in 2012 upon the isolation and characterisation of Cedar virus (CedPV), isolated from bat urine samples from a flying fox colony in Cedar Grove, South East Queensland. CedPV shows a remarkably similar genome organisation to HeV and NiV, antigenic cross-reactivity of the nucleocapsid protein between henipaviruses, and shares the same predominant entry receptor molecule, ephrin-B2 (Marsh et al. 2012). However, a critical difference between CedPV and HeV and NiV is that the CedPV P gene lacks coding capacity for the immune antagonising V protein, whilst the CedPV P protein shows an impaired capacity to bind and inhibit IFN signalling via signal transducer and activator of transcription (STAT)1 and STAT2 (Lieu et al. 2015). Accordingly, CedPV infection induces a robust type I interferon (IFN) response in human cells in vitro and does not cause clinical disease in ferret and guinea pig models of disease. Such findings highlight the importance of immune evasion in the context of henipavirus pathogenicity and demonstrate the diverse range of pathogenicity within the same genus.

1.2 Natural Reservoir and Other Novel Henipaviruses

In addition to these three viruses, the henipavirus genus is likely to be expanded in the future to accommodate the discovery and characterisation of emerging viruses from bats and other reservoirs. West African fruit bats harbour neutralising antibodies against HeV and NiV in particular, demonstrating a wider geographical range for henipaviruses not limited to pteropid bats (Hayman et al. 2008). Furthermore, a novel henipa-like virus, Mojiang paramyxovirus, was isolated from rats in the Yunnan Province of China in 2012 and may have caused fatal disease in three individuals (Wu et al. 2014). Alarmingly, a recent study looking at bat and human serum samples from Cameroon found that 3–4% of human samples were seropositive for henipaviruses, and that this was almost exclusively among individuals who reported butchering bat meat, providing the first evidence of human henipavirus spillover infections in Africa (Pernet et al. 2014).

2 Functional Genomics Analysis of Henipavirus Infection

2.1 Platforms for Functional Genomics and Challenges for Studying BSL-4 Pathogens

There are currently no licensed therapies to treat human cases of henipavirus infection. Therefore, gaining a deeper understanding of host pathways exploited by henipaviruses for infection may identify targets for new antiviral therapies. Viruses rely on the cell host machinery for completion of their infection cycle and therefore have adapted to interact with or exploit host molecules. Retroviruses, most DNA viruses, and many orthomyxoviruses replicate their genomes in the host nucleus. Conversely, most positive-sense single-stranded viruses such as picornaviruses and flaviviruses and negative-sense, single-stranded viruses such as filoviruses, rhabdoviruses, and paramyxoviruses are perceived as cytoplasmic viruses and therefore are believed to not have a nuclear stage in their life cycle, replicating their genome entirely in the cytoplasm (Lamb and Parks 2007). However, proteins of some of these viruses can traffic into nuclear compartments during infection (Peeples 1988; Yoshida et al. 1976; Ghildyal et al. 2003; Monaghan et al. 2014; Wang et al. 2010) and this movement is sometimes critical for efficient infection (Wang et al. 2010). This evidence indicates that the host nucleus may play a significant role in the infection cycle of henipaviruses and that the dynamics of virus–host interactions that occur in the nuclear compartments is an understudied area of molecular biology and virology. Furthermore, since important discoveries in cell biology often follow studies of how viruses exploit normal host machinery, investigations into these nuclear interactions may reveal interesting novel insights into the cell biology of the mammalian nucleus. With this in mind, functional genomics provides a powerful and unbiased approach to study these biological questions.

Functional genomics refers to the development and application of global (genome-wide or system-wide) experimental approaches to assess gene function by making use of the information and reagents provided by sequenced genomes (Hieter and Boguski 1997). A wide range of laboratory techniques can be considered as functional genomics, including genome interaction mapping (at the DNA level), microarrays, transcriptomics and serial analysis of gene expression (SAGE) (at the RNA level), yeast 2 hybrid systems and affinity chromatography and mass spectrometry (at the protein level) and loss-of-function studies such as mutational studies, RNA interference (RNAi) and clustered regularly interspaced short palindromic repeats (CRISPR) studies. Functional genomics has demonstrated much power in its ability to dissect the dynamic interplay between host and viral factors during a virus infection, paving the way for novel drug targets. For instance, a haploid genetic screen resulted in the discovery of the once elusive entry receptor for Ebola virus (Carette et al. 2011). There have been many full- or partial-genome RNAi screens of host–virus interactions, including orthomyxoviruses (Brass et al. 2009; Hao et al. 2008; Karlas et al. 2010; Konig et al. 2010; Shapira et al. 2009), retroviruses (Zhou et al. 2008; Konig et al. 2008; Brass et al. 2008) and flaviviruses (Ang et al. 2010; Sessions et al. 2009). Until recently, such information was lacking for henipaviruses, and perhaps surprisingly, for paramyxoviruses generally.

Functional genomics screens can be technically challenging, laborious and involve the use of robotics and advanced imaging equipment. Consequently there are technical and practical challenges to performing high-throughput screens at higher levels of containment. HeV and NiV are classified at BSL-4 agents due to their association with lethal human disease and the absence of preventive measures and effective treatments to combat infections. BSL-4 facilities feature additional precautions to protect workers from infections and prevent exposure, such as infectious work being conducted within class II biosafety cabinets, limited access by secure, locked doors, HEPA filtration of laboratory air, and additional primary containment (positive pressure air suits or class III biosafety cabinets). Due to these limitations, previous genome-wide screens for BSL-4 viruses used surrogate viruses, such as pseudotyped particles, and have been performed under BSL-2 conditions (Kouznetsova et al. 2015; Kleinfelter et al. 2015).

2.2 *Functional Genomics Studies on Henipaviruses*

Functional genomics have been employed to study henipavirus infection. For instance, the entry receptor of HeV and NiV, ephrin-B2, was identified by microarray analysis of infection-permissive and infection-resistant cell lines (Bonaparte et al. 2005). Transcriptomics and proteomics have been utilised to uncover key differences in cellular responses to HeV infection in HeV disease-susceptible (human) and disease-resistant (bat) cells, and suggest that activation of apoptosis pathways via the innate immune pathway may contribute to the tolerance of henipaviruses by flying foxes (Wynne et al. 2014). Here we largely focus on findings from two recent RNAi screens to identify protein-coding genes and host-encoded microRNAs impacting the henipavirus infection cycle in human cells. Not only can these findings be compared to published RNAi screens of host–virus interactions, but the identification of host genes required for infection (as opposed to those that are merely differentially expressed during infection) may deliver new targets for the development of antiviral therapies.

The large number of HeV incidents in Australia from 2011 to 2013 prompted researchers at our laboratory to establish the capability to perform genome-wide RNAi screens at BSL-4. Central to this work was the development of a recombinant HeV expressing the renilla luciferase construct, which allowed for high throughput and rapid measurement of virus infection (Marsh et al. 2013). This recombinant virus was shown to be lethal in the ferret model of henipavirus disease and exhibited a pathogenesis profile comparable to the wild-type virus. Functional genomics at high containment also required the establishment of protocols and/or safe work procedures for the operation and decontamination of liquid handling robots.

3 The Reliance of Henipavirus Infection on Nuclear and Nucleolar Proteins

3.1 Genome-Wide RNAi Screening

A genome-wide analysis of host protein-coding genes required for henipavirus infection involved a primary screen assaying 18,120 protein-coding genes, followed by a secondary deconvolution screen and a tertiary screen determining whether screen results obtained using recombinant HeV could be recapitulated using wild-type HeV and NiV (Deffrasnes et al. 2016). Applying a robust Z score normalisation method often used to interpret siRNA screen results (Birmingham et al. 2009; Zhang et al. 2006), 585 and 630 genes were identified that promoted or suppressed HeV infection, respectively, without adversely impacting cell numbers. At the completion of the primary screen, 200 proviral genes were selected based on rank for the secondary deconvolution screen. By this measure, 20 high- and 46 medium-confidence genes (>2 standard deviations from mean mock values for 4/4 or 3/4, or 2/4 siRNAs, respectively) were identified as being required for HeV infection. The apparent reliance of henipavirus infection on the nuclear or nucleolar host proteins was particularly striking, as over 40% of high confidence hits localise in the nucleus or nucleolus, with many involved in ribosome biogenesis (Table 1).

The nucleus is the site of gene expression and DNA transcription into mRNA, and houses the early steps of the RNAi pathway. The nucleus is separated from the cell cytoplasm by the nuclear envelop which contains nuclear pores and import/export proteins allowing the passage of small molecules such as mRNA. Nuclear import/export proteins such as XPO1 and KPNA3, which are required for

Table 1 Nuclear or nucleolar host protein-coding genes required for HeV infection

Entrez gene	Function
DDX10	Pre-rRNA cleavage and component of the U3 small subunit processome
EIF2S3	Recruitment of methionyl-tRNA to the 40S ribosomal subunit, initiation factor
ESF1	RPL & RPS family member, pre-rRNA processing
FBL	Pre-ribosome processing, chemical modification of pre-rRNA
GTPBP4	Ribosome subunit assembly
KPNA3	Nuclear export/import
IMP4	Pre-rRNA cleavage and component of the U3 small subunit processome
MRPL12	Structural constituent of the mitochondrial ribosome
POLR3E	rDNA transcription
PWP2	RNA binding and snoRNA binding
RPL13A	Pre-ribosome processing, modification of pre-rRNA
RPL7A	Component of the 60S ribosomal subunit
SP7	Transcriptional regulation
XPO1	Nuclear export

trafficking of larger molecules like proteins, were identified by RNAi screen as required for henipavirus infection (Deffrasnes et al. 2016).

The nucleolus is a highly dynamic structure and has increasingly been shown to play a critical role in virus–host interactions (Rawlinson and Moseley 2015; Xu et al. 2016). The nucleolus contains three regions composed of the fibrillar centre (FC) in the middle, surrounded by the dense fibrillary component (DFC) and the granular component (GC). This membrane-less structure contains a high concentration of proteins and RNAs and is the site of ribosomal RNA (rRNA) synthesis and ribosome production but is also a multifunctional structure in eukaryotic cells. Cell cycle progression, stress response, genetic silencing, regulation of apoptosis, cell migration and invasion are all functions associated with the nucleolus or partly regulated in this compartment (Rawlinson and Moseley 2015; Xu et al. 2016; Pederson 2010).

3.2 *Fibrillarin and Its Role in Henipavirus Infection*

Fibrillarin is the main nucleolar protein responsible for the chemical modification of ribosomal RNA (rRNA). This 34–38 kDa 2′-O-methyltransferase transfers methyl groups from its substrate, the S-adenosylmethionine (SAM), to the 2-hydroxyl groups of ribose target in rRNA. Fibrillarin has also been shown to methylate glutamine residue 104 of the human histone H2A, weakening its binding to the FACT (facilitator of chromatin transcription) complex and impacting chromatin remodelling and rDNA transcription by RNA Pol I (Tessarz et al. 2013), which points at an additional role for fibrillarin in ribosome biogenesis and translation.

Fibrillarin itself is methylated on several arginine residues by protein arginine N-methyltransferase 1 (PRMT1), which is thought to influence its activity (Rodriguez-Corona et al. 2015).

Expression levels of fibrillarin have been shown to be regulated by p53 through direct binding to fibrillarin intron 1. Abnormal levels of fibrillarin have been detected in p53-inactivated cancer cells and a decrease in p53 levels has been associated with an increase in fibrillarin expression, and conversely an increase in p53 expression results in decreased fibrillarin expression (Marcel et al. 2013). High levels of fibrillarin lead to changes in the rRNA methylation pattern, diminished translation fidelity and increase in IRES-mediated translation of some cancer genes. Moreover, ribosome biogenesis is often dysregulated and over-activated in cancer cells that have a decreased or absent p53 expression (Marcel et al. 2013).

In its N-terminal region, fibrillarin contains a glycine- and arginine-rich region (the GAR domain) enabling interaction with cellular and viral proteins, and acting as a nucleolar retention signal. Its C-terminal region (MTase) contains multiple RNA-binding domains, a catalytic site allowing for fibrillarin methyltransferase function, and is the site for NOP56/58 interaction. Fibrillarin is a part of at least one nucleolar ribonucleoprotein (snoRNP) complex comprising the NOP56, NOP58 and 15.5 K nucleolar proteins. X-ray data have suggested that the methylation of rRNA requires the formation of this complex with involvement of four fibrillarin molecules interacting with different regions of the target rRNAs. The yeast equivalent of fibrillarin, NOP1, has been more extensively studied than the human counterpart but fibrillarin is a well-conserved protein in most organisms, reinforcing the notion that all post-transcriptional processes involving fibrillarin such as chemical modification (methylation) of rRNA, pre-rRNA cleavage and ribosome assembly are essential for proper cellular functioning (Rodriguez-Corona et al. 2015).

In eukaryotes, ribosome biogenesis involves numerous nucleolar proteins and accessory factors, around 80 ribosomal proteins, many small nucleolar RNAs (snoRNAs), three RNA polymerases (RNA polymerase I, II and III) and four different species of rRNAs. The process of assembly of elongation-competent 80S ribosomes is divided into three major steps: (1) ribosomal DNA (rDNA) transcription into precursor rRNAs (pre-rRNAs), (2) processing of pre-RNAs into mature rRNAs, and then (3) assembly of rRNAs with ribosomal proteins into functional ribosomes. In the nucleolus, the RNA polymerase I (RNA Pol I) is responsible for transcribing the 18S, 5.8S and 28S rRNA from a single polycistronic pre-rRNA, while RNA pol III transcribes the 5S rRNA in the nucleus (Xue and Barna 2012). The pre-RNAs are then cleaved and modified during the pre-rRNA processing phase. All ribosomal proteins (RP) are transcribed in the cytoplasm by RNA Pol II and then translated before migrating to the nucleolus. These RP, along with nucleolar proteins such as fibrillarin and RPL13A, are responsible for modifying the rRNAs (ribose 2′-O-methylation, pseudouridylation, etc.) with the activity of more than 100 snoRNAs guiding the process in a site-specific manner. The main nucleolar protein involved in rRNA modification is fibrillarin, which methylates more than 100 sites essential for ribosome biogenesis and stability. Although these post-transcriptional modifications are crucial for ribosome functions, their roles are not yet fully understood. In eukaryotes, the large 60S subunit of ribosomes is made of the 5S, 5.8S, and 28S rRNA along with multiple large subunit ribosomal proteins (RPL), while the small 40S subunit is made of the 18S rRNA along with multiple small subunit ribosomal proteins (RPS). The two subunits are assembled in the nucleolus into the 80S ribosomes before being transferred into the cytoplasm.

Deffrasnes and colleagues showed that siRNA-mediated knockdown of fibrillarin expression dramatically reduced HeV protein production and viral genome replication but did not impact viral fusion, and that fibrillarin catalytic activity was essential to henipavirus infection. On the other hand, overexpression experiment did not lead to an increase in viral titers, suggesting that a simple reduction or increase in overall ribosome production is unlikely to explain the reliance of henipaviruses on fibrillarin activity (Deffrasnes et al. 2016).

3.3 *Modulation of Translation in Henipavirus Infection*

The requirement of fibrillarin and several other proteins from the ribosomal biogenesis pathway for henipavirus infection points a reliance on translation for efficient infection. However, while we tend to view ribosomes as homogenous, new studies reveal a more heterogeneous nature of ribosomes due to differences in the ribosomal proteins recruited, post-translational modifications of rRNA and rRNA composition. Moreover, ribosomal proteins have been found to have additional functions outside of their primary roles in ribosomes and to be involved in other nucleolar functions such as regulation of cell proliferation, tumorigenesis and DNA damage response (Xu et al. 2016; Xue and Barna 2012; Au and Jan 2014).

In eukaryotes, most messenger RNA (mRNA) harbour a 5′ 7-methylguanosine cap structure and a 3′ poly(A) tail, which are both required for canonical, cap-dependent translation. A cap-independent translation mechanism also utilised by a subset of host proteins is called Internal Ribosome Entry Site (IRES)-mediated translation. It is believed that most genes translated via an IRES are related to stress response, cell proliferation, cell death/survival, and that IRES-mediated translation happens when the canonical cap-dependent translation is inhibited either by the host reaction to environmental factors, damage, stress or infections. However, a group recently suggested that thousands of human genes are translated via this cap-independent mechanism, representing a 50-fold increase in the number of sequences previously associated with this translation pathway (Weingarten-Gabbay et al. 2016).

Recently a new type of translation has been described in vesicular stomatitis virus (VSV)-infected cells. This non-canonical cap-dependent protein translation involves the ribosomal protein RPL40 acting as a constituent of the large subunit of ribosomal complexes and suggests a novel ribosome-specialised translation initiation pathway benefiting viral mRNA translation (Lee et al. 2012). Translations of viral proteins from several other mononegaviruses, including the paramyxoviruses measles virus (MeV) and Newcastle disease virus (NDV), and a subset of cellular transcripts, are also RPL40-dependent.

How henipavirus mRNAs are translated is not fully understood. Whilst the RPL40-dependent form of cap-dependent translation remains to be characterised in detail, one could speculate that fibrillarin, like RPL40, acts a novel initiation factor for henipavirus mRNAs. The fact that depleting cells of fibrillarin did not impact synthesis of influenza A viral proteins (which occurs via the canonical cap-dependent pathway) suggests that henipavirus mRNA translation occurs via a non-canonical pathway, perhaps used by a subset of cellular transcripts. Such a concept would allow henipavirus protein synthesis to proceed in an environment where viruses may induce cellular translation shutdown in order to suppress host antiviral immune responses. There are several reports of paramyxoviruses blocking canonical translation pathways, including the MeV N protein binding to the eukaryotic initiation factor 3 (eIF3-p40) (Sato et al. 2007), whilst the P and V proteins of simian virus 5 (SV5) limit activation of the double-stranded RNA (dsRNA)-dependent protein kinase (PKR) to limit both host and viral protein translation (Gainey et al. 2008). Similar to SV5, siRNA-mediated depletion of PKR results in increased HeV growth (robust Z score 1.46), consistent with the notion that shutdown of host protein translation inhibits henipavirus infection.

If future studies do indeed demonstrate a role of fibrillarin in influencing the synthesis of ribosome subtypes required for viral protein translation, this may explain the targeting of fibrillarin by several viral proteins. Fibrillarin binds the HeV matrix (M) protein during the early stages of infection, whilst the HIV-Tat protein has been reported to bind fibrillarin and U3 snoRNA, both required for pre-rRNA processing, and this interaction reduces the pool of cytoplasmic ribosomes (Ponti et al. 2008). Intriguingly, the nucleoprotein of porcine reproductive and respiratory syndrome virus, the non-structural protein 1 (NS1) of a H3N2 influenza virus (Melen et al. 2012) and the non-structural protein 3b of the severe acute respiratory syndrome coronavirus (Yuan et al. 2005) all bind and co-localise with fibrillarin in the nucleolus; however, the reasons for this binding are yet to be determined.

4 Viral Targets Within the Host Cell Nucleus

4.1 Role of the M Protein in the Henipavirus in Infection Cycle

Many negative strand viruses encode viral proteins that localise in the nucleus and/or nucleolus at some point in their infection cycle [reviewed in (Rawlinson and Moseley 2015; Hiscox 2003; Oksayan et al. 2012; Flather and Semler 2015; Watkinson and Lee 2016)]. Within the *Paramyxoviridae*, nuclear localisation of matrix (M) protein has previously been described for NDV (Peeples 1988), Sendai virus (SeV) (Yoshida et al. 1976), human respiratory syncytial virus (Ghildyal et al. 2003), HeV (Monaghan et al. 2014) and NiV (Wang et al. 2010). During the early stages of henipavirus infection or when expressed ectopically (Monaghan et al. 2014; Wang et al. 2010), the HeV and NiV M proteins traffic through the nucleolus to the cytoplasm. It has been recently shown that nuclear traffic is required for the henipavirus M protein to coordinate viral budding. The henipavirus M protein is a structural protein that mediates viral assembly and budding (Liljeroos and Butcher 2012; Takimoto and Portner 2004; Eaton et al. 2007). Indeed, for both HeV-M and NiV-M, overexpression of these proteins alone is sufficient to trigger viral-like particles (VLPs) that bud into the supernatant. Wang and colleagues (2010) demonstrated that mutation of NiV-M nuclear localisation signals (NLS) or nuclear export signals (NES) blocks nuclear/cytoplasmic traffic and impairs viral budding. Furthermore, a highly conserved lysine residue in the NLS (K258) serves two functions: its positive charge mediates NiV-M nuclear import, while is also a potential site for monoubiquitination which regulates NiV-M nuclear export. Mutation of K258 or the treatment of cells with proteasome inhibitors such as MG132 inhibits both NiV-M budding and NiV infection. Consistent with this work, siRNA-mediated knockdown of the protein ubiquitin D (UBD) inhibits HeV and NiV infection (Deffrasnes et al. 2016).

4.2 The Impact of M-Binding Host Proteins on Hendra Virus Infection

This raises the question: do henipavirus M proteins traffic through the nucleolus for other reasons? The multi-faceted roles of paramyxovirus proteins in replication-specific roles and various cellular processes, particularly immune evasion, would suggest so. To explore whether M binds to host proteins associated with infection efficiency, results from the genome-wide RNAi HeV screen were cross-referenced against a proteomics study by Pentecost and colleagues cataloguing host proteins that bind HeV-M and NiV-M, among other paramyxovirus M proteins (Pentecost et al. 2015). That study revealed that the henipavirus M interactome spans hundreds of host proteins, with interactions with nuclear pore complex proteins, nuclear transport receptors and nucleolar proteins particularly prevalent. Interestingly, NiV-M and HeV-M interactomes show notable overlap to other paramyxovirus M proteins, including SeV and NDV, with over 60% of the proteins found in any single interactome also found in the interactomes of one or more of the other three viruses (Pentecost et al. 2015). Whilst the binding of fibrillarin to HeV-M was demonstrated by co-immunoprecipitation assays (Deffrasnes et al. 2016) and this was not observed by proteomics, interactions were observed between HeV-M and numerous nucleolar proteins such as NOP58 (Pentecost et al. 2015) which forms a complex with fibrillarin, supporting a functional interaction between fibrillarin and HeV-M.

The relative HeV growth (presented as robust Z scores) in cells depleted of the 389 HeV-M-binding host proteins is shown in Fig. 1a. Of the 327 candidates assayed, HeV-M binds to 22 protein-coding genes that have a large impact (robust Z score ≤ -2 or ≥ 2) on HeV infection, roughly evenly distributed between proviral (12) and antiviral (10) candidates. Designating all candidates genes with Z scores <0 as proviral and genes with Z scores >0 as antiviral, host proteins that bind HeV-M appears to be pro- and antiviral at approximately equal ratios with a slight enrichment of proviral genes (174 proviral candidates vs. 146 antiviral candidates).

An assessment of whether the relative abundance of HeV-M–host protein interactions indicated a likelihood of that host protein adopting a proviral or antiviral function was also carried out (Fig. 1b). The relative abundance of host proteins within the proteomics dataset is represented as the normalised spectral abundance factor (NSAF), with higher NSAF values presenting more abundant interactions. Plotting NSAF values against robust Z scores demonstrates that host proteins that bind HeV-M with high abundance (NSAFe5 scores between 250 and 938) were more proviral (11 candidates) than antiviral (4 candidates, Z score sums: proviral 13.9, antiviral 2.6). These candidates are listed in Table 2 and include several ribosomal proteins, further implicating M in host translation.

5 Host Proteins Targeted by Henipaviruses for Immune Evasion

5.1 Immune Modulating Function of P-Encoded Proteins

Akin to fibrillarin, the critical role of host molecules in henipavirus infection and pathogenesis can be inferred by their specific targeting by viral proteins. This is particularly true in the context of immune evasion, as the innate antiviral immune response is a known target for several henipavirus proteins.

The henipavirus genome contains six transcriptional units, *N*, *P*, *M*, *F*, *G* and *L*, coding for nine proteins (Eaton et al. 2007). The *P* gene alone codes for at least four of the proteins: P, W, V and C (Eaton et al. 2006). All four of these proteins are involved in modification of the immune response in the host cell, through inhibition of the type I interferon (IFN) responses [reviewed in (Audsley and Moseley 2013)]. Intracellular detection of pathogen-associated molecular patterns (PAMPs) is mediated by membrane-bound Toll-like receptors (TLRs) or cytoplasmic retinoic acid-inducible gene I (RIG-I)-like receptors (RLRs) and nucleotide-binding oligomerisation domain containing (NOD)-like receptors (NLRs). Engagement of these receptors with their agonists results in the activation of complex signalling pathways culminating in the production of cytokines and anti-microbial compounds. A critical component of this response is the type I IFN system, which induces a local antiviral state upon detection of viruses or intracellular bacteria or molecules associated with their replication (Schoggins and Rice 2011).

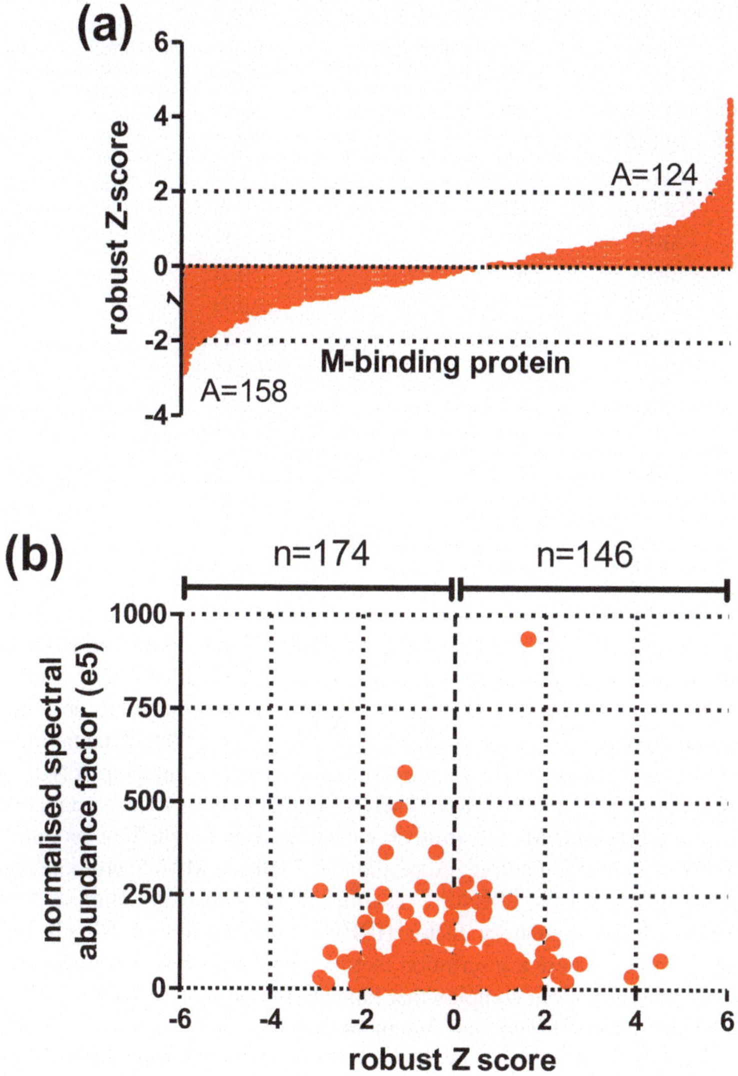

Fig. 1 The impact of HeV-M-binding host proteins on HeV infection. Cross-reference analysis of host proteins that bind HeV-M (Pentecost et al. 2015) and genome-wide analysis of host protein-coding genes associated with HeV infection (Deffrasnes et al. 2016). **a** Z scores resulting from growing HeV in cells depleted of HeV-M-binding proteins. Genes with Z scores <0 were designated proviral, while genes with Z scores >0 were designated antiviral. Values represent the sum of all the Z scores. It should be noted that 43 genes were excluded from analysis due to ambiguous gene identification listings in the proteomics study, whilst the silencing of 19 additional gene targets resulted in cell death that prevented the measurement of virus growth. **b** Plot of the Z score of HeV-M-binding proteins (x-axis) and relative abundance of HeV-M interactions, represented by normalised spectral abundance factor (y-axis)

Table 2 Robust Z score of HeV-M-binding host protein-coding genes

	Entrez gene ID	Relative abundance (NSAFe5)	Relative HeV infection (robust Z score)
Proviral	TUBAL3	580.35	−1.09
	RPL27A	482.31	−1.2
	RPL28	431.84	−1.12
	RPL38	422.59	−0.97
	ARF3	367.72	−1.5
	ARF1	367.72	−0.41
	ACTA1	274.63	−0.71
	NEDD8	273.90	−2.21
	RPL19	264.12	−2.92
	RPS27L	264.12	−0.22
	ACTA2	255.35	−1.57
Antiviral	TUBA8	938.83	1.61
	ARF4	287.60	0.26
	ACTC1	274.63	0.67
	ACTG2	255.35	0.11

Viral replication is typically detected by TLRs 3 and 7/8 in endosomal compartments (Alexopoulou et al. 2001; Lund et al. 2004), whilst RIG-I and/or melanoma differentiation-associated gene 5 (MDA5) recognise short or long viral dsRNA intermediates in the cytosol (Yoneyama et al. 2004; Triantafilou et al. 2012). TLR3 activates the TIR-domain-containing adapter-inducing IFN-β (TRIF) (Matsumoto et al. 2011), whilst RIG-I/MDA5 interact via their caspase recruitment domains (CARDs) with MAVS (mitochondrial activated signalling protein) (Seth et al. 2005) to induce signalling. Activation of TRIF or MAVS promotes recruitment of multiple cytosolic effectors, resulting in the phosphorylation and dimerisation of interferon regulatory factor (IRF) 3 or liberation of NF-κB from its inhibitory complex. These transcription factors then shuttle into the nucleus to form part of a large multiprotein complex that binds to the promoter region of IFN-β and initiates transcription (Honda and Taniguchi 2006).

The C-terminus of the HeV V protein binds and sequesters MDA5, thereby impairing IFN-β transcription in response to double-stranded RNA (Andrejeva et al. 2004). This binding appears to be conserved amongst most paramyxoviruses including NiV, SV5 and mumps virus (Childs et al. 2007). Intriguingly, RIG-I is not targeted by paramyxovirus V proteins, and perhaps consistent with this, the genome-wide RNAi screen suggested that depleting cells of MDA5 increased HeV infection (robust Z score 2.02), whilst targeting RIG-I had very little impact (Z score −0.37).

Similar to the NLR cytoplasmic antiviral immune responses, TLR3-dependent antiviral signalling is also inhibited by henipaviruses, with the W protein localising to the nucleus via the importin molecules KPNA3 and KPNA4 to block IRF3-responsive promoter activation by virus and intracellular dsRNA (Shaw et al.

2005). Transfecting NiV-W into cells in a dose-dependent manner sequesters inactive IRF3 in the nucleus, thus depleting the pool of available IRF3 for phosphorylation and activation. From the genome-wide screen, the impact of down-regulating TLR3 (Z score 1.16) and IRF3 (0.97) was a moderately antiviral phenotype.

5.2 Targeting of STAT by Henipaviruses

The best-characterised target of henipavirus immune evasion is the STAT proteins, critical signalling molecules in the context of type I IFN cytokine production conferring the antiviral state [reviewed in (Platanias 2005)]. The binding of type I IFN (IFN-α and IFN- β) and type II IFN (IFN-γ) to their respective receptor complexes leads to the phosphorylation and association of STAT1 and STAT2 heterodimers (for type I IFN signalling), or STAT1 homodimers (type II IFN). This prompts the formation of STAT1–STAT2–IRF9 (IFN-regulatory factor 9) complexes that translocate to the nucleus and bind IFN-stimulated response elements (ISREs) in DNA to initiate transcription of IFN-stimulated genes (ISGs). Whilst there are hundreds, potentially thousands of ISGs that collectively confer antiviral immunity, very few ISGs have been functionally characterised in the context of henipavirus infection. One ISG, cholesterol 25 hydroxylase (CH25H), inhibits infection by NiV and a range of other RNA viruses by blocking membrane fusion between host and viral membranes (Liu et al. 2013a). Consistent with this observation, CH25H blocked HeV infection in the genome-wide RNAi screen (robust Z score 1.05).

Henipaviruses, like other paramyxoviruses, generate multiple alternative mRNAs from the P gene locus—P, V and W (Thomas et al. 1988). A fourth protein, C, is generated by alternate translation initiation site selection from all these mRNAs and does not share sequence homology to the other proteins. The P, V, and W proteins share 407 amino acids in their N termini and all three proteins bind to STAT1 and STAT2 via this N-terminal region (Ciancanelli et al. 2009; Rodriguez et al. 2004). Virus–host interactions in this context prevent STAT1/2 phosphorylation and activation, and lead to their sequestration in high molecular weight complexes (Rodriguez et al. 2003; Rodriguez et al. 2002; Shaw et al. 2004). Interestingly, the siRNA-mediated inhibition of STAT1 increased HeV infection in the genome-wide screen, but inhibition of STAT2 did not (robust Z scores of 1.01 and −0.67). This preliminary observation suggests that STAT1 activity may have a greater impact on henipavirus infection than STAT2, and may implicate type II in antiviral immunity against henipaviruses.

5.3 Novel Function of the M Protein in Immune Evasion

Although the role of henipavirus P gene products in immune evasion is well-established, a recent study demonstrates the surprising ability of NiV-M to antagonise the antiviral type I IFN response (Bharaj et al. 2016). The study by Bharaj and colleagues shows that NiV-M binds to and targets the E3-ubiquitin ligase TRIM6 for degradation. TRIM6 catalyses the synthesis of unanchored polyubiquitin chains that are used as a substrate for the activation of IkB kinase-e (IKKe), which phosphorylates IRF3 and activates IRF3-dependent transcription of type I IFN, and TNF-α. TRIM6 targeting by NiV-M occurs in the cytoplasm via an unknown mechanism not involving the proteasome or the lysosome, and requires nuclear/cytoplasmic trafficking of NiV-M. Similar to viral budding, this function of M is dependant on nuclear traffic, as K258 mutants of NiV-M do not target TRIM6 for degradation. The study expands our understanding of immune antagonism and highlights the potential purpose of henipavirus M protein nuclear trafficking.

6 The Impact of Host-Encoded MicroRNAs on Henipavirus Infection

6.1 Role of MicroRNAs in Viral Infection

MicroRNAs (miRNAs) are a class of small ($\sim$21–22 b.p.), single-stranded non-coding RNA molecules (Fayyad-Kazan et al. 2014; Neel and Lebrun 2013; Skalsky and Cullen 2010) involved in post-transcriptional gene regulation. MiRNAs function by binding to complementary sequences typically located in 3′ untranslated region (3′ UTR) of specific mRNA targets (Fayyad-Kazan et al. 2014; Neel and Lebrun 2013; Skalsky and Cullen 2010; Liu et al. 2013a). Depending on the degree of complementarity, this generally results in the suppression or degradation of target mRNA, thereby preventing encoded proteins from being translated (Fayyad-Kazan et al. 2014; Neel and Lebrun 2013; Skalsky and Cullen 2010). Although far less frequent, miRNA binding may also cause an increase in target mRNA translation and thus up-regulation of protein expression (Vasudevan et al. 2007).

In terms of target complementarity, miRNAs do not require perfect base pairing (tenOever 2013). As a result, one miRNA has the potential to regulate a surprisingly broad network of genes (Skalsky and Cullen 2010; Zhang et al. 2013), with certain miRNAs found to have binding sites located on several hundred different mRNA sequences (Guo and Steitz 2014). Despite the potential for widespread impacts, studies have described the effects of miRNA gene regulation on protein expression levels as generally ‘subtle’ (tenOever 2013) or ‘typically relatively mild’ (Selbach et al. 2008). This is due to the fact that, in general, miRNAs do not entirely silence but rather moderately repress translation and, hence, effectively fine tune rather than knock out gene expression (Baek et al. 2008).

The role of miRNAs in the infection cycle of RNA viruses is becoming increasingly apparent. Certain miRNAs may promote virus replication by directly interacting with the viral genome or, alternatively, by down-regulating the expression of host genes that suppress virus infection (Skalsky and Cullen 2010; Roberts et al. 2011). Inhibiting specific 'proviral' miRNAs, therefore, may have a direct negative impact on the viral life cycle (Janssen et al. 2013) or alternatively render the intracellular environment unfavourable for virus replication (Stewart et al. 2013). In an example of the latter, miR-146a has been found to promote HeV infection by repressing ring finger protein 11, a negative regulator of NF-κB activity (Stewart et al. 2013). Furthermore, inhibiting miR-146a has been found to significantly reduce HeV replication in vitro (Stewart et al. 2013). On the other hand, miR-122 is an example of a miRNA that promotes hepatitis C virus (HCV) replication by directly interacting with the viral genome—this activity is the basis of the first miRNA inhibitor drug to enter phase II clinical trials (Janssen et al. 2013; Wilson and Sagan 2014).

6.2 Host-Encoded MicroRNAs and Henipavirus Infection

The functional genomics platform established as part of the screen of protein-coding genes associated with HeV infection was recently adapted to study the impact of host-encoded miRNAs on HeV growth (Foo et al. 2016). The screen involved the use of synthetic miRNA mimics and inhibitors targeting 834 microRNAs. Mimic and inhibitor screens identified 35 and 61 microRNAs, respectively, that promoted HeV infection, and 19 and 83 microRNAs, respectively, that inhibited virus infection. A major finding from this study was that all four members of the miR-181 family (-a to -d) promote infection by HeV and NiV. Infection promotion was primarily mediated via the ability of miR-181 to significantly enhance henipavirus-induced membrane fusion. Cell signalling receptors of ephrins, namely EphA5 and EphA7, were identified as novel negative regulators of henipavirus fusion. The expression of these receptors, as well as EphB4, was suppressed by miR-181 overexpression, suggesting that simultaneous inhibition of several Ephs by the miRNA contributes to enhanced infection and fusion. To our knowledge, this study represented the first evidence of a host-encoded miRNA promoting virus cell entry.

Previous studies have reported that members of the miR-181 family are involved in different aspects of immune regulation (Hutchison et al. 2013; Galicia et al. 2014; Zietara et al. 2013). Specifically, miR-181 has been found to play a central role in the regulation of B cell differentiation and T cell selection, maturation and sensitivity (Sun et al. 2014). For instance, induction of miR-181a has been found to occur at the CD4(+)–CD8(+) double-positive stage of T cell development, inhibiting the expression of CD69, Bcl-2 and T cell receptor—all involved in positive selection and T cell maturation (Neilson et al. 2007). In addition, miR-181c has been found to suppress CD4+ T cell activation by targeting interleukin 2 (IL-2) (Sun et al. 2014;

Xue et al. 2011). In addition, miR-181a expression levels have been shown to correlate with pro-inflammatory signals (e.g. IL-1β, IL-6 and TNF-α) in blood and various tissues of humans with chronic inflammation, as well as in the blood of LPS-treated mice (Xie et al. 2013). Consistent with the notion that miR-181 expression is immune-responsive, levels of miR-181 were up-regulated in the biofluids of ferrets and horses infected with HeV, suggesting that the host innate immune response may promote henipavirus spread and exacerbate disease severity.

The study of both miRNAs and protein-coding genes associated with HeV infection allows an assessment whether genes required for virus infection (i.e. proviral genes) are regulated by miRNAs that inhibit virus infection (i.e. antiviral microRNAs). Multiple members of the let-7 miRNA family inhibited HeV infection. There are 10 mature let-7 sequences in humans, with multiple roles described, including negative regulation of tumorigenesis (Shi et al. 2008; Esquela-Kerscher and Slack 2006). In a transcriptome-wide study in HeLa cells, genes significantly down-regulated by let-7b at either the mRNA level, protein level or both, included fourteen validated genes required for wild-type HeV infection, including *AKT1* (Selbach et al. 2008). Furthermore, six proviral genes contain putative let-7b binding sites in their 3′ UTR (*AKT1*, *C6Orf106*, *EIF2S3*, *HMGA1*, *IFITM3* and *SERPINH1*), as identified by DIANA-mirExTra (Alexiou et al. 2010). Collectively, these data suggest that let-7 miRNAs inhibit HeV by suppressing host proteins required for virus infection. Cross-referencing results from the protein-coding screen study showed that the majority of verified target genes for miR-181 and miR-17-92 miRNAs (proviral in the miRNA screen) were predominately antiviral, demonstrating a level of congruency between miRNA and protein-coding gene screens.

In contrast to let-7, all six members of the miRNA precursor miR-17 family (miR-17, -20a, -20b, -106a, -106b and -93), part of the oncogenic miR-17–92 polycistron, strongly promoted HeV infection. Interestingly, other miRNAs of the miR-17–92 cluster with distinct "seed" families (based on sequence identity at positions 2–7)—miR-18, miR-19 and miR-92) did not impact virus replication to a similar extent. The miR-17–92 cluster is a known oncogene locus—it is amplified in B cell lymphomas (Ota et al. 2004) and accelerates tumour development in a mouse B cell lymphoma model (He et al. 2005). Members of the miRNA precursor miR-17 family are expressed in almost all human tissues (Liang et al. 2007). In addition, miR-106a and -106b are expressed in peripheral blood mononuclear cells (PBMCs), platelets and exosomes derived from peripheral blood (Hunter et al. 2008).

7 Concluding Remarks

Henipaviruses are dangerous pathogens and control of disease caused by these viruses will critically rely on the development of new antiviral therapeutics and vaccination strategies. Currently, there is requirement for renewed research into the host immune responses to henipavirus infection and how competent immune responses may fight disease. A major challenge is to ascertain the molecular mechanisms of virus replication and immunity associated with protection to infection. The improved knowledge of functional genomics approaches and immune response to viral infection means that we now have the tools to further progress our understanding and knowledge. Nevertheless, this must be implemented to develop advanced infection control approaches.

References

Alexiou P et al (2010) The DIANA-mirExTra web server: from gene expression data to microRNA function. PLoS One 5(2):e9171

Alexopoulou L et al (2001) Recognition of double-stranded RNA and activation of NF-kappaB by Toll-like receptor 3. Nature 413(6857):732–738

Andrejeva J et al (2004) The V proteins of paramyxoviruses bind the IFN-inducible RNA helicase, mda-5, and inhibit its activation of the IFN-beta promoter. Proc Natl Acad Sci U S A 101 (49):17264–17269

Ang F et al (2010) Small interference RNA profiling reveals the essential role of human membrane trafficking genes in mediating the infectious entry of dengue virus. Virol J 7:24

Au HH, Jan E (2014) Novel viral translation strategies. Wiley Interdiscip Rev RNA 5(6):779–801

Audsley MD, Moseley GW (2013) Paramyxovirus evasion of innate immunity: diverse strategies for common targets. World J Virol 2(2):57–70

Baek D et al (2008) The impact of microRNAs on protein output. Nature 455(7209):64–71

Bharaj P et al (2016) The Matrix Protein of Nipah Virus Targets the E3-Ubiquitin Ligase TRIM6 to Inhibit the IKKepsilon Kinase-Mediated Type-I IFN Antiviral Response. PLoS Pathog 12 (9):e1005880

Birmingham A et al (2009) Statistical methods for analysis of high-throughput RNA interference screens. Nat Methods 6(8):569–575

Bonaparte MI et al (2005) Ephrin-B2 ligand is a functional receptor for Hendra virus and Nipah virus. Proc Natl Acad Sci U S A 102(30):10652–10657

Brass AL et al (2008) Identification of host proteins required for HIV infection through a functional genomic screen. Science 319(5865):921–926

Brass AL et al (2009) The IFITM proteins mediate cellular resistance to influenza A H1N1 virus, West Nile virus, and dengue virus. Cell 139(7):1243–1254

Carette JE et al (2011) Ebola virus entry requires the cholesterol transporter Niemann-Pick C1. Nature 477(7364):340–343

Childs K et al (2007) mda-5, but not RIG-I, is a common target for paramyxovirus V proteins. Virology 359(1):190–200

Ciancanelli MJ et al (2009) Nipah virus sequesters inactive STAT1 in the nucleus via a P gene-encoded mechanism. J Virol 83(16):7828–7841

Deffrasnes C et al (2016) Genome-wide siRNA Screening at Biosafety Level 4 Reveals a Crucial Role for Fibrillarin in Henipavirus Infection. PLoS Pathog 12(3):e1005478

Eaton BT et al (2006) Hendra and Nipah viruses: different and dangerous. Nat Rev Microbiol 4 (1):23–35

Eaton BT, Mackenzie JS Wang LF (2007) Henipaviruses. FIeld's Virology: Lippincott Williams & Wilkins

Esquela-Kerscher A, Slack FJ (2006) Oncomirs - microRNAs with a role in cancer. Nat Rev Cancer 6(4):259–269

Fayyad-Kazan H et al (2014) Downregulation of microRNA-24 and -181 parallels the upregulation of IFN-gamma secreted by activated human CD4 lymphocytes. Hum Immunol 75(7):677–685

Flather D, Semler BL (2015) Picornaviruses and nuclear functions: targeting a cellular compartment distinct from the replication site of a positive-strand RNA virus. Front Microbiol 6:594

Foo CH et al (2016) Dual microRNA Screens Reveal That the Immune-Responsive miR-181 Promotes Henipavirus Entry and Cell-Cell Fusion. PLoS Pathog 12(10):e1005974

From the Centers for Disease Control and Prevention (1999) Outbreak of Hendra-like virus–Malaysia and Singapore, 1998-1999. JAMA 281(19):1787–1788

Gainey MD et al (2008) Paramyxovirus-induced shutoff of host and viral protein synthesis: role of the P and V proteins in limiting PKR activation. J Virol 82(2):828–839

Galicia JC et al (2014) MiRNA-181a regulates Toll-like receptor agonist-induced inflammatory response in human fibroblasts. Genes Immun 15(5):333–337

Ghildyal R et al (2003) The matrix protein of Human respiratory syncytial virus localises to the nucleus of infected cells and inhibits transcription. Arch Virol 148(7):1419–1429

Guo YE, Steitz JA (2014) Virus meets host microRNA: the destroyer, the booster, the hijacker. Mol Cell Biol

Halpin K et al (2000) Isolation of Hendra virus from pteropid bats: a natural reservoir of Hendra virus. J Gen Virol 81(Pt 8):1927–1932

Hao L et al (2008) Drosophila RNAi screen identifies host genes important for influenza virus replication. Nature 454(7206):890–893

Hayman DT et al (2008) Evidence of henipavirus infection in West African fruit bats. PLoS One 3 (7):e2739

He L et al (2005) A microRNA polycistron as a potential human oncogene. Nature 435 (7043):828–833

Hieter P, Boguski M (1997) Functional genomics: it's all how you read it. Science 278(5338): 601–602

Hiscox JA (2003) The interaction of animal cytoplasmic RNA viruses with the nucleus to facilitate replication. Virus Res 95(1–2):13–22

Honda K, Taniguchi T (2006) IRFs: master regulators of signalling by Toll-like receptors and cytosolic pattern-recognition receptors. Nat Rev Immunol 6(9):644–658

Hossain MJ et al (2008) Clinical presentation of nipah virus infection in Bangladesh. Clin Infect Dis 46(7):977–984

Hunter MP et al (2008) Detection of microRNA expression in human peripheral blood microvesicles. PLoS One 3(11):e3694

Hutchison ER et al (2013) Evidence for miR-181 involvement in neuroinflammatory responses of astrocytes. Glia 61(7):1018–1028

Janssen HL et al (2013) Treatment of HCV infection by targeting microRNA. N Engl J Med 368 (18):1685–1694

Karlas A et al (2010) Genome-wide RNAi screen identifies human host factors crucial for influenza virus replication. Nature 463(7282):818–822

Kleinfelter LM et al (2015) Haploid Genetic Screen Reveals a Profound and Direct Dependence on Cholesterol for Hantavirus Membrane Fusion. MBio 6(4)

Konig R et al (2008) Global analysis of host-pathogen interactions that regulate early-stage HIV-1 replication. Cell 135(1):49–60

Konig R et al (2010) Human host factors required for influenza virus replication. Nature 463 (7282):813–817

Kouznetsova J et al (2015) Identification of 53 compounds that block Ebola virus-like particle entry via a repurposing screen of approved drugs. Emerg Microbes Infect 3(12):e84

Lamb R, Parks G (2007) Paramyxoviridae: The Viruses and Their Replication. FIeld's Virology: Lippincott Williams & Wilkins

Lee AS, Burdeinick-Kerr R, Whelan SP (2012) A ribosome-specialized translation initiation pathway is required for cap-dependent translation of vesicular stomatitis virus mRNAs. Proc Natl Acad Sci U S A 110(1):324–329

Liang Y et al (2007) Characterization of microRNA expression profiles in normal human tissues. BMC Genomics 8:166

Lieu KG et al (2015) The non-pathogenic Henipavirus Cedar paramyxovirus phosphoprotein has a compromised ability to target STAT1 and STAT2. Antiviral Res. 124:69–76
Liljeroos L, Butcher SJ (2012) Matrix proteins as centralized organizers of negative-sense RNA virions. Front Biosci (Landmark Ed). 18:696–715
Liu J et al (2013) Mechanism of T cell regulation by microRNAs. Cancer Biol Med 10(3):131–137
Liu SY et al (2013a) Interferon-inducible cholesterol-25-hydroxylase broadly inhibits viral entry by production of 25-hydroxycholesterol. Immunity. 38(1):92–105
Luby SP et al (2009a) Recurrent zoonotic transmission of Nipah virus into humans, Bangladesh, 2001–2007. Emerg Infect Dis 15(8):1229–1235
Luby SP, Gurley ES, Hossain MJ (2009b) Transmission of human infection with Nipah virus. Clin Infect Dis 49(11):1743–1748
Lund JM et al (2004) Recognition of single-stranded RNA viruses by Toll-like receptor 7. Proc Natl Acad Sci U S A 101(15):5598–5603
Marcel V et al (2013) p53 acts as a safeguard of translational control by regulating fibrillarin and rRNA methylation in cancer. Cancer Cell. 24(3):318–330
Marsh GA et al (2012) Cedar virus: a novel Henipavirus isolated from Australian bats. PLoS Pathog. 8(8):e1002836
Marsh GA et al (2013) Recombinant Hendra viruses expressing a reporter gene retain pathogenicity in ferrets. Virol J 10:95
Matsumoto M, Oshiumi H, Seya T (2011) Antiviral responses induced by the TLR3 pathway. Rev Med Virol 21(2):67–77
Melen K et al (2012) Influenza A H3N2 subtype virus NS1 protein targets into the nucleus and binds primarily via its C-terminal NLS2/NoLS to nucleolin and fibrillarin. Virol J 9:167
Middleton D et al (2014) Hendra virus vaccine, a one health approach to protecting horse, human, and environmental health. Emerg Infect Dis 20(3):372–379
Mohd Nor MN, Gan CH, Ong BL (2000) Nipah virus infection of pigs in peninsular Malaysia. Rev Sci Tech 19(1):160–165
Monaghan P et al (2014) Detailed morphological characterisation of Hendra virus infection of different cell types using super-resolution and conventional imaging. Virol J 11:200
Murray K et al (1995) A morbillivirus that caused fatal disease in horses and humans. Science 268 (5207):94–97
Neel JC, Lebrun JJ (2013) Activin and TGFbeta regulate expression of the microRNA-181 family to promote cell migration and invasion in breast cancer cells. Cell Signal 25(7):1556–1566
Neilson JR et al (2007) Dynamic regulation of miRNA expression in ordered stages of cellular development. Genes Dev 21(5):578–589
Oksayan S et al (2012) Subcellular trafficking in rhabdovirus infection and immune evasion: a novel target for therapeutics. Infect Disord Drug Targets 12(1):38–58
Ota A et al (2004) Identification and characterization of a novel gene, C13orf25, as a target for 13q31-q32 amplification in malignant lymphoma. Cancer Res 64(9):3087–3095
Pederson, T., The nucleus introduced. Cold Spring Harb Perspect Biol, 2010. 3(5)
Peeples ME (1988) Differential detergent treatment allows immunofluorescent localization of the Newcastle disease virus matrix protein within the nucleus of infected cells. Virology 162 (1):255–259
Pentecost M et al (2015) Evidence for ubiquitin-regulated nuclear and subnuclear trafficking among Paramyxovirinae matrix proteins. PLoS Pathog 11(3):e1004739
Pernet O et al (2014) Evidence for henipavirus spillover into human populations in Africa. Nat Commun 5:5342
Platanias LC (2005) Mechanisms of type-I- and type-II-interferon-mediated signalling. Nat Rev Immunol 5(5):375–386
Ponti D et al (2008) The HIV Tat protein affects processing of ribosomal RNA precursor. BMC Cell Biol 9:32
Rawlinson SM, Moseley GW (2015) The nucleolar interface of RNA viruses. Cell Microbiol 17 (8):1108–1120

Roberts AP, Lewis AP, Jopling CL (2011) The role of microRNAs in viral infection. Prog Mol Biol Transl Sci 102:101–139
Rodriguez JJ, Cruz CD, Horvath CM (2004) Identification of the nuclear export signal and STAT-binding domains of the Nipah virus V protein reveals mechanisms underlying interferon evasion. J Virol 78(10):5358–5367
Rodriguez JJ, Parisien JP, Horvath CM (2002) Nipah virus V protein evades alpha and gamma interferons by preventing STAT1 and STAT2 activation and nuclear accumulation. J Virol 76 (22):11476–11483
Rodriguez JJ, Wang LF, Horvath CM (2003) Hendra virus V protein inhibits interferon signaling by preventing STAT1 and STAT2 nuclear accumulation. J Virol 77(21):11842–11845
Rodriguez-Corona U et al (2015) Fibrillarin from Archaea to human. Biol Cell 107(6):159–174
Sato H et al (2007) Measles virus N protein inhibits host translation by binding to eIF3-p40. J Virol 81(21):11569–11576
Schoggins JW, Rice CM (2011) Interferon-stimulated genes and their antiviral effector functions. Curr Opin Virol 1(6):519–525
Selbach M et al (2008) Widespread changes in protein synthesis induced by microRNAs. Nature 455(7209):58–63
Sessions OM et al (2009) Discovery of insect and human dengue virus host factors. Nature 458 (7241):1047–1050
Seth RB et al (2005) Identification and characterization of MAVS, a mitochondrial antiviral signaling protein that activates NF-kappaB and IRF 3. Cell 122(5):669–682
Shapira SD et al (2009) A physical and regulatory map of host-influenza interactions reveals pathways in H1N1 infection. Cell 139(7):1255–1267
Shaw ML et al (2004) Nipah virus V and W proteins have a common STAT1-binding domain yet inhibit STAT1 activation from the cytoplasmic and nuclear compartments, respectively. J Virol 78(11):5633–5641
Shaw ML et al (2005) Nuclear localization of the Nipah virus W protein allows for inhibition of both virus- and toll-like receptor 3-triggered signaling pathways. J Virol 79(10):6078–6088
Shi XB (2008) C.G. Tepper, and R.W. deVere White, Cancerous miRNAs and their regulation. Cell Cycle 7(11):1529–1538
Skalsky RL, Cullen BR (2010) Viruses, microRNAs, and host interactions. Annu Rev Microbiol 64:123–141
Stewart CR et al (2013) Promotion of Hendra virus replication by microRNA 146a. J Virol 87 (7):3782–3791
Sun X, Sit A, Feinberg MW (2014) Role of miR-181 family in regulating vascular inflammation and immunity. Trends Cardiovasc Med 24(3):105–112
Takimoto T, Portner A (2004) Molecular mechanism of paramyxovirus budding. Virus Res 106 (2):133–145
tenOever BR (2013) RNA viruses and the host micro RNA machinery. Nat Rev Microbiol 11(3):169–180
Tessarz P et al (2013) Glutamine methylation in histone H2A is an RNA-polymerase-I-dedicated modification. Nature 505(7484):564–568
Thomas SM, Lamb RA, Paterson RG (1988) Two mRNAs that differ by two nontemplated nucleotides encode the amino coterminal proteins P and V of the paramyxovirus SV5. Cell 54 (6):891–902
Triantafilou K et al (2012) Visualisation of direct interaction of MDA5 and the dsRNA replicative intermediate form of positive strand RNA viruses. J Cell Sci 125(Pt 20):4761–4769
Update: outbreak of Nipah virus–Malaysia and Singapore (1999) MMWR Morb Mortal Wkly Rep 48(16):335-337
Vasudevan S, Tong Y, Steitz JA (2007) Switching from repression to activation: microRNAs can up-regulate translation. Science 318(5858):1931–1934
Wang YE et al (2010) Ubiquitin-regulated nuclear-cytoplasmic trafficking of the Nipah virus matrix protein is important for viral budding. PLoS Pathog 6(11):e1001186

Watkinson RE, Lee B (2016) Nipah virus matrix protein: expert hacker of cellular machines. FEBS Lett 590(15):2494–2511

Weingarten-Gabbay S et al (2016) Comparative genetics. Systematic discovery of cap-independent translation sequences in human and viral genomes. Science 351(6270)

Wilson JA, Sagan SM (2014) Hepatitis C virus and human miR-122: insights from the bench to the clinic. Curr Opin Virol 7C:11–18

Wu Z et al (2014) Novel Henipa-like virus, Mojiang Paramyxovirus, in rats, China, 2012. Emerg Infect Dis 20(6):1064–1066

Wynne JW et al (2014) Proteomics informed by transcriptomics reveals Hendra virus sensitizes bat cells to TRAIL-mediated apoptosis. Genome Biol 15(11):532

Xie W et al (2013) miR-181a and inflammation: miRNA homeostasis response to inflammatory stimuli in vivo. Biochem Biophys Res Commun 430(2):647–652

Xu X, Xiong X, Sun Y (2016) The role of ribosomal proteins in the regulation of cell proliferation, tumorigenesis, and genomic integrity. Sci China Life Sci 59(7):656–672

Xue Q et al (2011) Human activated CD4(+) T lymphocytes increase IL-2 expression by downregulating microRNA-181c. Mol Immunol 48(4):592–599

Xue S, Barna M (2012) Specialized ribosomes: a new frontier in gene regulation and organismal biology. Nat Rev Mol Cell Biol 13(6):355–369

Yob JM et al (2001) Nipah virus infection in bats (order Chiroptera) in peninsular Malaysia. Emerg Infect Dis 7(3):439–441

Yoneyama M et al (2004) The RNA helicase RIG-I has an essential function in double-stranded RNA-induced innate antiviral responses. Nat Immunol 5(7):730–737

Yoshida T et al (1976) Membrane (M) protein of HVJ (Sendai virus): its role in virus assembly. Virology 71(1):143–161

Young PL et al (1996) Serologic evidence for the presence in Pteropus bats of a paramyxovirus related to equine morbillivirus. Emerg Infect Dis 2(3):239–240

Yuan X et al (2005) Nucleolar localization of non-structural protein 3b, a protein specifically encoded by the severe acute respiratory syndrome coronavirus. Virus Res 114(1–2):70–79

Zhang XD et al (2006) Robust statistical methods for hit selection in RNA interference high-throughput screening experiments. Pharmacogenomics 7(3):299–309

Zhang Y, Wang Z, Gemeinhart RA (2013) Progress in microRNA delivery. J Control Release 172 (3):962–974

Zhou H et al (2008) Genome-scale RNAi screen for host factors required for HIV replication. Cell Host Microbe 4(5):495–504

Zietara N et al (2013) Critical role for miR-181a/b-1 in agonist selection of invariant natural killer T cells. Proc Natl Acad Sci U S A 110(18):7407–7412

Roles of Non-coding RNAs in Respiratory Syncytial Virus (RSV) Infection

Ralph A. Tripp and Abhijeet A. Bakre

Abstract Analysis of host gene expression profiles following viral infections of target cells/tissues can reveal crucial insights into the host: virus interaction and enables the development of novel therapeutics and prophylactics. Regions of the host genome that do not code for protein, encode structural, and functional non-coding RNAs that are important not only in regulation of host gene expression but also may impact viral replication. This review summarizes the role of host non-coding RNAs during replication of multiple respiratory viruses with a focus on Respiratory Syncytial Virus (RSV), an important pediatric pathogen. This review highlights the current state of knowledge and understanding regarding the function (s) of ncRNAs for respiratory viral infection and host immunity in general.

Contents

R.A. Tripp · A.A. Bakre (✉)
Department of Infectious Diseases, University of Georgia, Athens, GA 30602, USA
e-mail: bakre@uga.edu

Current Topics in Microbiology and Immunology (2018) 419:215–242
DOI 10.1007/82_2017_32

Published Online: 24 June 2017

1 RSV

1.1 Etiology and Management

Respiratory Syncytial Virus (RSV) is a *Pneumovirus* belonging to the family *Paramyxoviridae*. A and B genotypes circulate globally and are transmitted via respiratory droplets or fomites. A typical RSV season extends from November to March but viruses can exhibit annual/biennial cycles of transmission based on geography (Mlinaric-Galinovic et al. 2009). RSV infects the very young and old and infection rates approach >90% of children by the age of 2 (Hall 2012; Nair et al. 2010). Clinical and in vitro data suggest that RSV can remain in a variety of cell types including macrophages, dendritic cells, and innervating neurons (Agoti et al. 2015; Bramley et al. 1999; Dakhama et al. 1997; Estripeaut et al. 2008; Gaona et al. 2014; Hegele et al. 1994; Konig et al. 1996; Mejias et al. 2008; Ostler et al. 2001; Piedimonte 2013; Piedimonte and Perez 2014; Rivera-Toledo et al. 2015; Schwarze et al. 2004; Sikkel et al. 2008; Valdovinos and Gomez 2003; Zheng et al. 2015).

Typical RSV infections are mild and restricted to the upper respiratory tract leading to symptoms like runny nose, low appetite, coughing, and sneezing. Lower respiratory tract infections resulting from a combination of viral and host factors can lead to RSV bronchiolitis, i.e., inflammation of the airway epithelium leading to difficulty in breathing, wheezing, and cyanosis and may need mechanical ventilation in neonates. Clinical manifestations of RSV disease are very similar to other viral respiratory diseases complicating diagnosis and clinical management. The dynamic interactions between viral factors and the host drive the immune response to RSV. Host factors that contribute to lower respiratory tract illness (LRTI) include preterm birth, age <2 months, low vitamin D receptor levels in the mother, congenital pulmonary/cardiac/neuromuscular disorders, immunodeficiency, defects in cellular immunity, pollution, and single nucleotide polymorphisms (SNPs) in Toll-like receptor 4 (TLR4)/Surfactant protein A and D (SP-A, SP-D) (Ramet et al. 2011; Zeng et al. 2011), in cytokines and chemokines (IFN, IL-6, IL-10, TGF-β, CCR5) (Choi et al. 2013), and others host features have been attributed. The two major viral surface factors are also known to modulate disease severity, i.e., the RSV G and F proteins involved in attachment, viral egress, and modulation of the host immune response (Moore et al. 2013; Oshansky et al. 2010).

The effects linked to RSV infection are just beginning to be fully elucidated. For example, studies have shown that RSV infection alters the immune response to secondary bacterial infections (Thorburn et al. 2006), and there is a strong correlation between severity of RSV infection in infancy and development of respiratory disorders such as asthma in adult life (Bacharier et al. 2012; Holt 2015; Sigurs et al. 1995, 2000, 2010; Feldman et al. 2015). Despite six decades of RSV research, clinical management of RSV diseases is symptomatic; no effective antivirals or vaccines are available despite extensive studies. A better understanding of the molecular events at the host-virus interface will enable a systems biology perspective on the virulence and host immune pathways that lead to RSV disease.

1.2 RSV Genome

RSV is a pleomorphic virus and infectious virus may exist as either as spherical viruses, or as filamentous extensions (McCurdy and Graham 2003). RSV has a ~15 kb long negative sense single-stranded RNA (ssRNA) genome that encodes 10 genes and 11 proteins in the order 3′-NS1/2-N-P-M-SH-G-F-M2-1/2-L-5′ (Broadbent et al. 2015). Each gene segment is flanked by a non-transcribed intergenic spacer region consisting of a gene start and gene stop sequence that helps the viral RNA-dependent RNA polymerase (RdRP) catalyze gene transcription/genome replication. Gene transcription in all paramyxoviruses including RSV follows a decreasing gradient, genes at the 3′ end are transcribed and translated at a higher level compared to those at the 5′ end owing to the propensity of the RdRP to fall off between gene segments and restart at the 3′ end. Among the 10 genes, three surface proteins RSV glycoprotein (G), fusion protein (F), and small hydrophobic protein (SH) perform myriad functions ranging from viral attachment and adhesion to dampening of host immune responses. A summary of RSV genes and their known functions is given in Table 1.

RSV infection is initiated by attachment of the virus to cell surface receptors (glycosaminoglycans (McCurdy and Graham 2003; Broadbent et al. 2015; Boyoglu-Barnum et al. 2013, 2015; Caidi et al. 2012; Connors et al. 1991; Haynes et al. 2009; Li et al. 2006)/Nucleolin (Moore et al. 2008)/CX3CR1) (Feldman et al. 1999, 2000) and surface G protein. Attachment is followed by fusion of the host and viral lipid membranes and release of the viral ribonucleoprotein (vRNP) complex into the host cell cytosol. Viral genomes released are then transcribed and translated into individual gene products in a decreasing 3′→5′ gradient. The negative sense ssRNA genome is also copied into a positive sense complementary RNA (cRNA) which acts as a template for production of more negative sense vRNA. vRNAs associate with newly translated viral proteins (N, P, and M) to form viral ribonucleoprotein particles (vRNPs) which associate with lipid rafts in the membrane and bud out as progeny virions.

2 Organization and Classification of Non-coding RNAs in the Human Genome

A large proportion of eukaryotic genomes contain regions that do not encode any proteins and were considered to be non-coding "junk" DNA (Lander et al. 2001). These junk DNA regions encode structural and functional RNAs at fourfold excess

Table 1 Overview of RSV gene function

RSV gene product	Function	References.
Surface proteins		
Glycoprotein (G)	Attachment, fractalkine mimic, antibody decoy; TLR 2 antagonist; Essential for in vivo viral replication; induces SOCS1/3 to decrease IFN production	Oshansky et al. (2010), Boyoglu-Barnum et al. (2013, 2015), Caidi et al. (2012), Connors et al. (1991), Haynes et al. (2009), Li et al. (2006), Moore et al. (2008), Feldman et al. (1999, 2000), Shields et al. (2003), Harcourt et al. (2004, 2006), Haynes et al. (2003), Tripp et al. (2001, 2003), Zhang et al. (2010)
Fusion protein (F)	Penetration into cells, TLR4 agonist, Fusion of cells to form syncytia; Essential for viral entry	Connors et al. (1991), Feldman et al. (2000), Hallak et al. (2007), Mastrangelo and Hegele (2013)
Small Hydrophobic (SH)	inhibition of apoptosis; Putative viroporin	Akerlind-Stopner et al. (1993), Collins and Mottet (1993), Fuentes et al. (2007), Gan et al. (2012), Heminway et al. (1994), Karron et al. (1997), Kochva et al. (2003), Li et al. (2014), Rixon et al. (2004, 2005), Russell et al. (2015), Taylor et al. (2014), Techaarpornkul et al. (2001), Triantafilou et al. (2013)
Intrinsic		
Non-structural 1 (NS1) Non-structural 2 (NS2)	Not part of the virion; Suppress innate immune response; Block interferon signaling; inhibit apoptosis; activate PI3 Kinase; trigger epithelial cell shedding; modulate miRNA expression; inhibition of Th subset differentiation; Epithelial cell shedding	Bakre et al. (2015), Thornburg et al. (2012)
Matrix (M)	vRNP assembly; viral transcription/replication; maturation of viral filaments	Mitra et al. (2012), Ghildyal et al. (2006)
Nucleocapsid (N)	RNA binding; dual function polymerase transcription factor; Shifts between viral transcription and replication	
Phosphoprotein (P)	Mediates N recognition by Polymerase (L); Essential for viral transcription and replication; Chaperone; interacts with M2-1; regulates viral uncoating	Asenjo et al. (2006, 2008)

(continued)

Table 1 (continued)

RSV gene product	Function	References.
M2-1/M2-2	Zinc binding; vRNA binding; Induction of NF-κβ	Cai et al. (2015)
Polymerase (L)	Viral gene transcription; genome replication; mRNA guanylyation; association with lipid rafts	Crowe et al. (1996), Fix et al. (2011), Grosfeld et al. (1995), Laganas et al. (2015), Liuzzi et al. (2005), Luongo et al. (2009), McDonald et al. (2004), Sourimant et al. (2015), Stec et al. (1991), Tang et al. (2002)

over protein coding transcripts with a high degree of conservation across the eukaryotic evolutionary tree (Ponting and Belgard 2010). The present classification recognizes at least 26 different categories of ncRNAs (Consortium 2015) with variable abundance and more being continually discovered. The major classes include long non-coding RNAs (lncRNAs), microRNAs (miRNAs), piwi associated RNAs (piRNAs), and structural RNAs such as ribosomal RNAs (rRNAs), transfer RNAs (tRNAs). The role(s) of many ncRNAs in normal development as well as disease remain to be explored. This review will focus on the roles of microRNAs and transfer RNA fragments during RSV replication since these are the two well-studied classes as of now.

2.1 microRNAs (miRNAs)

The human genome encodes ~2500 genes that encode 21–23 nt long small non-coding RNAs known as miRNAs. miRNAs are extremely conserved along the eukaryotic evolutionary tree, co-evolving with eukaryotes and less than 12 miRNAs having been lost from the evolution from deuterostomes to mammals (Berezikov et al. 2007; Grimson et al. 2008). This conservation of miRNA diversity also extends to miRNA function and structure with miRNAs across divergent species exhibiting similar structural organization and target specificity which is elaborated upon in the following sections. miRNAs are transcribed from intergenic or intronic loci. Most ancient miRNAs were intergenic; subsequent exaptation has led to nearly 50% miRNAs being intron encoded thus allowing co-transcription with host genes (Campo-Paysaa et al. 2011). miRNAs may also be encoded by transfer RNAs (tRNAs) (Pederson 2010), snoRNAs (Ender et al. 2008), direct repeats in transposons (Piriyapongsa et al. 2007), ribosomal RNA (Chak et al. 2015; Castellano and Stebbing 2013; Yoshikawa and Fujii 2016), and independent miRNA transcription units-termed mirtrons (Berezikov et al. 2007; Ruby et al. 2007).

2.1.1 miRNA Biogenesis

For the intronic or intergenic miRNAs, transcription is thus RNA Polymerase II (RNAPOlII) (Lee et al. 2004) mediated but miRNAs encoded in rRNAs/tRNAs/direct repeats may be transcribed by other RNA polymerases (Borchert et al. 2006). The primary miRNA transcript, called a pri-miRNA, has a characteristic stem loop structure which is processed in the nucleus by a microprocessor complex consisting of two proteins Drosha and DGCR8. The microprocessor complex cleaves the pri-miRNA into a pre-miRNA which is then exported out of the nucleus is an Exportin-5 mediated process into the cytosol. Pre-miRNAs are further processed by another RNAse III enzyme Dicer to produce a 21–23 nt long double stranded miRNA complex, one strand of which is thermodynamically more stable and is referred to as the guide strand while the other strand is less stable and referred to as the passenger strand. The involvement of these cellular proteins in miRNA biogenesis argues that miRNAs would be produced only in nucleated cells. Interestingly however, miRNAs are also abundant in non-nucleated cells such as RBCs (Sangokoya et al. 2010; Hamilton 2010) with distinct expression linked to blood storage (Kannan and Atreya 2010) or disease state (Chen et al. 2008).

2.1.2 Mechanism of miRNA Action

The first six nucleotides at the 5′ of the miRNA guide strand are known as the "seed sequence" and can bind to a complementary "miRNA recognition element" (MRE) in the 3′ UTR of a target transcript and delay/block translation or cause transcript decay. Since the region of complementarity between the seed site and the MRE is usually only 6 nts, each miRNA can potentially regulate the expression of hundreds of genes. miRNAs are hence considered molecular rheostats that dampen a global response rather than targeting a single gene. Indeed, computational and experimental studies predict/validate that 74–92% of the transcriptome in 4 model species is miRNA regulated (Miranda et al. 2006). Though most miRNAs are believed to act via binding to the 3′ UTR, several miRNAs can also bind to promoters (Zardo et al. 2012), coding regions or 5′ UTRs within genes (Miranda et al. 2006) and alter gene expression.

2.1.3 Theater of miRNA Function

Owing to their somewhat promiscuous mode of action, miRNAs can regulate several cellular pathways involved in normal homeostasis as well as disease. In addition to their intracellular role, miRNAs also function across cell/tissue types. miRNAs are exported in vesicular bodies called exosomes from multiple cell types. Exosomes are 30–150 nm diameter vesicles that are produced by invagination of the plasma membrane followed by fusion with endosomes to form multi-vesicular bodies that contain a small fraction of the cytoplasm and associated cellular cargo. Exosomes are produced by all cell types and are believed to be important intercellular signaling conduits, though the exact mechanisms are poorly understood and probably are cell/tissue type specific. Owing to their small diameter, the typical exosomal cargo consists of small RNAs, peptides, smaller transcripts, and proteins and metabolites. A couple of studies have shown preferential loading of exosomes with 3′ UTRs (Batagov and Kurochkin 2013) though the implication of this has yet to be determined. Exported exosomes can fuse with cells from the same or different tissue types and deliver their cargo inducing a paracrine signaling cascade similar to hormones. Analysis of exosomal contents from different biofluids has shown that exosome associated miRNA profiles correlate with disease state or development (Rani 2014; Kulkarni et al. 2016; Hornick et al. 2015; Bi et al. 2015; Madhavan et al. 2015; McKiernan et al. 2016; Ho et al. 2014; Kosanovic and Jankovic 2014; Benito-Martin et al. 2013; Chen et al. 2013; Conde-Vancells et al. 2010).

2.2 *Transfer RNA Fragments (tRFs)*

Transfer RNAs (tRNAs) are central to translation of mRNA to proteins. Humans encode 513 nuclear and 22 mitochondrial tRNA genes that incorporate 61 different amino acids during translation (Chan and Lowe 2016). The number of tRNA genes varies among individuals (Iben and Maraia 2014) and these exhibit tissue and cell type differences in expression (Dittmar et al. 2006). The cellular "tRNAome" is in a state of constant flux with tRNA biosynthesis complemented with tRNA quality control and degradation in response to cellular stimuli or stress signals (de Nadal et al. 2011), although tRNA has a half-life range from hours to days (Hopper 2013) in contrast to mRNAs which are short-lived.

Mature tRNAs are produced from transcription of tRNA genes by RNA polymerase III complex containing transcription factor TFIIC and TFIIIB. Initial tRNA transcripts contain a 5′ and 3′ trailer sequences that are removed by RNAse P and Z, respectively (Skowronek et al. 2014) in conjunction with adaptor protein La (Maraia and Lamichhane 2011). The 3′CCA trinucleotide is added to this transcript (Wolfe et al. 1996). Primary transcripts are processed via splicing, undergo numerous modifications (Kirchner and Ignatova 2015) followed by aminoacyl transferase (aaRS) catalyzed amino acylation before being exported out of the nucleus into the cytosol (Kirchner and Ignatova 2015). Mature tRNAs may also traffic to the nucleus in a retrograde fashion for recycling to cytoplasm (Whitney et al. 2007).

Transfer RNAs are specifically cleaved during cellular stress (Gebetsberger and Polacek 2013) including viral infection into fragments that originate from the 5′ end or the 3′ end (3′tRFs) or the anti-codon loop to generate transfer RNA fragments (tRFs) known, respectively, as 5′tRFs, 3′tRFs, or tRF-1. Functionally, tRFs can have multiple activities akin to short interfering RNAs (siRNAs) degrading target transcripts (Maute et al. 2013), but can also regulate ribosomal loading and protein chain elongation (Sobala and Hutvagner 2013). The molecular pathways of tRF biogenesis, the role and function are poorly understood presently, but are believed to result from cellular stress induced during nutrient starvation to recycle essential nitrogen and phosphate needed for cell survival (Huang et al. 2015). Intriguingly, tRNA cleavage is not accompanied with decline in levels of full length tRNAs (Thompson and Parker 2009; Saikia et al. 2012).Transfer RNA cleavage to generate tRFs may be mediated by the TRAMP pathway in the nucleus that degrades pre-tRNA molecules in the nucleus (Maraia and Lamichhane 2011; LaCava et al. 2005; Kadaba et al. 2004; Anderson 2005; Wang et al. 2008), or via cytosolic degradation of mature tRNAs via the Rapid tRNA decay (RTD) pathway. The TRAMP pathway consists of a polyadenylase Trf4 (topoisomerase 1- related 4), a RNA helicase Mtr4p (mRNA transport regulator 4 protein) and Air2 (Arginine methyltransferase interacting RING finger protein 2) which interacts with Rrp6, a 3′ exoribonuclease of the nuclear exosome. The RTD pathway involves Met22 (Methionine requiring protein 22) (Chernyakov et al. 2008) and cytosolic 5′–3′ exonucleases such as Rat 1 (Ribonucleic acid trafficking protein 1) (Chernyakov et al. 2008), exoribonuclease 1 (Xrn1) (Chernyakov et al. 2008; Watanabe et al. 2013), endonucleases ELAC2 (Lee et al. 2009), Dicer (Haussecker et al. 2010; Cole et al. 2009; Babiarz et al. 2008), and Angiogenin (ANG) (Yamasaki et al. 2009). The enzyme ANG is a member of the RNAse A family endonucleases in humans (Yamasaki et al. 2009), localized in the nuclei of cells (Cooper et al. 2015) and involved in ribosomal RNA (rRNA) transcription. Expression of Angiogenin is cell density independent and drives cellular proliferation (Cooper et al. 2015). ANG recognizes tRNAs with CA in their anti-codon loop (UCA- Serine, CCA-Proline, ACA-Threonine, GCA-Alanine) which constitute a small proportion of the tRNAome and cleaves them preferentially into 5′ and 3′ tRFs (Czech et al. 2013). However, since all tRNAs also carry a CCA motif in their 3′ end aminoacyl acceptor site, ANG can non-selectively remove this CCA motif from the tRNA 3′ end and cause a global shutdown of translation in a reversible process (Czech et al. 2013). The role of ANG and other nucleases in tRF biogenesis are still emerging.

3 Non-coding RNA Function in RSV Replication

3.1 miRNA Regulation of Host Gene Expression

RSV replication affects host gene expression at early and late time-points during infection. There is good evidence that these global increases in the host transcription are however accompanied by concomitant decreases in the nuclear proteome both in cell lines as well as primary human bronchoepithelial cells infected with RSV, suggesting that RSV downregulates host protein translation.

It was shown that miRNA deregulation occurred in human bronchial epithelial cells infected with a RSV construct bearing a green fluorescent protein (GFP). In these studies, the RSV strain (rgRSV) led to maximal miR-221 repression and computational analysis shortlisted Nerve growth factor (NGF), a neurotrophic factor important in regulating apoptotic cell death as a major miR-221 target. miR-221 repression was accompanied with high expression of NGF in infected cells. To validate if miR-221 regulated NGF, mimics that induce miR-221 incorporation into the RISC complex were transfected into uninfected bronchial epithelial cells. Flow cytometric analysis of NGF expression following miR-221 mimic transfection showed a substantial knockdown of NGF expression relative to negative control. Additionally, miR-221 mimic transfection reduced the abundance of apoptotic cells in infected cells suggesting that miR-221 downregulation is pro-viral. These preliminary data showed that miRNAs have a role in gene regulation during viral infection but did not shed any insights on the mechanisms involved. Moreover, analysis of NGF 5′ and 3′ UTRs as well as coding regions using a variety of miRNA target finding algorithms failed to demonstrate a sequence complementarity between miR-221 and NGF suggesting that miR-221 downregulation of NGF was an off-target effect. Analysis of miR-221 over expression has been shown to suppress tumor suppressor genes Bcl 2 binding component 3 (*Bbc3/Puma*) (Zhang et al. 2010), Bcl2 interacting protein *Bim* (Terasawa et al. 2009) Phosphatase and Tension homolog (PTEN) (Zhang et al. 2010), and transcription factor Forkhead Box O3 (Fox03a) (Hamada et al. 2012) which can lead to induction of NGF (Terasawa et al. 2009).

Recently, it was demonstrated that RSV infection of respiratory epithelial and dendritic cells (DCs) led to a cell type-specific alteration in miRNA expression (Thornburg et al. 2012). The study showed that RSV infection upregulated let-7b in DCs, and let-7i and miR-30b in respiratory epithelial cells in a replication dependent manner (Thornburg et al. 2012). Further, it was shown that non-structural genes, NS1 and NS2, inhibited upregulation of let-7i and miR-30b in epithelial cells suggesting that these non-structural genes can modulate the host antiviral response via deregulation of miRNAs. These data demonstrate the intricate interactions between viral genes and host regulatory mechanisms mediated by small non-coding RNAs. To elucidate not only the profiles of miRNAs deregulated during RSV infection, but to better understand the mechanisms involved, the expression of host miRNAs was examined at 24 h pi (Bakre et al. 2012). Five miRNAs (let-7, miR-24, miR-26b, miR-520a, and miR-337) were found to be induced, while two miRNAs (miR-198 and -595) were repressed in A549 cells relative to mock. To determine the function of these differentially expressed miRNAs on host gene expression, the combined data from previous host gene expression profiles with miRNA target predictions were examined, which helped to identify a set of genes and miRNAs that showed inverse correlations. Twelve of the shortlisted genes SOCS3, CCND1, SMOX, HOXA1, TNFAIP3, ELF4, DYRK2, CCL7, PLAUR, VLDLR, GLRX3, and SERPING1) were then analyzed using luciferase-UTR assays. Upon transfection with these luciferase-UTR constructs, miRNA inhibitor treatment increased luciferase expression for five (SOCS3, CCND1, ELF4, DYRK2, and CCL7) genes while mimic treatment repressed these genes further relative to non-targeting and mock controls. These data showed that these five genes were genuine targets of miRNAs let-7f and miR-24 (Fig. 1). Moreover, we showed that DYRK2 which is

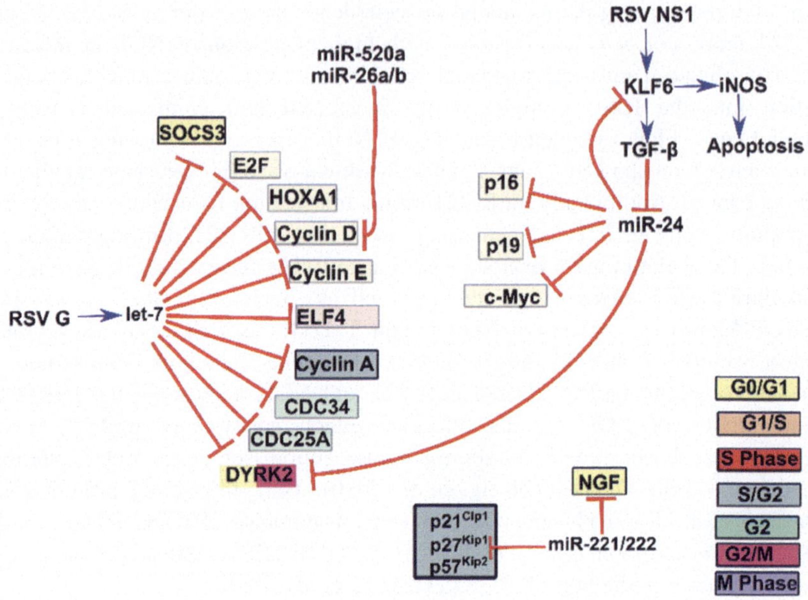

Fig. 1 Involvement of miRNAs in regulating different stages of cell cycle during RSV infection. *Red line* indicates repression. *Purple arrows* indicate induction. Gene names are grouped according to stage of cell cycle

independently regulated by let-7 and miR-24 miRNAs also showed increased induction upon concomitant inhibition of let-7f and miR-24 suggesting that these two miRNAs can work cooperatively on shared target genes.

3.2 Viral Regulators of Host miRNA Expression

The role of viral genes in miRNA deregulation is poorly understood. Analysis of let-7f and miR-24 expression following infection of A549 cells with wild type or viruses lacking the G protein showed that let-7f induction was markedly reduced in viruses lacking the G protein. Conversely, treatment of cells with purified RSV G protein led to important induction of miRNA let-7f suggesting a direct role for this gene in miRNA induction (Bakre et al. 2012). These data are important since the RSV G protein has been shown to have important roles in molecular mimicry and immunomodulation (Tripp et al. 2001). The carboxy terminal ectodomain of RSV G protein contains two mucin-like domains separated by a non-glycosylated central conserved region (CCR). RSV G CCR contains a CX3C motif formed from four key cysteine residues in this region (a.a 173–186) that form a cysteine noose via disulfide bonding (Gorman et al. 1997). The CX3C motif mimics an identical motif found in fractalkine (CX3CL1/FKN), interacting with the fractalkine receptor CX3CR1 instead (Tripp et al. 2001). CX3CR1 has been recently identified as a RSV co-receptor (Johnson et al. 2015; Chirkova et al. 2015) and this molecular mimicry between RSV G CX3C and CX3CR1 results in G protein-induced lymphocyte chemotaxis which can be inhibited by anti-G protein antibody (Tripp et al. 2001). Disrupting the CX3C-CX3CR1 interaction has important immunological outcomes; it reduces TNFα and type I IFN production in vitro (Chirkova et al. 2013), alters $CX3CR1^{+}$ cell trafficking, cytokine, chemokine, and substance P expression (Harcourt et al. 2006; Tripp et al. 2003), reduces viral load, bronchoalveolar lavage (BAL) leukocyte infiltration (Haynes et al. 2009) and immune cell trafficking, and IFNα and IL-4 levels in BAL of CX3CR1-deficient mice infected with RSV (Johnson et al. 2012). Anti-RSV G protein prophylaxis or treatment reduces pulmonary mucus production, inhibits infection-induced airway dysfunction, and reduces pulmonary cell infiltration in response to RSV infection (Boyoglu-Barnum et al. 2015; Haynes et al. 2009; Chirkova et al. 2013). Moreover, our preliminary data suggest that the organization of the CX3C motif alters miRNA deregulation. Mutations in the CX3C motif alter the profile and tempo of miRNA expression during infection and can thus affect downstream signaling pathways involved in viral replication and or host immune response.

Our lab has observed that the RSV NS1 protein can suppress miRNA expression via induction of transcription factor KLF6 and Tumor Growth Factor beta (TGF-β) (Bakre et al. 2015). Expression of miRNAs let-7 and miR-24 was significantly induced during infection with viruses lacking NS1 gene. Transient NS1 overexpression led to induction of KLF6, TGF-β, and concomitant silencing of miR-24 expression. Conversely, silencing KLF6 repressed TGF-β and induced miR-24 suggesting that in a wild type infection RSV NS1 can suppress miR-24 activity. Analysis of miR-24 targets identifies numerous cell cycle targets as well as cytokines and suggests that RSV modifies these to alter host cell cycle dynamics (Fig. 2). Since miRNAs may regulate multiple gene networks, it is important to understand if miRNA activity is essential for viral replication. Transfection of A549 cells with miRNA specific inhibitors followed by infection led to repression of viral replication suggesting that the miRNAs investigated have pro-viral function.

3.3 *miRNA Biomarkers of RSV Disease*

miRNA expression patterns and tempo can affect both disease states and progression. Examining the miRNA expression profiles in nasal mucosa specimens from a small group of RSV-positive and -negative infants, and comparing the expression profiles using quantitative PCR, it was shown that miRNA biomarkers could be observed between age and gender but not for individual miRNA expression. However, a considerable downregulation of some miRNAs were observed specifically for miR-34b, miR-34c, miR-125b, miR-29c, miR-125a, miR-429, and miR-27b, and upregulation of miR-155, miR-31, miR-203a, miR-16, and let-7d in RSV-positive infants relative to healthy controls (Inchley et al. 2015). The expression of miR-125a and miR-429 were appreciably different between infants with mild and severe disease. These data suggest that miRNA expression profiles may be able to discern between disease state and or severity. The differentially expressed miRNAs were predicted to target multiple pathways such as NF-κβ signaling, regulation of apoptosis, cellular proliferation, maturation of DCs, polarization of T helper 1 (Th1) cells, upregulation of Toll-like receptor 4 (TLR4), repression of inflammatory innate responses, induction of nitric oxide production (Inchley et al. 2015) suggesting that a miRNA response to RSV infection can have both diagnostic and prognostic value if the miRNAs are repeatable and reliable.

While these miRNA findings suggest there are gene pathways under post-transcriptional regulation during RSV infection, the impact of miRNAs on regulating host and possibly viral gene regulation is global. For example, a number of miRNAs have been identified that showed differential expression upon RSV infection, have focused on deregulation caused by miR-221 (Othumpangat et al. 2012). Many of these miRNAs have isoforms transcribed from the 5′ and 3′ end arms and it is not clear which mature miRNAs were deregulated. However, pathway prediction analysis of these miRNAs using DIANA miRPath v.3 (Vlachos et al. 2015) shows that these miRNAs alone or together affect multiple pathways. Similarly, in our recent microarray analysis of miRNA expression following RSV infection from our lab (Bakre et al. 2012), targets for miRNA regulation were chosen based on their differential expression during RSV infection (Martinez et al. 2007). These data likely represent a small subset of host genes affected during RSV infection, and it is possible we have failed to identify genes/pathways that may have important roles in mRNA regulation of RSV replication and disease (Fig. 2).

3.4 Biogenesis and Function of transfer RNA Fragments During RSV Infection

It has been demonstrated that RSV infection led to a significant increase in three 5′-tRFs, Gly-GCC, Glu-CTC, and Cys-GCA (Wang et al. 2013). Time-course studies showed that these tRFs were detected as early as 6 h pi and were abundant at 24 h pi. Expression of Glu-CTC was replication dependent and tRFs were located predominantly in the cytosol suggesting that mechanisms of tRF biogenesis probably involved the RTD pathway. Anti-sense inhibition of the Glu-CTC reduced viral replication while overexpression of tRF-Glu-CTC led to increased viral replication in respiratory epithelial cells. Knockdown of Angiogenin (ANG) but not Dicer/Drosha/ELAC2/RNAseL reduced Glu-CTC production validating ANG as the cytosolic endonuclease responsible for Glu-CTC production.

In a related study, it was demonstrated that the 3′ end of tRF-Glu-CTC functions as a siRNA targeting the 3′ UTR of apolipoprotein ER 2 (APOER2) and reducing

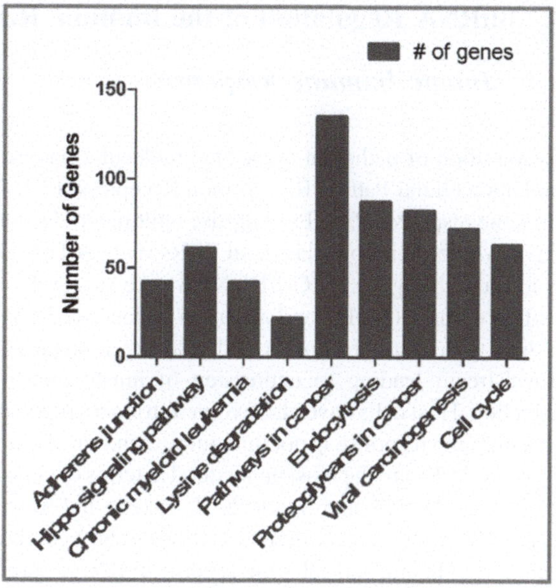

Fig. 2 Potential miRNA regulated pathways during RSV replication. Potential pathways regulated by differentially expressed miRs identified in (Bakre et al. 2012; Othumpangat et al. 2012; Inchley et al. 2015) were analyzed using DIANA miRPath v.3. Top 10 statistically significant ($p < 0.05$) pathways are shown

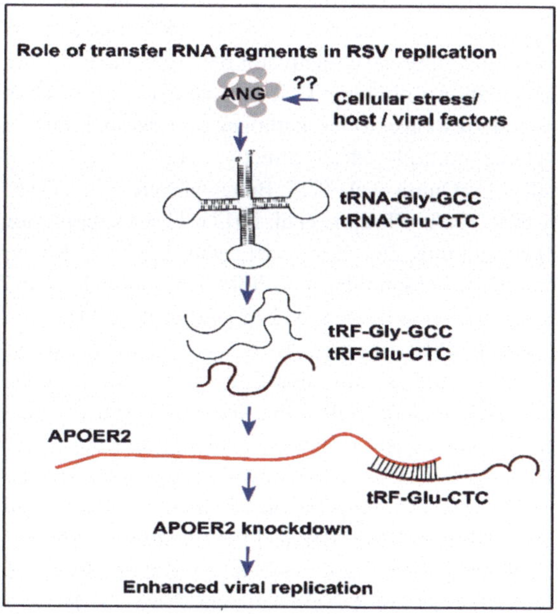

Fig. 3 Role of transfer RNA fragments (tRFs) in RSV replication. Cellular stress or host/viral factors induce Angiogenin (ANG) activity leading to 5′tRF production and silencing of APOER2 and increased viral replication

APOER2 expression, an antiviral factor (Deng et al. 2015). APOER2 is a low density lipoprotein receptor (LDLR) that is involved in receptor mediated endocytosis of ligands for lysosomal degradation and as a receptor for cholesterol transport apolipoprotein E (Fig. 3).

4 miRNA Regulation of the Immune Response

4.1 *Innate Immune Response*

Recognition of pathogen-associated molecular patterns (PAMPs) using cell surface and intracellular Pattern Recognition Receptors (PRRs) is the cornerstone of the host immune response. PAMPs comprise multiple molecular classes and eukaryotic hosts have evolved multiple classes of PRRs such as Toll-Like receptors (TLRs), retinoic acid inducible gene (RIG-I/DDX-58)-like receptors (RLRs), nucleotide oligomerization domain (NOD), and absent in melanoma 2 (AIM2)-like receptors, as well as C-like type lectin receptors and scavenger receptors (SRs) to detect and signal downstream leading to expression of innate effector molecules that respond to infection. Host cells have also evolved stringent regulatory mechanisms to ensure that this immune response is not unregulated and miRNAs have an important role in this process. TLRs are the most understood in terms of signaling and expression of effector molecules (O'Neill and Bowie 2007; Takeda and Akira 2004; Li and Shi 2013) and it is now understood that different cell types and ligands in the PAMP-TLR interaction can lead to ligand and cell type-specific miRNA expression and regulation (Taganov et al. 2006; Cremer et al. 2009; Eis et al. 2005; Tili et al. 2007; Moschos et al. 2007; Liu et al. 2009; Bazzoni et al. 2009; Jennewein et al. 2010). Similarly, miR-132 expression is induced by cyclic AMP response element binding protein and post-transcriptional co-activator p300 (Nahid et al. 2011; Lagos et al. 2010). miRNA deregulation following TLR engagement is a mechanism to hone and amplify the immune response to the pathogen and dampen host antagonistic activity. MiRNAs regulate multiple PRRs directly (TLR2 repression by miR-105, miR-146a, and miR-143 (Quinn et al. 2013; Benakanakere et al. 2009; Guo et al. 2013), upregulation by miR-19a/b (Philippe et al. 2012), TLR3 suppression by miR-223 (Johnnidis et al. 2008) and miR-26a (Benveniste et al. 1987), TLR4 suppression by miRNAs let-7e/-7i, miR-223 (Johnnidis et al. 2008; Androulidaki et al. 2009; O'Hara et al. 2010), and TLR4 induction by miR-511 (Tserel et al. 2011).

PAMP-PRR engagement is modulated via various adaptor and effector molecules and studies show that these molecules are under miRNA regulation as well. Examples include miR-146a and miR-146b regulation of IRAK1/IRAK2 kinase and TRAF6 ligases (Taganov et al. 2006; Hou et al. 2009; Li et al. 2013; Lin et al. 2013; Curtale et al. 2013), miR-29 suppression of TRAF4 expression (Ahmed et al. 2013), miR-155 suppression of multiple NF-κβ pathway members (Ceppi et al. 2009; Tili et al. 2007), miR-155 regulation of p38 MAPK signaling, TAB2 activity (Ceppi et al. 2009), and MyD88 (Tang et al. 2010) activity along with miR-149 (Xu et al. 2014) and miR-203 (Wei et al. 2013). MiR-145 regulates MyD88 adaptor-like protein TIRAP (Starczynowski et al. 2010) while miR-200b/c (Wendlandt et al. 2012) and miR-21(Chen et al. 2013) regulate MyD88 directly. MiR-346 targets BTK, a key kinase involved in TLR4, TLR7, TLR8, and TLR9 signaling pathways (Alsaleh et al. 2009; Horwood et al. 2003).

Signaling cascades emanating from PAMP-PRR interaction culminate in expression of cytokines and other anti-microbial moieties. miRNAs can regulate key transcription factors, e.g., miR-329 targets NF-κβp65 (Garg et al. 2013), miR-17-5p and miR-20a and miR-223 target STAT3 (Zhang et al. 2011; Chen et al. 2012), miR-155 targets CCAAT/enhancer binding protein-β (CEBP) (Worm et al. 2009), FOXP3, and E26 (Kohlhaas et al. 2009; Quinn et al. 2014) while miR-132 targets p300 (Lagos et al. 2010) and directly control the expression of interferon stimulated genes (ISGs) (Li and Shi 2013; Martinez and Walhout 2009; Filipowicz et al. 2008; Olivieri et al. 2013; O'Neill et al. 2011; Nahid et al. 2011; Ma et al. 2011; Zhou et al. 2011; Coll and O'Neill 2010). miR-9 (Bazzoni et al. 2009) and miR-210 (Qi et al. 2012) have been shown to directly target the NF-κβ1 transcript. Additionally, miR-4661 suppresses IFN-α (Li et al. 2012), miR-26a, miR-34a miR-145 and let-7b suppress IFN-β (Zhou et al. 2011; Witwer et al. 2010), miR-125b and miR-187 suppress tumor necrosis factor (TNF) (Tili et al. 2007; Rossato et al. 2012), miR-16, miR-365, miR-142-3p suppresses IL-6 (Zhou et al. 2011; Xu et al. 2011; Sun et al. 2011), miR-106a/b suppress IL-10 (Sharma et al. 2009), miR-21 suppresses IL-12 (Lu et al. 2009) and miR-29 suppresses IFN-γ (Ma et al. 2011). This is affected via binding to AU-rich elements (AREs) in their 3′ UTRs or by destabilizing the interactions with other RNA binding proteins. MiR-16 and miR-221 accelerate TNF degradation by destabilizing tristetraprolin (TTP) (Jing et al. 2005), an important RNA binding protein while miR-579 and miR125b reduce TNF translation by binding to TTP (Tili et al. 2007; El Gazzar and McCall 2010).

miRNAs can also regulate signaling induced upon cytokine binding to its cognate receptor. MiR-155 suppresses SOCS1 (Wang et al. 2010), miR-146a regulates Notch1, a positive regulator of IL-12p70 (Bai et al. 2012), miR-98 and let-7 miRs regulate cytokine inducible Src homology 2 (CIS) and SOCS4 proteins (Hu et al. 2010). Figure 4 summarizes our present understanding of these mechanisms.

4.2 Adaptive Immune Response

miRNAs also regulate the adaptive response to infection by regulating the development, maturation, and function of the cellular arm of immunity. These include T, B, and antigen presenting cells (APCs). For example, miRNA-155 regulates T- and B-cell development from hematopoietic stem cells (HSCs) in the bone marrow (Cobb et al. 2005; Muljo et al. 2005; Rodriguez et al. 2007). Similarly, miR-125b controls the size of the HSC compartment in mice (Ooi et al. 2010; Surdziel et al. 2011), and is repressed upon lineage commitment (Kirigin et al. 2012; O'Connell et al. 2010). Additionally, miR-181a and miR-150 exhibit characteristic deregulation (Henao-Mejia et al. 2013); inhibition of miR-181a impairs T-cell selection (Curtale and Citarella 2013). miR-125b regulates key T-cell developmental cytokines such as Interferon-γ (IFNγ), IL-2 receptor beta (IL2RB), IL-10 receptor alpha (IL10Rα), and the interferon repressor, Blimp-1 in naïve CD4 + T cells (Rossi et al. 2011), while miRNA-182 regulates T helper cell proliferation via the transcription factor Forkhead box O1 (Foxo1) (Bronevetsky et al. 2013). These data show that miRNAs and potentially other non-coding RNAs have a huge impact on the host response to infection.

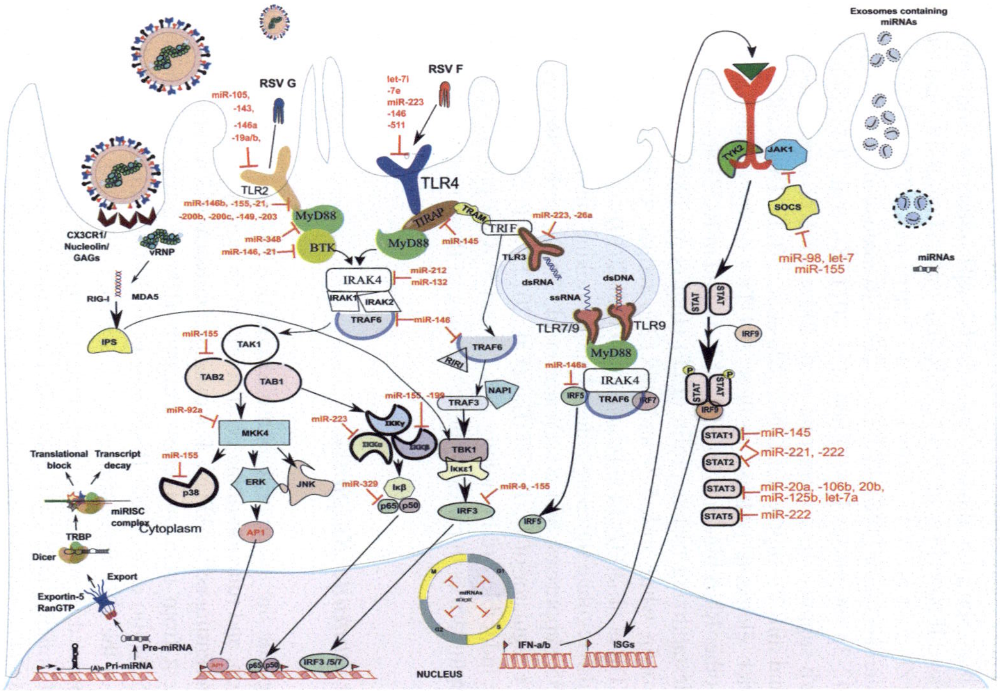

Fig. 4 MiRNA regulation of the immune response. An overview of miRNA biogenesis and the regulation of the innate immune response during RSV infection is shown. *Black* arrows indicate induction while *red lines* denote inhibition of gene function/activity. Only prominent nodes are shown

5 Future Directions

Non-coding RNAs have a significant impact on host gene expression both during normal physiological conditions as well as during infection and stress. Present data suggest that the functions of ncRNAs are poorly understood and further research into the roles of ncRNAs is needed to help accelerated identification and development of anti-RSV prophylactics and therapeutics. Investigations into other ncRNA classes and impact of RSV infection are needed to gain mechanistic insights into the host-virus interface.

References

Agoti CN et al (2015) Successive respiratory syncytial virus epidemics in local populations arise from multiple variant introductions, providing insights into virus persistence. J Virol 89 (22):11630–11642

Ahmed F et al (2013) Tumor necrosis factor receptor associated factor-4: an adapter protein overexpressed in metastatic prostate cancer is regulated by microRNA-29a. Oncol Rep 30 (6):2963–2968

Akerlind-Stopner B et al (1993) Antibody responses of children to the C-terminal peptide of the SH protein of respiratory syncytial virus and the immunological characterization of this protein. J Med Virol 40(2):112–120

Alsaleh G et al (2009) Bruton's tyrosine kinase is involved in miR-346-related regulation of IL-18 release by lipopolysaccharide-activated rheumatoid fibroblast-like synoviocytes. J Immunol 182(8):5088–5097

Anderson JT (2005) RNA turnover: unexpected consequences of being tailed. Curr Biol 15(16): R635–R638

Androulidaki A et al (2009) The kinase Akt1 controls macrophage response to lipopolysaccharide by regulating microRNAs. Immunity 31(2):220–231

Asenjo A, Calvo E, Villanueva N (2006) Phosphorylation of human respiratory syncytial virus P protein at threonine 108 controls its interaction with the M2-1 protein in the viral RNA polymerase complex. J Gen Virol 87(Pt 12):3637–3642

Asenjo A, Gonzalez-Armas JC, Villanueva N (2008) Phosphorylation of human respiratory syncytial virus P protein at serine 54 regulates viral uncoating. Virology 380(1):26–33

Babiarz JE et al (2008) Mouse ES cells express endogenous shRNAs, siRNAs, and other Microprocessor-independent dicer-dependent small RNAs. Genes Dev 22(20):2773–2785

Bai Y et al (2012) Integrin CD11b negatively regulates TLR9-triggered dendritic cell cross-priming by upregulating microRNA-146a. J Immunol 188(11):5293–5302

Bakre A et al (2012) Respiratory syncytial virus modifies microRNAs regulating host genes that affect virus replication. J Gen Virol 93(Pt 11):2346–2356

Bakre A et al (2015) Human respiratory syncytial virus non-structural protein NS1 modifies miR-24 expression via transforming growth factor-beta. J Gen Virol 96(11):3179–3191

Batagov AO, Kurochkin IV (2013) Exosomes secreted by human cells transport largely mRNA fragments that are enriched in the 3′-untranslated regions. Biol Direct 8:12

Bazzoni F et al (2009) Induction and regulatory function of miR-9 in human monocytes and neutrophils exposed to proinflammatory signals. Proc Natl Acad Sci USA 106(13):5282–5287

Berezikov E et al (2007) Mammalian mirtron genes. Mol Cell 28(2):328–336

Borchert GM, Lanier W, Davidson BL (2006) RNA polymerase III transcribes human microRNAs. Nat Struct Mol Biol 13(12):1097–1101

Boyoglu-Barnum S et al (2013) A respiratory syncytial virus (RSV) anti-G protein F(ab')2 monoclonal antibody suppresses mucous production and breathing effort in RSV rA2-line19F-infected BALB/c mice. J Virol 87(20):10955–10967

Boyoglu-Barnum S et al (2015) An anti-G protein monoclonal antibody treats RSV disease more effectively than an anti-F monoclonal antibody in BALB/c mice. Virology 483:117–125

Broadbent L et al (2015) Respiratory syncytial virus, an ongoing medical dilemma: an expert commentary on respiratory syncytial virus prophylactic and therapeutic pharmaceuticals currently in clinical trials. Influenza Other Respir Viruses 9(4):169–178

Bacharier LB et al. (2012) Determinants of asthma after severe respiratory syncytial virus bronchiolitis. J Allergy Clin Immunol 130(1): 91–100 (e3)

Bramley AM et al (1999) Effects of respiratory syncytial virus persistence on airway responsiveness and inflammation in guinea-pigs. Eur Respir J 14(5):1061–1067

Benakanakere MR et al (2009) Modulation of TLR2 protein expression by miR-105 in human oral keratinocytes. J Biol Chem 284(34):23107–23115

Benito-Martin A et al (2013) Osteoprotegerin in exosome-like vesicles from human cultured tubular cells and urine. PLoS One 8(8):e72387

Benveniste EN, Herman PK, Whitaker JN (1987) Myelin basic protein-specific RNA levels in interleukin-2-stimulated oligodendrocytes. J Neurochem 49(4):1274–1279

Bi S et al (2015) Correlation between serum exosome derived miR-208a and acute coronary syndrome. Int J Clin Exp Med 8(3):4275–4280

Bronevetsky Y et al (2013) T cell activation induces proteasomal degradation of Argonaute and rapid remodeling of the microRNA repertoire. J Exp Med 210(2):417–432

Cai H et al (2015) Zinc binding activity of human metapneumovirus M2-1 protein is indispensable for viral replication and pathogenesis in vivo. J Virol 89(12):6391–6405

Caidi H et al (2012) Combination therapy using monoclonal antibodies against respiratory syncytial virus (RSV) G glycoprotein protects from RSV disease in BALB/c mice. PLoS One 7 (12):e51485

Campo-Paysaa F et al (2011) microRNA complements in deuterostomes: origin and evolution of microRNAs. Evol Dev 13(1):15–27

Castellano L, Stebbing J (2013) Deep sequencing of small RNAs identifies canonical and non-canonical miRNA and endogenous siRNAs in mammalian somatic tissues. Nucleic Acids Res 41(5):3339–3351

Ceppi M et al (2009) MicroRNA-155 modulates the interleukin-1 signaling pathway in activated human monocyte-derived dendritic cells. Proc Natl Acad Sci USA 106(8):2735–2740

Chak LL et al (2015) A deeply conserved, noncanonical miRNA hosted by ribosomal DNA. RNA 21(3):375–384

Chan PP, Lowe TM (2016) GtRNAdb 2.0: an expanded database of transfer RNA genes identified in complete and draft genomes. Nucleic Acids Res 44(D1):D184–D189

Chen SY et al (2008) The genomic analysis of erythrocyte microRNA expression in sickle cell diseases. PLoS One 3(6):e2360

Chen Q et al (2012) Inducible microRNA-223 down-regulation promotes TLR-triggered IL-6 and IL-1beta production in macrophages by targeting STAT3. PLoS One 7(8):e42971

Chen CY, Hogan MC, Ward CJ (2013a) Purification of exosome-like vesicles from urine. Methods Enzymol 524:225–241

Chen Y et al (2013b) HCV-induced miR-21 contributes to evasion of host immune system by targeting MyD88 and IRAK1. PLoS Pathog 9(4):e1003248

Chernyakov I et al (2008) Degradation of several hypomodified mature tRNA species in Saccharomyces cerevisiae is mediated by Met22 and the 5′–3′ exonucleases Rat1 and Xrn1. Genes Dev 22(10):1369–1380

Chirkova T et al (2013) Respiratory syncytial virus G protein CX3C motif impairs human airway epithelial and immune cell responses. J Virol 87(24):13466–13479

Chirkova T et al (2015) CX3CR1 is an important surface molecule for respiratory syncytial virus infection in human airway epithelial cells. J Gen Virol 96(9):2543–2556

Choi EH, Lee HJ, Chanock SJ (2013) Human genetics and respiratory syncytial virus disease: current findings and future approaches. Curr Top Microbiol Immunol 372:121–137

Cobb BS et al (2005) T cell lineage choice and differentiation in the absence of the RNase III enzyme Dicer. J Exp Med 201(9):1367–1373

Cole C et al (2009) Filtering of deep sequencing data reveals the existence of abundant Dicer-dependent small RNAs derived from tRNAs. RNA 15(12):2147–2160

Collins PL, Mottet G (1993) Membrane orientation and oligomerization of the small hydrophobic protein of human respiratory syncytial virus. J Gen Virol 74(Pt 7):1445–1450

Coll RC, O'Neill LA (2010) New insights into the regulation of signalling by toll-like receptors and nod-like receptors. J Innate Immun 2(5):406–421

Conde-Vancells J et al (2010) Candidate biomarkers in exosome-like vesicles purified from rat and mouse urine samples. Proteomics Clin Appl 4(4):416–425

Connors M et al (1991) Respiratory syncytial virus (RSV) F, G, M2 (22 K), and N proteins each induce resistance to RSV challenge, but resistance induced by M2 and N proteins is relatively short-lived. J Virol 65(3):1634–1637

Cooper DA et al (2015) RNase L targets distinct sites in influenza A virus RNAs. J Virol 89 (5):2764–2776

Cremer TJ et al (2009) MiR-155 induction by F. novicida but not the virulent F. tularensis results in SHIP down-regulation and enhanced pro-inflammatory cytokine response. PLoS One 4:8508

Crowe JE Jr et al (1996) Acquisition of the ts phenotype by a chemically mutagenized cold-passaged human respiratory syncytial virus vaccine candidate results from the acquisition of a single mutation in the polymerase (L) gene. Virus Genes 13(3):269–273

Curtale G, Citarella F (2013) Dynamic nature of noncoding RNA regulation of adaptive immune response. Int J Mol Sci 14(9):17347–17377

Curtale G et al (2013) Negative regulation of Toll-like receptor 4 signaling by IL-10-dependent microRNA-146b. Proc Natl Acad Sci USA 110(28):11499–11504

Czech A et al (2013) Reversible and rapid transfer-RNA deactivation as a mechanism of translational repression in stress. PLoS Genet 9(8):e1003767

Consortium RN (2015) RNAcentral: an international database of ncRNA sequences. Nucleic Acids Res 43(Database issue): D123–9

Dakhama A, Vitalis TZ, Hegele RG (1997) Persistence of respiratory syncytial virus (RSV) infection and development of RSV-specific IgG1 response in a guinea-pig model of acute bronchiolitis. Eur Respir J 10(1):20–26

de Nadal E, Ammerer G, Posas F (2011) Controlling gene expression in response to stress. Nat Rev Genet 12(12):833–845

Deng J et al (2015) Respiratory syncytial virus utilizes a tRNA fragment to suppress antiviral responses through a novel targeting mechanism. Mol Ther 23(10):1622–1629

Dittmar KA, Goodenbour JM, Pan T (2006) Tissue-specific differences in human transfer RNA expression. PLoS Genet 2(12):e221

Eis PS et al (2005) Accumulation of miR-155 and BIC RNA in human B cell lymphomas. Proc Natl Acad Sci USA 102:3627–3632

Ender C et al (2008) A human snoRNA with microRNA-like functions. Mol Cell 32(4):519–528

Estripeaut D et al (2008) Respiratory syncytial virus persistence in the lungs correlates with airway hyperreactivity in the mouse model. J Infect Dis 198(10):1435–1443

El Gazzar M, McCall CE (2010) MicroRNAs distinguish translational from transcriptional silencing during endotoxin tolerance. J Biol Chem 285(27):20940–20951

Feldman SA, Hendry RM, Beeler JA (1999) Identification of a linear heparin binding domain for human respiratory syncytial virus attachment glycoprotein G. J Virol 73(8):6610–6617

Feldman SA, Audet S, Beeler JA (2000) The fusion glycoprotein of human respiratory syncytial virus facilitates virus attachment and infectivity via an interaction with cellular heparan sulfate. J Virol 74(14):6442–6447

Feldman AS et al (2015) Toward primary prevention of asthma. Reviewing the evidence for early-life respiratory viral infections as modifiable risk factors to prevent childhood asthma. Am J Respir Crit Care Med 191(1):34–44

Fix J et al (2011) The insertion of fluorescent proteins in a variable region of respiratory syncytial virus L polymerase results in fluorescent and functional enzymes but with reduced activities. Open Virol J 5:103–108

Filipowicz W, Bhattacharyya SN, Sonenberg N (2008) Mechanisms of post-transcriptional regulation by microRNAs: are the answers in sight? Nat Rev Genet 9(2):102–114

Fuentes S et al (2007) Function of the respiratory syncytial virus small hydrophobic protein. J Virol 81(15):8361–8366

Gaona J et al (2014) Respiratory syncytial virus persistence in macrophages upregulates Fcgamma receptors expression. Viruses 6(2):624–639

Garg M, Potter JA, Abrahams VM (2013) Identification of microRNAs that regulate TLR2-mediated trophoblast apoptosis and inhibition of IL-6 mRNA. PLoS One 8(10):e77249

Gebetsberger J, Polacek N (2013) Slicing tRNAs to boost functional ncRNA diversity. RNA Biol 10(12):1798–1806

Gorman JJ et al (1997) Determination of the disulfide bond arrangement of human respiratory syncytial virus attachment (G) protein by matrix-assisted laser desorption/ionization time-of-flight mass spectrometry. Protein Sci 6(6):1308–1315

Guo H et al (2013) The regulation of Toll-like receptor 2 by miR-143 suppresses the invasion and migration of a subset of human colorectal carcinoma cells. Mol Cancer 12:77

Gan SW et al (2012) The small hydrophobic protein of the human respiratory syncytial virus forms pentameric ion channels. J Biol Chem 287(29):24671–24689

Ghildyal R, Ho A, Jans DA (2006) Central role of the respiratory syncytial virus matrix protein in infection. FEMS Microbiol Rev 30(5):692–705

Grimson A et al (2008) Early origins and evolution of microRNAs and Piwi-interacting RNAs in animals. Nature 455(7217):1193–1197

Grosfeld H, Hill MG, Collins PL (1995) RNA replication by respiratory syncytial virus (RSV) is directed by the N, P, and L proteins; transcription also occurs under these conditions but requires RSV superinfection for efficient synthesis of full-length mRNA. J Virol 69(9):5677–5686

Hall CB (2012) The burgeoning burden of respiratory syncytial virus among children. Infect Disord Drug Targets 12(2):92–97

Hallak LK, Kwilas SA, Peeples ME (2007) Interaction between respiratory syncytial virus and glycosaminoglycans, including heparan sulfate. Methods Mol Biol 379:15–34

Hamada N et al (2012) MicroRNA expression profiling of NGF-treated PC12 cells revealed a critical role for miR-221 in neuronal differentiation. Neurochem Int 60(8):743–750

Hamilton AJ (2010) MicroRNA in erythrocytes. Biochem Soc Trans 38(Pt 1):229–231

Harcourt JL, Karron RA, Tripp RA (2004) Anti-G protein antibody responses to respiratory syncytial virus infection or vaccination are associated with inhibition of G protein CX3C-CX3CR1 binding and leukocyte chemotaxis. J Infect Dis 190(11):1936–1940

Harcourt J et al (2006) Respiratory syncytial virus G protein and G protein CX3C motif adversely affect CX3CR1 + T cell responses. J Immunol 176(3):1600–1608

Haussecker D et al (2010) Human tRNA-derived small RNAs in the global regulation of RNA silencing. RNA 16(4):673–695

Haynes LM et al (2003) Enhanced disease and pulmonary eosinophilia associated with formalin-inactivated respiratory syncytial virus vaccination are linked to G glycoprotein CX3C-CX3CR1 interaction and expression of substance P. J Virol 77(18):9831–9844

Haynes LM et al (2009) Therapeutic monoclonal antibody treatment targeting respiratory syncytial virus (RSV) G protein mediates viral clearance and reduces the pathogenesis of RSV infection in BALB/c mice. J Infect Dis 200(3):439–447

Hegele RG et al (1994) Persistence of respiratory syncytial virus genome and protein after acute bronchiolitis in guinea pigs. Chest 105(6):1848–1854

Heminway BR et al (1994) Analysis of respiratory syncytial virus F, G, and SH proteins in cell fusion. Virology 200(2):801–805

Henao-Mejia J et al (2013) The microRNA miR-181 is a critical cellular metabolic rheostat essential for NKT cell ontogenesis and lymphocyte development and homeostasis. Immunity 38(5):984–997
Holt PG (2015) The mechanism or mechanisms driving atopic asthma initiation: the infant respiratory microbiome moves to center stage. J Allergy Clin Immunol 136(1):15–22
Ho DH et al (2014) Increased DJ-1 in urine exosome of Korean males with Parkinson's disease. Biomed Res Int 2014:704678
Hopper AK (2013) Transfer RNA post-transcriptional processing, turnover, and subcellular dynamics in the yeast Saccharomyces cerevisiae. Genetics 194(1):43–67
Hornick NI et al (2015) Serum exosome MicroRNA as a minimally-invasive early biomarker of AML. Sci Rep 5:11295
Horwood NJ et al (2003) Bruton's tyrosine kinase is required for lipopolysaccharide-induced tumor necrosis factor alpha production. J Exp Med 197(12):1603–1611
Hou J et al (2009) MicroRNA-146a feedback inhibits RIG-I-dependent Type I IFN production in macrophages by targeting TRAF6, IRAK1, and IRAK2. J Immunol 183(3):2150–2158
Hu G et al (2010) MicroRNA-98 and let-7 regulate expression of suppressor of cytokine signaling 4 in biliary epithelial cells in response to Cryptosporidium parvum infection. J Infect Dis 202 (1):125–135
Huang H et al (2015) Bulk RNA degradation by nitrogen starvation-induced autophagy in yeast. EMBO J 34(2):154–168
Iben JR, Maraia RJ (2014) tRNA gene copy number variation in humans. Gene 536(2):376–384
Inchley CS et al (2015) Nasal mucosal microRNA expression in children with respiratory syncytial virus infection. BMC Infect Dis 15:150
Jennewein C et al (2010) MicroRNA-27b contributes to lipopolysaccharide-mediated peroxisome proliferator-activated receptor gamma (PPARgamma) mRNA destabilization. J Biol Chem 285 (16):11846–11853
Jing Q et al (2005) Involvement of microRNA in AU-rich element-mediated mRNA instability. Cell 120(5):623–634
Johnnidis JB et al (2008) Regulation of progenitor cell proliferation and granulocyte function by microRNA-223. Nature 451(7182):1125–1129
Johnson CH et al (2012) Effect of chemokine receptor CX3CR1 deficiency in a murine model of respiratory syncytial virus infection. Comp Med 62(1):14–20
Johnson SM et al (2015) Respiratory syncytial virus uses CX3CR1 as a receptor on primary human airway epithelial cultures. PLoS Pathog 11(12):e1005318
Konig B et al (1996) IL-8 release from human neutrophils by the respiratory syncytial virus is independent of viral replication. J Leukoc Biol 60(2):253–260
Kadaba S et al (2004) Nuclear surveillance and degradation of hypomodified initiator tRNAMet in S. cerevisiae. Genes Dev 18(11):1227–1240
Kannan M, Atreya C (2010) Differential profiling of human red blood cells during storage for 52 selected microRNAs. Transfusion 50(7):1581–1588
Karron RA et al (1997) Respiratory syncytial virus (RSV) SH and G proteins are not essential for viral replication in vitro: clinical evaluation and molecular characterization of a cold-passaged, attenuated RSV subgroup B mutant. Proc Natl Acad Sci USA 94(25):13961–13966
Kirchner S, Ignatova Z (2015) Emerging roles of tRNA in adaptive translation, signalling dynamics and disease. Nat Rev Genet 16(2):98–112
Kirigin FF et al (2012) Dynamic microRNA gene transcription and processing during T cell development. J Immunol 188(7):3257–3267
Kochva U, Leonov H, Arkin IT (2003) Modeling the structure of the respiratory syncytial virus small hydrophobic protein by silent-mutation analysis of global searching molecular dynamics. Protein Sci 12(12):2668–2674
Kohlhaas S et al (2009) Cutting edge: the Foxp3 target miR-155 contributes to the development of regulatory T cells. J Immunol 182(5):2578–2582

Kosanovic M, Jankovic M (2014) Isolation of urinary extracellular vesicles from Tamm- Horsfall protein-depleted urine and their application in the development of a lectin-exosome-binding assay. Biotechniques 57(3):143–149
Kulkarni S et al. (2016) Identifying urinary and serum exosome biomarkers for radiation exposure using a data dependent acquisition and SWATH-MS combined workflow. Int J Radiat Oncol Biol Phys 2016 Nov 1. 96(3):566–77. doi: 10.1016/j.ijrobp.2016.06.008
LaCava J et al (2005) RNA degradation by the exosome is promoted by a nuclear polyadenylation complex. Cell 121(5):713–724
Lagos D et al (2010) miR-132 regulates antiviral innate immunity through suppression of the p300 transcriptional co-activator. Nat Cell Biol 12(5):513–519
Laganas VA et al (2015) Characterization of novel respiratory syncytial virus inhibitors identified by high throughput screen. Antiviral Res 115:71–74
Lander ES et al (2001) Initial sequencing and analysis of the human genome. Nature 409 (6822):860–921
Lee Y et al (2004) MicroRNA genes are transcribed by RNA polymerase II. EMBO J 23 (20):4051–4060
Lee YS et al (2009) A novel class of small RNAs: tRNA-derived RNA fragments (tRFs). Genes Dev 23(22):2639–2649
Li Y, Shi X (2013) MicroRNAs in the regulation of TLR and RIG-I pathways. Cell Mol Immunol 10(1):65–71
Li Y et al (2012) MicroRNA-466 l inhibits antiviral innate immune response by targeting interferon-alpha. Cell Mol Immunol 9(6):497–502
Li S et al (2013) MicroRNA-146a represses mycobacteria-induced inflammatory response and facilitates bacterial replication via targeting IRAK-1 and TRAF-6. PLoS One 8(12):e81438
Lin L et al (2013) Type I IFN inhibits innate IL-10 production in macrophages through histone deacetylase 11 by downregulating microRNA-145. J Immunol 191(7):3896–3904
Liu G et al (2009) miR-147, a microRNA that is induced upon Toll-like receptor stimulation, regulates murine macrophage inflammatory responses. Proc Natl Acad Sci USA 106 (37):15819–15824
Lu TX, Munitz A, Rothenberg ME (2009) MicroRNA-21 is up-regulated in allergic airway inflammation and regulates IL-12p35 expression. J Immunol 182(8):4994–5002
Li XQ et al (2006) Respiratory syncytial virus (RSV) infects neuronal cells and processes that innervate the lung by a process involving RSV G protein. J Virol 80(1):537–540
Li Y et al (2014) Inhibition of the human respiratory syncytial virus small hydrophobic protein and structural variations in a bicelle environment. J Virol 88(20):11899–11914
Liuzzi M et al (2005) Inhibitors of respiratory syncytial virus replication target cotranscriptional mRNA guanylylation by viral RNA-dependent RNA polymerase. J Virol 79(20):13105–13115
Luongo C et al (2009) Codon stabilization analysis of the "248" temperature sensitive mutation for increased phenotypic stability of respiratory syncytial virus vaccine candidates. Vaccine 27 (41):5667–5676
Ma X et al (2011a) MicroRNAs in NF-kappaB signaling. J Mol Cell Biol 3(3):159–166
Ma F et al (2011b) The microRNA miR-29 controls innate and adaptive immune responses to intracellular bacterial infection by targeting interferon-gamma. Nat Immunol 12(9):861–869
Madhavan B et al (2015) Combined evaluation of a panel of protein and miRNA serum-exosome biomarkers for pancreatic cancer diagnosis increases sensitivity and specificity. Int J Cancer 136(11):2616–2627
Maraia RJ, Lamichhane TN (2011) 3′ processing of eukaryotic precursor tRNAs. Wiley Interdiscip Rev RNA 2(3):362–375
Martinez NJ, Walhout AJ (2009) The interplay between transcription factors and microRNAs in genome-scale regulatory networks. BioEssays 31(4):435–445
Martinez I et al (2007) Distinct gene subsets are induced at different time points after human respiratory syncytial virus infection of A549 cells. J Gen Virol 88(Pt 2):570–581
Maute RL et al (2013) tRNA-derived microRNA modulates proliferation and the DNA damage response and is down-regulated in B cell lymphoma. Proc Natl Acad Sci USA 110(4):1404–1409

McKiernan J et al (2016) A novel urine exosome gene expression assay to predict high-grade prostate cancer at initial biopsy. JAMA Oncol 2(7):882–889

Moschos SA et al (2007) Expression profiling in vivo demonstrates rapid changes in lung microRNA levels following lipopolysaccharide-induced inflammation but not in the anti-inflammatory action of glucocorticoids. BMC Genom 8:240

Mastrangelo P, Hegele RG (2013) RSV fusion: time for a new model. Viruses 5(3):873–885

McCurdy LH, Graham BS (2003) Role of plasma membrane lipid microdomains in respiratory syncytial virus filament formation. J Virol 77(3):1747–1756

McDonald TP et al (2004) Evidence that the respiratory syncytial virus polymerase complex associates with lipid rafts in virus-infected cells: a proteomic analysis. Virology 330(1):147–157

Miranda KC et al (2006) A pattern-based method for the identification of MicroRNA binding sites and their corresponding heteroduplexes. Cell 126(6):1203–1217

Mitra R et al (2012) The human respiratory syncytial virus matrix protein is required for maturation of viral filaments. J Virol 86(8):4432–4443

Moore EC, Barber J, Tripp RA (2008) Respiratory syncytial virus (RSV) attachment and nonstructural proteins modify the type I interferon response associated with suppressor of cytokine signaling (SOCS) proteins and IFN-stimulated gene-15 (ISG15). Virol J 5:116

Mejias A et al (2008) Respiratory syncytial virus persistence: evidence in the mouse model. Pediatr Infect Dis J 27(10 Suppl):S60–S62

Mlinaric-Galinovic G et al (2009) Does the viral subtype influence the biennial cycle of respiratory syncytial virus? Virol J 6(1):1–7

Moore ML, Stokes KL, Hartert TV (2013) The impact of viral genotype on pathogenesis and disease severity: respiratory syncytial virus and human rhinoviruses. Curr Opin Immunol 25 (6):761–768

Muljo SA et al (2005) Aberrant T cell differentiation in the absence of Dicer. J Exp Med 202 (2):261–269

Nair H et al (2010) Global burden of acute lower respiratory infections due to respiratory syncytial virus in young children: a systematic review and meta-analysis. Lancet 375(9725):1545–1555

Nahid MA et al (2011a) Polymicrobial infection with periodontal pathogens specifically enhances microRNA miR-146a in ApoE-/- mice during experimental periodontal disease. Infect Immun 79(4):1597–1605

Nahid MA, Satoh M, Chan EK (2011b) MicroRNA in TLR signaling and endotoxin tolerance. Cell Mol Immunol 8(5):388–403

O'Hara SP et al (2010) NFkappaB p50-CCAAT/enhancer-binding protein beta (C/EBPbeta)-mediated transcriptional repression of microRNA let-7i following microbial infection. J Biol Chem 285(1):216–225

O'Neill LA, Bowie AG (2007) The family of five: TIR-domain-containing adaptors in Toll-like receptor signalling. Nat Rev Immunol 7(5):353–564

O'Connell RM et al (2010) MicroRNAs enriched in hematopoietic stem cells differentially regulate long-term hematopoietic output. Proc Natl Acad Sci USA 107(32):14235–14240

Olivieri F et al (2013) Toll like receptor signaling in "inflammaging": microRNA as new players. Immun Ageing 10(1):11

O'Neill LA, Sheedy FJ, McCoy CE (2011) MicroRNAs: the fine-tuners of Toll-like receptor signalling. Nat Rev Immunol 11(3):163–175

Ooi AG et al (2010) MicroRNA-125b expands hematopoietic stem cells and enriches for the lymphoid-balanced and lymphoid-biased subsets. Proc Natl Acad Sci USA 107(50):21505–21510

Oshansky CM et al (2010) Respiratory syncytial virus F and G proteins induce interleukin 1alpha, CC, and CXC chemokine responses by normal human bronchoepithelial cells. J Infect Dis 201 (8):1201–1207

Ostler T et al (2001) Long-term persistence and reactivation of T cell memory in the lung of mice infected with respiratory syncytial virus. Eur J Immunol 31(9):2574–2582

Othumpangat S, Walton C, Piedimonte G (2012) MicroRNA-221 modulates RSV replication in human bronchial epithelium by targeting NGF expression. PLoS One 7(1):e30030

Philippe L et al (2012) TLR2 expression is regulated by microRNA miR-19 in rheumatoid fibroblast-like synoviocytes. J Immunol 188(1):454–461

Pederson T (2010) Regulatory RNAs derived from transfer RNA? RNA 16(10):1865–1869

Piriyapongsa J, Marino-Ramirez L, Jordan IK (2007) Origin and evolution of human microRNAs from transposable elements. Genetics 176(2):1323–1337

Ponting CP, Belgard TG (2010) Transcribed dark matter: meaning or myth? Hum Mol Genet 19 (R2):R162–R168

Piedimonte G (2013) Respiratory syncytial virus and asthma: speed-dating or long-term relationship? Curr Opin Pediatr 25(3):344–349

Piedimonte G, Perez MK (2014) Alternative mechanisms for respiratory syncytial virus (RSV) infection and persistence: could RSV be transmitted through the placenta and persist into developing fetal lungs? Curr Opin Pharmacol 16:82–88

Qi J et al (2012) microRNA-210 negatively regulates LPS-induced production of proinflammatory cytokines by targeting NF-kappaB1 in murine macrophages. FEBS Lett 586(8):1201–1207

Quinn EM et al (2013) MicroRNA-146a is upregulated by and negatively regulates TLR2 signaling. PLoS One 8(4):e62232

Quinn SR et al (2014) The role of Ets2 transcription factor in the induction of microRNA-155 (miR-155) by lipopolysaccharide and its targeting by interleukin-10. J Biol Chem 289 (7):4316–4325

Rani S (2014) MicroRNA profiling of exosomes isolated from biofluids and conditioned media. Methods Mol Biol 1182:131–144

Rixon HW et al (2004) The small hydrophobic (SH) protein accumulates within lipid-raft structures of the Golgi complex during respiratory syncytial virus infection. J Gen Virol 85(Pt 5):1153–1165

Rixon HW et al (2005) The respiratory syncytial virus small hydrophobic protein is phosphorylated via a mitogen-activated protein kinase p38-dependent tyrosine kinase activity during virus infection. J Gen Virol 86(Pt 2):375–384

Ruby JG, Jan CH, Bartel DP (2007) Intronic microRNA precursors that bypass Drosha processing. Nature 448(7149):83–86

Russell RF et al (2015) Partial attenuation of respiratory syncytial virus with a deletion of a small hydrophobic gene is associated with elevated Interleukin-1beta responses. J Virol 89 (17):8974–8981

Rodriguez A et al (2007) Requirement of bic/microRNA-155 for normal immune function. Science 316(5824):608–611

Rossato M et al (2012) IL-10-induced microRNA-187 negatively regulates TNF-alpha, IL-6, and IL-12p40 production in TLR4-stimulated monocytes. Proc Natl Acad Sci USA 109(45): E3101–E3110

Rossi RL et al (2011) Distinct microRNA signatures in human lymphocyte subsets and enforcement of the naive state in CD4 + T cells by the microRNA miR-125b. Nat Immunol 12 (8):796–803

Ramet M, Korppi M, Hallman M (2011) Pattern recognition receptors and genetic risk for rsv infection: value for clinical decision-making? Pediatr Pulmonol 46(2):101–110

Rivera-Toledo E, Torres-Gonzalez L, Gomez B (2015) Respiratory syncytial virus persistence in murine macrophages impairs IFN-beta response but not synthesis. Viruses 7(10):5361–5374

Saikia M et al (2012) Genome-wide identification and quantitative analysis of cleaved tRNA fragments induced by cellular stress. J Biol Chem 287(51):42708–42725

Sharma A et al (2009) Posttranscriptional regulation of interleukin-10 expression by hsa-miR-106a. Proc Natl Acad Sci USA 106(14):5761–5766

Skowronek E et al (2014) tRNA 3′ processing in yeast involves tRNase Z, Rex1, and Rrp6. RNA 20(1):115–130

Sobala A, Hutvagner G (2013) Small RNAs derived from the 5′ end of tRNA can inhibit protein translation in human cells. RNA Biol 10(4):553–563

Starczynowski DT et al (2010) Identification of miR-145 and miR-146a as mediators of the 5q- syndrome phenotype. Nat Med 16(1):49–58
Sun Y et al (2011) Targeting of microRNA-142-3p in dendritic cells regulates endotoxin-induced mortality. Blood 117(23):6172–6183
Surdziel E et al (2011) Enforced expression of miR-125b affects myelopoiesis by targeting multiple signaling pathways. Blood 117(16):4338–4348
Sangokoya C, LaMonte G, Chi JT (2010) Isolation and characterization of microRNAs of human mature erythrocytes. Methods Mol Biol 667:193–203
Shields B et al (2003) Multiple heparin binding domains of respiratory syncytial virus G mediate binding to mammalian cells. Arch Virol 148(10):1987–2003
Sigurs N et al (1995) Asthma and immunoglobulin E antibodies after respiratory syncytial virus bronchiolitis: a prospective cohort study with matched controls. Pediatrics 95(4):500–505
Sigurs N et al (2000) Respiratory syncytial virus bronchiolitis in infancy is an important risk factor for asthma and allergy at age 7. Am J Respir Crit Care Med 161(5):1501–1507
Sigurs N et al (2010) Asthma and allergy patterns over 18 years after severe RSV bronchiolitis in the first year of life. Thorax 65(12):1045–1052
Sourimant J et al (2015) Fine mapping and characterization of the L-polymerase-binding domain of the respiratory syncytial virus phosphoprotein. J Virol 89(8):4421–4433
Stec DS, Hill MG 3rd, Collins PL (1991) Sequence analysis of the polymerase L gene of human respiratory syncytial virus and predicted phylogeny of nonsegmented negative-strand viruses. Virology 183(1):273–287
Schwarze J et al (2004) Latency and persistence of respiratory syncytial virus despite T cell immunity. Am J Respir Crit Care Med 169(7):801–805
Sikkel MB et al (2008) Respiratory syncytial virus persistence in chronic obstructive pulmonary disease. Pediatr Infect Dis J 27(10 Suppl):S63–S70
Taganov KD et al (2006a) NF-kappaB-dependent induction of microRNA miR-146, an inhibitor targeted to signaling proteins of innate immune responses. Proc Natl Acad Sci USA 103:12481–12486
Taganov KD et al (2006b) NF-kappaB-dependent induction of microRNA miR-146, an inhibitor targeted to signaling proteins of innate immune responses. Proc Natl Acad Sci USA 103 (33):12481–12486
Takeda K, Akira S (2004) TLR signaling pathways. Semin Immunol 16(1):3–9
Tang B et al (2010) Identification of MyD88 as a novel target of miR-155, involved in negative regulation of Helicobacter pylori-induced inflammation. FEBS Lett 584(8):1481–1486
Terasawa K et al (2009) Sustained activation of ERK1/2 by NGF induces microRNA-221 and 222 in PC12 cells. FEBS J 276(12):3269–3276
Thompson DM, Parker R (2009) Stressing out over tRNA cleavage. Cell 138(2):215–219
Tili E et al (2007a) Modulation of miR-155 and miR-125b levels following lipopolysaccharide/TNF-alpha stimulation and their possible roles in regulating the response to endotoxin shock. J Immunol 179:5082–5089
Tili E et al (2007b) Modulation of miR-155 and miR-125b levels following lipopolysaccharide/TNF-alpha stimulation and their possible roles in regulating the response to endotoxin shock. J Immunol 179(8):5082–5089
Tserel L et al (2011) MicroRNA expression profiles of human blood monocyte-derived dendritic cells and macrophages reveal miR-511 as putative positive regulator of Toll-like receptor 4. J Biol Chem 286(30):26487–26495
Tang RS et al (2002) Clustered charge-to-alanine mutagenesis of human respiratory syncytial virus L polymerase generates temperature-sensitive viruses. Virology 302(1):207–216
Taylor G et al (2014) Recombinant bovine respiratory syncytial virus with deletion of the SH gene induces increased apoptosis and pro-inflammatory cytokines in vitro, and is attenuated and induces protective immunity in calves. J Gen Virol 95(Pt 6):1244–1254
Techaarpornkul S, Barretto N, Peeples ME (2001) Functional analysis of recombinant respiratory syncytial virus deletion mutants lacking the small hydrophobic and/or attachment glycoprotein gene. J Virol 75(15):6825–6834

Thornburg NJ, Hayward SL, Crowe JE Jr (2012) Respiratory syncytial virus regulates human microRNAs by using mechanisms involving beta interferon and NF-kB. MBio 3(6) pii: e00220-12. doi: 10.1128/mBio.00220-12
Triantafilou K et al (2013) Human respiratory syncytial virus viroporin SH: a viral recognition pathway used by the host to signal inflammasome activation. Thorax 68(1):66–75
Tripp RA et al (2001) CX3C chemokine mimicry by respiratory syncytial virus G glycoprotein. Nat Immunol 2(8):732–738
Tripp RA et al (2003) The G glycoprotein of respiratory syncytial virus depresses respiratory rates through the CX3C motif and substance P. J Virol 77(11):6580–6584
Thorburn K et al (2006) High incidence of pulmonary bacterial co-infection in children with severe respiratory syncytial virus (RSV) bronchiolitis. Thorax 61(7):611–615
Valdovinos MR, Gomez B (2003) Establishment of respiratory syncytial virus persistence in cell lines: association with defective interfering particles. Intervirology 46(3):190–198
Vlachos IS et al (2015) DIANA-miRPath v3.0: deciphering microRNA function with experimental support. Nucleic Acids Res 43(W1):W460–W466
Wang X et al (2008) Degradation of hypomodified tRNA(iMet) in vivo involves RNA-dependent ATPase activity of the DExH helicase Mtr4p. RNA 14(1):107–116
Wang P et al (2010) Inducible microRNA-155 feedback promotes type I IFN signaling in antiviral innate immunity by targeting suppressor of cytokine signaling 1. J Immunol 185(10):6226–6233
Wang Q et al (2013) Identification and functional characterization of tRNA-derived RNA fragments (tRFs) in respiratory syncytial virus infection. Mol Ther 21(2):368–379
Watanabe K et al (2013) Degradation of initiator tRNAMet by Xrn1/2 via its accumulation in the nucleus of heat-treated HeLa cells. Nucleic Acids Res 41(8):4671–4685
Wei J et al (2013) MyD88 as a target of microRNA-203 in regulation of lipopolysaccharide or Bacille Calmette-Guerin induced inflammatory response of macrophage RAW264.7 cells. Mol Immunol 55(3–4):303–309
Wendlandt EB et al (2012) The role of microRNAs miR-200b and miR-200c in TLR4 signaling and NF-kappaB activation. Innate Immun 18(6):846–855
Whitney ML et al (2007) Rapid and reversible nuclear accumulation of cytoplasmic tRNA in response to nutrient availability. Mol Biol Cell 18(7):2678–2686
Witwer KW et al (2010) MicroRNA regulation of IFN-beta protein expression: rapid and sensitive modulation of the innate immune response. J Immunol 184(5):2369–2376
Wolfe CL, Hopper AK, Martin NC (1996) Mechanisms leading to and the consequences of altering the normal distribution of ATP(CTP):tRNA nucleotidyltransferase in yeast. J Biol Chem 271(9):4679–4686
Worm J et al (2009) Silencing of microRNA-155 in mice during acute inflammatory response leads to derepression of c/ebp Beta and down-regulation of G-CSF. Nucleic Acids Res 37 (17):5784–5792
Xu Z et al (2011) miR-365, a novel negative regulator of interleukin-6 gene expression, is cooperatively regulated by Sp1 and NF-kappaB. J Biol Chem 286(24):21401–21412
Xu G et al (2014) MicroRNA-149 negatively regulates TLR-triggered inflammatory response in macrophages by targeting MyD88. J Cell Biochem 115(5):919–927
Yamasaki S et al (2009) Angiogenin cleaves tRNA and promotes stress-induced translational repression. J Cell Biol 185(1):35–42
Yoshikawa M, Fujii YR (2016) Human ribosomal RNA-derived resident microRNAs as the transmitter of information upon the cytoplasmic cancer stress. Biomed Res Int 2016:7562085
Zeng R et al (2011) The role of cytokines and chemokines in severe respiratory syncytial virus infection and subsequent asthma. Cytokine 53(1):1–7
Zheng J et al (2015) A respiratory syncytial virus persistent-infected cell line system reveals the involvement of SOCS1 in the innate antiviral response. Virol Sin 30(3):190–199
Zhang W et al (2010a) Vaccination to induce antibodies blocking the CX3C-CX3CR1 interaction of respiratory syncytial virus G protein reduces pulmonary inflammation and virus replication in mice. J Virol 84(2):1148–1157

Zhang C et al (2010b) PUMA is a novel target of miR-221/222 in human epithelial cancers. Int J Oncol 37(6):1621–1626

Zhang M et al (2011) Both miR-17-5p and miR-20a alleviate suppressive potential of myeloid-derived suppressor cells by modulating STAT3 expression. J Immunol 186 (8):4716–4724

Zhou R, O'Hara SP, Chen XM (2011) MicroRNA regulation of innate immune responses in epithelial cells. Cell Mol Immunol 8(5):371–379

Zardo G et al (2012) Polycombs and microRNA-223 regulate human granulopoiesis by transcriptional control of target gene expression. Blood 119(17):4034–4046

Roles of Non-coding RNAs During Herpesvirus Infection

Meaghan H. Hancock and Rebecca L. Skalsky

Abstract Non-coding RNAs (ncRNAs) play essential roles in multiple aspects of the life cycles of herpesviruses and contribute to lifelong persistence of herpesviruses within their respective hosts. In this chapter, we discuss the types of ncRNAs produced by the different herpesvirus families during infection, some of the cellular ncRNAs manipulated by these viruses, and the overall contributions of ncRNAs to the viral life cycle, influence on the host environment, and pathogenesis.

Abbreviations

KSHV	Kaposi's sarcoma associated herpesvirus
HVS	Herpesvirus saimiri
HCMV	Human Cytomegalovirus
RhCMV	Rhesus Cytomegalovirus
MCMV	Mouse Cytomegalovirus
RCMV	Rat Cytomegalovirus
EBV	Epstein Barr virus
LCV	Lymphocryptovirus
HSV	Herpes simplex virus
miRNA	MicroRNA
snoRNA	Small nucleolar RNA
sisRNA	Stable intronic sequence RNA
ncRNA	Non-coding RNA
EBERs	EBV-encoded RNAs
BARTs	BamHI A rightward transcripts
BHRF1	BamHI right forward 1
PAN	Polyadenylated nuclear RNA
LAT	Latency-associated transcript
HSURs	HVS Sm class U RNAs

M.H. Hancock · R.L. Skalsky (✉)
Vaccine and Gene Therapy Institute at Oregon Health and Science University, Beaverton, OR, USA
e-mail: skalsky@ohsu.edu

Current Topics in Microbiology and Immunology (2018) 419:243–280
DOI 10.1007/82_2017_31

Published Online: 04 July 2017

UTR	Untranslated region
LMP	Latent membrane protein
RTA	Replication and transcription activator protein
EBNA	EBV nuclear antigen
RNA	Ribonucleic acid
RISC	RNA-induced silencing complex
PAR-CLIP	Photoactivatable ribonucleoside enhanced cross-linking and immunoprecipitation
HITS-CLIP	High throughput sequencing of RNA isolated by cross-linking and immunoprecipitation
ChIRP	Chromatin isolation by RNA purification
CHART	Capture hybridization analysis of RNA targets
TF	Transcription factor

Contents

1 Introduction

Herpesviruses are large, enveloped, double-stranded DNA viruses that establish life-long, persistent infections in their hosts. Classified into three families (alpha, beta, and gamma) based on sequence homology and genome structure, all herpesviruses transition between two complex, distinct states during their life cycles: lytic and latent replication. The ability to successfully navigate between these phases all the while effectively combating host anti-viral innate and adaptive immune responses is facilitated in part by viral non-coding RNAs (ncRNAs). Viral genomes are compact in size, thus herpesviruses maximize use of their limited genetic space by encoding RNAs with multi-functional roles. Furthermore, many viral ncRNAs are non-immunogenic and can be expressed in times of host shut-off when protein translation is inhibited. These unique features of viral ncRNAs provide multiple methods for strategically manipulating gene expression within the internal and external cellular environment to effectively support the herpesvirus life cycle. While many herpesvirus ncRNAs continue to be defined and functionally characterized, numerous studies have highlighted the important and diverse roles that these molecules play in regulating viral replication, gene expression, and virus persistence.

2 Non-coding RNAs Made by Herpesviruses

Non-coding RNAs are defined as functional RNAs that lack protein-coding potential and do not associate with translation factors such as ribosomes. In humans, a surprisingly significant portion of the genome (>75%) is transcribed, yet only ~2% of these transcripts are protein coding (Kapranov et al. 2007). The remaining transcripts represent ncRNAs which include highly abundant ribosomal RNAs (rRNAs), transfer RNAs (tRNAs), and regulatory RNAs such as microRNAs (miRNAs), piwiRNAs (piRNAs), small nucleolar RNAs (snoRNAs) and long ncRNAs. Many of these classes of ncRNAs have been observed in herpesviruses which are subject to the same control networks that regulate cellular gene expression, and produce viral transcripts that share structural features with cellular RNAs. For protein-coding viral mRNAs, this includes a 5′ methylguanosine cap, 3′ polyadenylated tail, and cis elements that regulate translation. The ncRNAs are more diverse in structure, size, and function. Based on size, the herpesvirus ncRNAs fall into two general classes: long ncRNAs (>150–200 nt) and small ncRNAs (~18–100 nt). Figure 1a shows the genomic locations of several ncRNAs encoded by human herpesviruses representative of the three families.

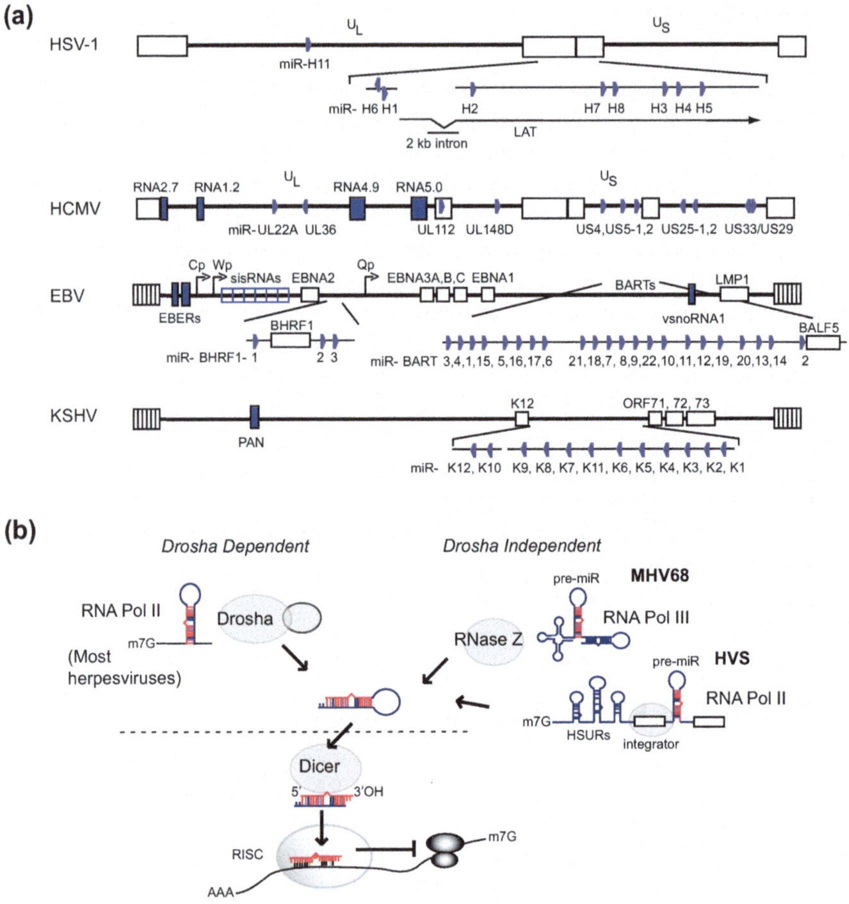

Fig. 1 Non-coding RNAs expressed by herpesviruses. **a** Genomic locations (not to scale) of large and small ncRNAs (*highlighted*) encoded by the alpha, beta, and gamma-herpesviruses. Shown are representative human herpesviruses from each family and approximate locations of viral ncRNAs. miRNA precursors are represented by *arrows* that indicate the direction in which they are transcribed. For HSV-1, only miRNAs reported to be RISC-associated are shown (Flores 2013). **b** Drosha dependent (canonical) and Drosha independent miRNA biogenesis. The majority of herpesvirus miRNAs are generated via the canonical pathway. MHV68 and HVS bypass Drosha/DGCR8 cleavage and utilize alternative methods to generate pre-miRNAs that are co-transcribed with other viral ncRNAs

2.1 Herpesvirus-Encoded MiRNAs: The Short of It

The majority of viral small ncRNAs are ~22 nt cytoplasmic, single-stranded miRNAs that direct activity of the cellular RNA interference machinery and contribute to the silencing of target RNAs. Over 300 herpesvirus miRNAs have been identified within the past 12 years, and at least seven of the nine known human herpesviruses have been shown to encode miRNAs (Pfeffer et al. 2004, 2005; Samols et al. 2005; Cai et al. 2005; Grey et al. 2005; Umbach et al. 2008; Chen et al. 2010; Tuddenham et al. 2012; Tang et al. 2008); to date, no miRNAs have been identified for varicella zoster virus (VZV) (Umbach et al. 2009), the causative agent of chicken pox and shingles, or human herpesvirus 7 (HHV7) which, along with HHV6B, is associated with roseola in young children. miRNAs and other small ncRNAs may yet be identified for these viruses as additional RNA sequencing studies are performed. Both the alpha and gamma-herpesvirus miRNAs are organized in clusters (Fig. 1a) within genomic loci that are transcriptionally active during latent infection; many of these viral miRNAs are processed from stable introns of larger ncRNAs as described below. In contrast, beta-herpesvirus miRNAs, such as the human cytomegalovirus (HCMV) miRNAs, are dispersed throughout the viral genome and many are encoded within protein-coding regions that are transcriptionally active during acute stages of infection.

MiRNAs are potent post-transcriptional regulators of gene expression that are essential players in herpesvirus biology. miRNAs act by guiding the RNA induced silencing complex (RISC) to sequence complementary sites on target RNAs-primarily in the 3′ UTRs of mRNAs, which subsequently interferes with protein expression and leads to destabilization of the target RNA (Bartel 2004, 2009). Over 1800 miRNAs are encoded within the human genome and are predicted to regulate upwards of 30–50% of protein-coding transcripts (Bartel 2009). Consequently, miRNAs regulate multiple biological processes, including cell differentiation, proliferation, apoptosis, stress responses, and the development of immunological responses. Aberrant miRNA activity, such as the loss of miRNA binding sites in 3′ UTRs from alternative splicing or polyadenylation, has been observed in many diseases including cancer (Mayr and Bartel 2009; Sandberg et al. 2008).

Pfeffer et al. (2004) identified the first viral miRNAs in 2004 by sequencing size fractionated, small RNAs (18–24 nt) from B cells infected with the human gamma-herpesvirus, Epstein Barr virus (EBV). EBV is a highly successful pathogen and persistently infects >90% of adults worldwide. While primary infection is asymptomatic or induces infectious mononucleosis, in immunocompromised individuals, the virus is linked to malignancies such as Burkitt's lymphoma, non-Hodgkin's lymphoma, and nasopharyngeal carcinoma. EBV encodes a total of 25 precursor miRNAs (pre-miRNAs), three within the BHRF1 region and two large miRNA clusters within the BamHI rightward transcripts (BART) region (Cai et al. 2005; Chen et al. 2010; Grundhoff and Sullivan 2011; Pfeffer et al. 2004, 2005; Skalsky and Cullen 2015) (Fig. 1a). The BART miRNAs, transcribed from the BART P1 and P2 promoters, are detectable at varying levels throughout infection while the BHRF1 miRNAs are expressed during specific latency programs when the latent C or W promoters are active (Amoroso et al. 2011; Edwards et al. 2008; Xing and Kieff 2011). Homologs of the EBV miRNAs can be found in several Old World non-human primate (NHP) lymphocryptoviruses (LCV), including rhesus LCV and herpesvirus papio, which infects baboons (Cai et al. 2006; Riley et al. 2010; Skalsky et al. 2014; Walz et al. 2010). rLCV currently holds the record for most viral miRNAs; 70 potential miRNAs can be generated from the 35 pre-miRNAs (Riley et al. 2010; Skalsky et al. 2014; Walz et al. 2010).

A second lymphotrophic human gamma-herpesvirus, Kaposi's sarcoma-associated herpesvirus (KSHV), encodes 12 pre-miRNAs within the latency-associated region. KSHV is linked to KS and lymphoproliferative diseases such as primary effusion lymphoma (PEL). 10 of the KSHV miRNAs are expressed from an intron mapping between vFLIP (ORF71) and K12 while the other two are encoded within K12 (Cai et al. 2005; Pfeffer et al. 2005; Samols et al. 2005). Rhesus rhadinovirus (RRV) and other NHP rhadinoviruses encode positional homologs of the KSHV miRNAs and at least one of the NHP virus miRNAs has sequence homology to KSHV miR-K10a (Schafer et al. 2007; Umbach et al. 2009; Walz et al. 2010; Bruce et al. 2013; Skalsky et al. 2016). Small RNA sequencing studies on KSHV infected B cell lymphoma lines revealed that viral miRNAs account for a large portion ($\sim$30–70%) of the total miRNA population (Gottwein et al. 2011; Umbach et al. 2010), and are therefore expected to exert significant impacts on host gene expression.

Neurotropic alpha-herpesviruses such as Herpes simplex viruses 1 and 2 (HSV-1 and HSV-2), the rhesus macaque herpes B virus, and the highly infectious pseudorabies virus (PRV) of swine, encode a locus known as the latency associated transcript (LAT) which remains transcriptionally active during latency. No viral proteins are produced; rather, large, capped and polyadenylated transcripts accumulate to high levels and are each spliced to liberate a stable intron (Tycowski et al. 2015). The unstable spliced LAT exon is cleaved by the miRNA biogenesis machinery to liberate multiple miRNAs. HSV-1 and HSV-2 establish latent infection in the trigeminal or sacral dorsal root ganglia neurons, respectively. Disease can arise from viral reactivation and usually presents as lesions in the form of cold sores for HSV-1 or genital sores for HSV-2. During lytic HSV-1 or HSV-2 infection, miRNAs from LAT and other genomic loci can be detected (Jurak et al. 2014; Umbach et al. 2008; Tang et al. 2008). A total of 17 miRNAs have been reported for HSV-1; however, only 9 of these (namely, miR-H1 to miR-H8 and miR-H11) were demonstrated to be actively loaded into RISC following in vitro HSV-1 17syn + infection of neuronal cell lines (Flores et al. 2013). Expression levels of HSV-1 miRNAs vary significantly, and inconsistencies in viral miRNA detection in many studies are likely due to the lack of tractable models for HSV-1 infection.

Human cytomegalovirus (HCMV), a ubiquitous virus that is generally asymptomatic in immunocompetent individuals but can cause severe disease in immunocompromised individuals such as neonates and transplant recipients (Britt 2008), encodes 11 pre-miRNAs scattered throughout its genome. All HCMV miRNAs are detected during lytic infection of fibroblasts (Dunn et al. 2005; Grey et al. 2005; Stark et al. 2012) however only a subset of the miRNAs are detectable in in vitro latency models (Fu et al. 2014; Lau et al. 2016; Meshesha et al. 2016; Shen et al. 2014). Mouse CMV (MCMV) and rat CMV (RCMV) encode 18 and 14 pre-miRNAs, respectively, which, like HCMV, are scattered throughout the genome but bear no homology to the primate CMV miRNAs (Buck et al. 2007; Meyer et al. 2011).

2.2 Biogenesis of Viral MicroRNAs

Most herpesvirus miRNAs are processed by the canonical host miRNA biogenesis pathway, whereby long pri-miRNAs are transcribed by RNA Pol II and cleaved by nuclear RNAse III Drosha along with its co-factor DGCR8 (DiGeorge critical region 8) (Fig. 1b) (Ambros 2004; Bartel 2004, 2009; Ha and Kim 2014). Drosha cleavage releases a ~60 nt precursor miRNA (pre-miRNA) hairpin containing a 5′ mono-phosphate and 2 nt 3′ overhang. Pre-miRNAs are then exported to the cytoplasm through the nuclear pore complex via Exportin-5 (XPO5) and cleaved by a second RNAse III enzyme, Dicer, to liberate ~22 nt miRNA:miRNA* duplexes. One duplex strand is bound by an Argonaute protein and incorporated into RISC as the mature miRNA. Viral miRNA biogenesis is dependent solely on the cellular machinery and to date, there are no viral factors known to be required for the generation of herpesvirus miRNAs. Single nucleotide polymorphisms (SNPs) in several herpesvirus pre-miRNA sequences, particularly for two KSHV miRNAs, miR-K12-5 and miR-K12-9, have been shown to alter miRNA processing events by interfering with Drosha and/or Dicer cleavage (Han et al. 2013; Gottwein et al. 2006). Additionally, post-transcriptional modifications of pri- or pre-miRNAs, such as adenosine to inosine (A-to-I) editing by the ADAR1 enzyme, can alter miRNA biogenesis. A-to-I editing of the EBV primary miR-BART6 transcript abrogates Drosha cleavage, release of the pre-miRNA, and thus alters total expression levels of the mature miR-BART6 (Iizasa et al. 2010). The clinical significance of SNPs and post-transcriptional modifications in human gamma-herpesvirus miRNAs has not been fully evaluated, although it is thought these may contribute to increased pathogenesis risk or disease states by altering viral miRNA expression levels (Iizasa et al. 2010; Han et al. 2013; Marshall et al. 2010).

Not all herpesviruses exploit Drosha for miRNA biogenesis. At least two gamma-herpesviruses have been shown to generate miRNAs in a Drosha- and DGCR8-independent manner. The 14 murine MHV68 pre-miRNAs are generated from eight RNA Pol III tRNA-pre-miRNA chimera transcripts to yield a total of 28 mature miRNAs. These MHV68 tRNA-miRNA-encoding RNAs (TMERs) are cleaved by cellular tRNAseZ (ELAC1) and other as yet unknown nucleases to separate the pre-miRNAs from the tRNA prior to export into the cytoplasm and canonical Dicer cleavage (Bogerd et al. 2010; Diebel et al. 2010; Tycowski et al. 2015). Many MHV68 TMERs contain one tRNA and two pre-miRNAs, and mature miRNAs generated from the pre-miRNAs closest to the tRNAs are more abundant. The second pre-miRNA terminates in a 3′ polyU tract, a remnant of RNA Pol III transcription.

Pre-miRNAs of the oncogenic New World monkey virus herpesvirus saimiri (HVS) are co-transcribed with the HVS Sm class U RNAs (HSURs) by RNA Pol II (Croen et al. 1987). The resulting multi-hairpin transcripts contain cis elements, including Sm binding regions and uridine-rich, small nuclear RNA (snRNA) 3′ signal elements (3′ box) that contribute to further processing. The host Integrator complex, consisting of ~12 proteins in humans that are responsible for snRNA biogenesis, recognizes the 3′ box and cleaves the transcripts, thereby separating the 5′ HSURs from the downstream viral pre-miRNAs (Croen et al. 1987). This cleavage event generates the 5′ ends of the pre-miRNAs; another 3′ box-like sequence element recognized by the Integrator complex is needed to generate the 3′ ends of the pre-miRNAs (Xie et al. 2015). Released pre-miRNAs are subsequently exported into the cytoplasm to join the canonical miRNA biogenesis pathway (Fig. 1b). Elucidating these non-canonical pathways for MHV68 and HVS miRNA processing has contributed to our overall understanding of miRNA biogenesis in eukaryotes.

2.3 *Other Herpesvirus Small NcRNAs*

In addition to miRNAs, herpesviruses express a number of other small ncRNAs, although the regulatory functions for many are still unclear. Encoded 100 nt downstream of EBV miR-BART2 is a 65 nt viral snoRNA (v-snoRNA1) (Hutzinger et al. 2009) (Fig. 1a). V-snoRNA1 contains sequence and structural elements similar to canonical C/D box cellular snoRNAs that localize to the nucleolus, interact with Nop65, Nop58, and fibrillarin to form functional snoRNA protein (snoRNP) complexes, and chemically modify RNAs. A smaller 24-nt RNA with miRNA-like characteristics may be processed from v-snoRNA1 and can been detected by Northern blot in both EBV + B and epithelial cells (Hutzinger et al. 2009; Lung et al. 2013). The 24 nt v-snoRNA1-derived RNA was reported to bind to a region in the EBV BALF5 3′ UTR and induce cleavage of the BALF5 mRNA (Hutzinger et al. 2009). How this interaction might occur is unclear since the v-snoRNA1 is not detectable in RISC immunoprecipitation experiments (Riley et al. 2012; Skalsky et al. 2012).

Identified in RNA-seq datasets from latently infected B cells, the EBV stable intronic sequence RNA 1 (sisRNA-1) is a highly abundant, 81 nt RNA derived from a stable intron that arises from transcription across the five to eight W repeats (Moss and Steitz 2013). These W repeats are included as the 5′ ends of long transcripts encoding the EBV nuclear antigens (EBNAs); some of these EBNA transcripts might have additional ncRNA capabilities (Concha et al. 2012; Moss and Steitz 2013). Functions for EBV sisRNA-1 are unknown; however, the RNA contains a short U-rich hairpin loop and additional motifs that may interact with proteins (Moss and Steitz 2013). RNA-seq experiments have unveiled hundreds of other novel EBV transcripts and stable viral introns produced during lytic infection that appear to lack protein coding potential and may have regulatory functions (Cao et al. 2015; Concha et al. 2012; O'Grady et al. 2014); at least some of these novel EBV ncRNAs may sequester miRNAs and alter viral gene expression (Cao et al. 2015).

Two HSV-1 small ncRNAs [sRNA1 (65 nt) and sRNA2 (36 nt)] arise from within the first 1.5 kb of the stable LAT intron during productive HSV-1 infection of neuronal cultures (Perng et al. 1994) and in latently infected mouse trigeminal ganglia (Stevens et al. 1987). LAT sRNA2 inhibits the expression of ICP4 in transient co-transfection experiments, and both sRNAs contribute to blocking productive infection of a LAT-null virus. Additionally, both sRNAs cooperate to partially block cold-shock induced apoptosis in neuro-2A cells (Stevens et al. 1987). Thus, the functions of these sRNAs may contribute to the overall requirement of the LAT region in latency and reactivation of HSV-1.

Varying in size from 75–143 nt in length, the seven HSURs of HVS are abundantly expressed in latently infected marmoset T cells (10^3-10^4 copies/cell) [reviewed in (Tycowski et al. 2015)]. The HSURs share some sequence similarities to Sm-class small nuclear RNAs (snRNAs). These non-coding RNAs are dispensable for the T cell transformation phenotype of HVS, but remain highly conserved in other New World primate gamma-herpesviruses, suggesting other important functions for the virus. At least one of the HSURs can sequester a host miRNA (discussed below) that contributes to T cell development (Cazalla et al. 2010).

2.4 The Long(er) Non-coding RNAs

Some of the most abundant viral transcripts in herpesvirus-infected cells are lncRNAs. HCMV expresses four major ncRNAs (RNA2.7, RNA1.2, RNA4.9, and RNA5.0) (Fig. 1a) that account for greater than 65% of polyadenylated transcripts in infected human fibroblasts in vitro (Gatherer et al. 2011). RNA2.7 (also known as β2.7), the best-studied HCMV lncRNA, is expressed with early kinetics and accounts for more than 20% of polyadenylated lytic viral transcripts (McDonough and Spector 1983; McDonough et al. 1985). Additionally, RNA 2.7 has been detected both in in vitro latency models and in seropositive donors (Rossetto et al. 2013). RNA2.7 is encoded within the inverted repeats flanking the unique long (U_L) segment of the viral genome, has a cytoplasmic localization (Gawn and Greaves 2002), and may encode several short polypeptides (Stern-Ginossar et al. 2012). Another HCMV lncRNA, RNA4.9, has also been detected in seropositive donors and in in vitro models of HCMV latency (Rossetto et al. 2013). RNA4.9 initiates within oriLyt, terminates just downstream of UL69 (Gatherer et al. 2011) and may encode a polypeptide (Stern-Ginossar et al. 2012). RNA5.0 has been identified as an AT-rich stable intron encoded between the UL105 and UL111a loci. Mutation of the 5′ splice donor site does not affect viral replication in fibroblasts (Kulesza and Shenk 2004). The MCMV homolog of RNA 5.0 is also dispensable for replication in fibroblasts but may be a virulence factor in vivo (Kulesza and Shenk 2006). MCMV may also encode other long non-coding RNAs that await characterization (Juranic Lisnic et al. 2013). Finally, HCMV RNA 1.2 is encoded within the terminal repeats, may also encode a polypeptide (Stern-Ginossar et al. 2012), but has not been studied extensively (Gatherer et al. 2011; Hutchinson and Tocci 1986).

EBV expresses two nuclear RNA pol III transcripts, EBER1 (167 nt) and EBER2 (172 nt), that accumulate to ~10^5–10^6 copies per cell and are the most abundant viral transcripts in EBV-infected cells (Lerner et al. 1981; Tycowski et al. 2015). The EBERs are detectable throughout EBV infection and therefore are often used for diagnostics. Originally identified as interactors of the La protein, which binds the RNA pol III polyU tracts of the EBERs, both EBER1 and EBER2 form extensively base-paired secondary structures to scaffold RNA binding proteins. EBER1 relocates the ribosomal protein L22 to the nucleus (Fok et al. 2006) and also binds several cellular hnRNPs including AUF1 (Lee et al. 2012), which alters the stability of AU-rich RNAs. Recently identified, novel functions for EBER2 (Lee et al. 2015) are discussed below and illustrated in Fig. 2.

During lytic replication, the 1.1 kb Kaposi-sarcoma associated herpesvirus (KSHV) polyadenylated nuclear RNA (PAN) makes up >75% of the total polyadenylated transcripts in KSHV infected cells (Arias et al. 2014; Conrad 2016). Recent studies show that PAN may be virion associated and accumulates in the nucleus at high levels (>500,000 copies/cell) within a few hours after infection (Rossetto and Pari 2012; Arias et al. 2014; Conrad 2016). PAN contains a 79 nt triple helical expression and nuclear retention element (ENE) that contributes to its stable secondary structure (Fig. 2a) and facilitates nuclear localization (Devi-Rao et al. 1994; Conrad and Steitz 2005). Sequence analysis of other rhadinovirus

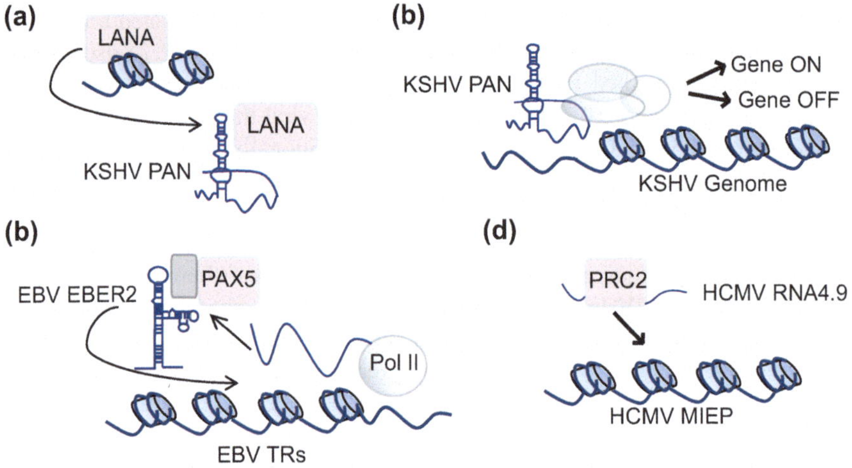

Fig. 2 Viral long ncRNA regulation of gene expression. **a** KSHV PAN RNA can serve as a decoy by sequestering LANA from binding to and repressing KSHV lytic promoters **b** KSHV PAN can also guide various cellular factors (i.e. demethylases such as JMJD3 and UTX or methylases such as Ezh2) to sites on the KSHV genome which consequently alter histone modifications that mediate transcriptional activation or repression. **c** EBV EBER2 forms a complex with cellular PAX5 and an unknown bridging factor to regulate transcription from the terminal repeat (TR) regions of the EBV genome. Nascent TR transcripts pair with EBER2 to stabilize the interaction. **d** HCMV RNA4.9 interacts with the PRC2 complex and may bind regions of the viral genome to regulate the major immediate early promoter

genomes such as rhesus rhadinovirus (RRV), japanese macaque rhadinovirus (JMRV), and retroperitoneal fibromatosis-associated herpesvirus (RFHV), indicate the presence of PAN homologs (Estep et al. 2013; Tycowski et al. 2012).

Particularly for EBV and KSHV, recent transcriptome-wide approaches to characterize viral gene expression patterns have uncovered several novel lncRNAs present during lytic infection, and similar to HCMV lncRNAs, many initiate near lytic origins. EBV BHLF1 is an abundant 2.5 kb ori-LytL associated transcript that has been gaining recognition as a potential viral lncRNA since the transcript seems to lack protein coding potential due to several stop codons (Arvey et al. 2012; Krug 2013); BHLF1 may contribute to control of lytic DNA replication through regulation of the Cp and Wp promoters (Rennekamp and Lieberman 2011). Encoded within the BamHI locus adjacent to ori-LytR, the EBV LF3 transcript also contains multiple stop codons, suggesting function as a ncRNA (Lin et al. 2013). Other EBV ncRNAs include the alternatively spliced, polyadenylated nuclear BART lncRNAs that negatively regulate expression of several host genes (Marquitz et al. 2014), a second ~2.8 kb sisRNA (sisRNA-2) generated from the larger introns of W repeat transcripts that has no characterized function (Moss and Steitz 2013), and bi-directional oriP transcripts that remain nuclear, are subject to ADAR hyper-editing, and may fold into long hairpin structures (Cao et al. 2015). GapmeR knockdown of oriPtL activity in EBV + Burkitt's lymphoma cell lines indicates a role in enhancing lytic replication (Cao et al. 2015). KSHV similarly transcribes several non-coding regions of the viral genome during lytic reactivation that are also adjacent to an oriLyt, which may allow replication factors access to the viral DNA (Wang et al. 2008; Chandriani et al. 2010).

HSV LAT (latency-associated transcript) is the most abundantly transcribed gene in infected neurons and is encoded anti-sense to ICP0 (Kim et al. 2015). LAT encodes an 8.3 kb transcript that is spliced to yield a stable 2 kb intron that can be further processed into a stable 1.5 kb intron (Javier et al. 1988) that is found at >10^4 copies/cell. The role of the 1.5 kb stable intron in latency establishment, maintenance and reactivation has been extensively studied [reviewed in (Scarpini et al. 2001)] and is discussed below.

3 Functions of Viral LncRNAs

Much of what we know about lncRNA functions comes from studies to characterize the >60,000 cellular lncRNAs encoded within the human genome (Iyer et al. 2015). Many cellular lncRNAs are transcribed by RNA pol II and are subject to splicing and polyadenylation akin to cellular mRNAs (Derrien et al. 2012). More common to lncRNAs is their nuclear localization, unique secondary or tertiary structures, and interactions with other RNAs, DNA, or proteins for activity (Cech and Steitz 2014). Cellular lncRNAs have been shown to regulate gene expression through chromatin remodeling or interactions with splicing machinery; some act as molecular scaffolds to bring effector components into close proximity while others contribute to RNA stability or post-transcriptional or post-translational modifications [reviewed in (Cech and Steitz 2014)]. Together, these studies have revealed the diversity of lncRNA functions, and thus, we would expect viral lncRNAs to harbor comparable or perhaps, even more unique roles.

3.1 Roles in Metabolic Processes and Apoptosis

Functions for the HCMV lncRNAs during viral infection have not been extensively studied. Reeves et al. (2007) examined the interacting partners of HCMV RNA2.7 and identified the mitochondrial Complex I subunit NADH Ubiquinone Oxidoreductase. RNA2.7 blocked the relocalization of cellular GRIM-19, a subunit of Complex I required for its assembly and function, in response to the Complex I inhibitor rotenone. The maintenance of Complex I function allows for the stabilization of mitochondrial membrane potential and continued production of ATP. Although deletion of RNA 2.7 from the viral genome does not greatly affect viral growth in fibroblasts (McSharry et al. 2003; Reeves et al. 2007), there is an observed growth disadvantage in the presence of rotenone or low glucose growth medium (Reeves et al. 2007). Continued ATP production is likely critical for efficient lytic replication and HCMV mediates this, at least in part, through the function of its lncRNA.

The HSV stable intron LAT plays an important role in blocking apoptosis in infected neurons, although the exact mechanism remains unknown (Marshall et al. 2010; Ray et al. 2012). Caspase 8 and 9-induced apoptosis is blocked by LAT (Hansen et al. 2010a) and caspase 3 activation in response to Granzyme B can also be inhibited by LAT (Jiang et al. 2011). Two sRNAs encoded within the first 1.5 kb of the LAT region may contribute to this phenotype (Shen et al. 2014). Given that HSV establishes latency in long-lived neuronal cells, blocking apoptosis is likely a critical function of LAT during the lifespan of the host.

3.2 Viral Long NcRNAs Modify Viral Gene Expression

Several cellular and viral lncRNAs have been found associated with chromatin remodeling factors (Khalil et al. 2009; Rossetto and Pari 2012) and are thought to act as molecular scaffolds to allow for histone modifications in the epigenetic control of viral or cellular gene expression (Chinen and Tani 2012; Saxena and Carninci 2011). Following trafficking of the viral DNA into the nucleus, herpesvirus genomes become nucleosome-associated and take on highly structured, organized chromatin architecture (Toth et al. 2010). Histone modifications compact the chromatin to repress transcription, which is a key step in silencing viral gene expression for latency.

There are several examples of herpesvirus lncRNAs that directly contribute to transcriptional control of viral immediate early promoters. In latently infected CD14 + monocytes, HCMV RNA4.9 interacts with components of the polycomb repressor complex PRC2 (Fig. 2d) and may be involved in the enrichment of H3K27me3 on histone tails at the site of the HCMV major immediate early promoter (MIEP) (Rossetto et al. 2013). PRC complexes are also involved in the formation of heterochromatin and histone marks on KSHV genomes, and may be titrated away from the viral genome by KSHV PAN to enhance lytic replication [reviewed in (Campbell et al. 2014)]. Recent ChIRP (chromatin isolation by RNA purification) experiments on KSHV infected B cells indicate that KSHV PAN lncRNA interacts with PRC2 components such as the lysine methylase Ezh2 (Rossetto et al. 2013). PAN also interacts with demethylases JMJD3 and UTX that act on the ORF50 promoter to drive expression of the viral transactivator RTA (Fig. 2b) (Rossetto and Pari 2012). It was originally determined that HSV LAT contributes to the heterochromatinization of lytic gene promoters during latency through the induction of lysine 9 methylation on histone H3 (Wang et al. 2005). Additionally, facultative heterochromatin (H3K9 and H3K27 trimethylations) on lytic gene promoters during latency is reduced in small animal models of HSV-1 using LAT-deficient viruses (Amelio et al. 2006; Cliffe et al. 2009; Kwiatkowski et al. 2009). The mechanism of LAT-induced heterochromatinization of the lytic viral gene promoters is unclear; however, LAT may be involved in recruitment of chromatin remodeling complexes (Kwiatkowski et al. 2009; Cliffe et al. 2013).

In vivo studies suggest that PAN can also interact with the KSHV latency associated nuclear antigen (LANA). LANA has many functions in the maintenance of latent viral genomes, and has been shown to repress transcription from lytic viral promoters including the ORF50 promoter (Lan et al. 2004). In vitro, the PAN/LANA interaction has been mapped to LANA's DNA binding domain which inhibits interactions with histone H3 (Campbell et al. 2014). One current model is that PAN sequesters LANA from binding lytic viral promoters (Fig. 2a).

Recent work from the Steitz group demonstrated that the EBV nuclear EBER2 physically interacts with paired box protein 5 (PAX5), a transcription factor involved in B cell differentiation (Li et al. 2011). Capture hybridization analysis of RNA targets (CHART) was used to identify areas of DNA where EBER2 localized. This analysis showed that EBER2 can bind to the EBV terminal repeats at sites that overlapped with the previously reported PAX5 binding sites (Arvey et al. 2012). Subsequent experiments demonstrated that EBER2 and PAX5 interact to form a complex that includes paraspeckle components SFPQ, NONO, and RBM14 to help regulate LMP1 and LMP2A RNA expression (Lee et al. 2015, 2016) (Fig. 2c). Consistent with this model, deletion of EBER2 from the virus leads to a ~5-fold increase in LMP2A transcript levels in lymphoblastoid cell lines (LCLs); surprisingly, no significant changes were observed at the protein level (Gregorovic et al. 2015).

4 Small Silencers of Viral Gene Expression

Viral miRNAs are considered key contributors in mediating the balance between lytic and latent replication [reviewed in (Grey 2015)]. One way in which viral miRNAs can modulate this balance is through the direct silencing of lytic transcripts, which is important for establishing and/or maintaining latency. EBV miR-BART2-5p, which is encoded anti-sense to the viral DNA polymerase catalytic subunit BALF5, has perfect complementarity to the BALF5 3′ UTR and induces site-specific endonucleolytic cleavage of the BALF5 transcript during lytic replication (Barth et al. 2008; Pfeffer et al. 2004). HCMV miR-UL112, miR-US33, miR-US5-1, and miR-US5-2 are each encoded on strands opposite of viral protein coding genes, and therefore postulated to inhibit viral gene expression. HCMV miR-UL112 is anti-sense to UL114 (Pfeffer et al. 2005), which encodes a uracil DNA glycosylase involved in DNA replication, and inhibits UL114 protein expression by direct binding to the UL114 transcript (Stern-Ginossar et al. 2009). Surprisingly, despite full complementarity between miR-UL112 and the UL114 mRNA, there is no evidence that miR-UL112 acts like a siRNA to induce cleavage of the UL114 transcript. miR-US5-1 and miR-US5-2 are both anti-sense to US7 and act cooperatively through two perfect binding sites and one imperfect binding site to inhibit US7 protein expression (Tirabassi et al. 2011). While US7 functions are uncharacterized, rhesus CMV encodes homologs of US7 and miR-US5-2 (Tirabassi et al. 2011; Hancock et al. 2012), indicating that these interactions are important to the CMV life cycle.

Herpesvirus miRNAs further contribute to the lytic/latent balance by restricting immediate early viral transactivators (Grey et al. 2007; Umbach et al. 2008). HCMV miR-UL112 negatively regulates the major immediate early transcript IE72 (UL123, IE1) as well as UL112/113, which is involved in viral DNA replication, and UL120/121, which presumably encodes a membrane protein (Grey et al. 2007). Mutation of miR-UL112 within the context of full-length virus or the miR-UL112 binding site within the UL123 3′ UTR leads to an increase in IE1 expression (Murphy et al. 2008). Additional studies demonstrate that post-transcriptional regulation of major viral lytic transactivators may be a conserved function of herpesvirus miRNAs. miR-H2-3p of HSV-1 is encoded antisense to ICP0 and thus silences protein expression from the transcript (Umbach et al. 2008). HSV-2 encodes a positional homolog, miR-III, that likewise targets the HSV-2 ICP0 mRNA (Umbach et al. 2010; Tang et al. 2008). Both miR-H2-3p and miR-III have been shown to decrease ICP0 protein expression in latency models (Umbach et al. 2008; Flores et al. 2013). Additionally, miR-H6 of HSV-1 targets the IE transcriptional activator ICP4 (Umbach et al. 2008; Tang et al. 2008, 2011). HSV-1 miR-H4 (Flores et al. 2013) and HSV-2 miR-I and miR-II (Tang et al. 2008, 2011) overlap with the coding sequence and 5′ UTR of the virulence factor ICP34.5 and have been shown to decrease protein expression. There is no experimental evidence for viral targets of the PRV miRNAs and no miRNAs are directly complementary to

PRV transcripts. However, in silico analysis suggests that the ICP0 and ICP4 homologs in PRV may also be PRV miRNA targets.

For human gamma-herpesviruses, studies of direct viral miRNA control over viral transactivators have not been so straightforward. KSHV miR-K9* was reported to target the ORF50 RNA, thus inhibiting expression of the immediate-early transactivator RTA and preventing lytic cycle entry during KSHV latent infection (Bellare and Ganem 2009); however, these observations have not been confirmed and other studies indicate the interaction may be indirect (Lei et al. 2010; Lu et al. 2010). EBV miR-BART20-5p has been reported to target transcripts of two immediate early transactivators, EBV BZLF1 and BRLF1, which share a 3′ UTR (Jung et al. 2014). These interactions may also be indirect, as RISC-capture experiments on EBV + cells have failed to confirm miR-BART20-5p binding sites within the BZLF1 or BHRF1 RNAs (Kang et al. 2015; Riley et al. 2012; Skalsky et al. 2014). Analysis of RISC-associated sites in wild-type EBV-infected LCLs indicate that EBV BALF2, BNRF1, BHRF1 and LMP1 transcripts are regulated by viral miRNAs (Skalsky et al. 2014). Viral miRNA targeting of LMP1 and BHRF1 3′ UTRs have been confirmed by several groups, and functional consequences of LMP1 regulation have been examined with regard to the activation of NF-kB signaling pathways (Lo et al. 2007; Riley et al. 2012; Skalsky et al. 2012, 2014). EBV LMP2A and EBNA2 are also subject to viral miRNA regulation (Riley et al. 2012; Skalsky et al. 2012, 2014; Lung et al. 2009; Kang et al. 2015), but like LMP1 and BHRF1, do not act as transactivators of the lytic cascade.

4.1 Host Silencers of Viral Gene Expression

Not only do herpesviruses utilize their own miRNAs to aid in latency establishment, but several examples exist whereby cellular miRNAs can directly limit herpesvirus gene expression in specific cell types. This was first demonstrated for EBV latent membrane protein 1 (LMP1) which is negatively regulated by the myc-regulated miR-17 seed family (includes miR-17, miR-20a, miR-106a/b) through a conserved binding site within the LMP1 3′ UTR (Riley et al. 2012; Skalsky et al. 2012, 2014). The functional consequences of this interaction have not been fully evaluated; however, an attractive hypothesis is that the timing and levels of LMP1 transcripts together with kinetics of miR-17 expression following de novo B cell infection may play an important part in transitioning through EBV-mediated B cell growth and differentiation programs (Forte and Luftig 2011; Skalsky and Cullen 2015).

Use of cell-type specific host miRNAs to silence lytic transcripts is an emerging theme for alpha- and beta-herpesviruses. The neuronal-specific miRNA miR-138 was shown to target the ICP0 transcript of HSV-1 (Pan et al. 2014). When the miR-138 target site within the ICP0 3′ UTR was mutated, the resultant virus displayed enhanced expression of lytic transcripts both in neuronal cultures and following corneal inoculation in mice resulting in increased mortality. Likewise,

HCMV utilizes members of the miR-200 family to target the 3′ UTR of UL122 (IE2) (O'Connor et al. 2014). miR-200 family members are expressed highest in undifferentiated cells of the myeloid lineage, where HCMV establishes latency. Using in vitro systems of HCMV latency, the authors show that a mutant lacking the miR-200 binding site in the 3′ UTR of UL122 does not efficiently establish latency as measured by IE1 expression and production of extracellular virions. Thus, both HSV-1 and HCMV may have evolved to co-opt cellular miRNAs for targeting their major viral transactivators in order to aid in latency establishment.

5 Viral MiRNA Regulation of Transcription Factors

A broader way that viral miRNAs indirectly control the timing and/or level of viral gene expression is through the targeting of host transcription factors (TFs). For example, one way KSHV miR-K3 may inhibit lytic replication is through the targeting of cellular CAAT box TF nuclear factor I/B (NFIB) which binds the KSHV ORF50 promoter region and activates RTA expression (Lu et al. 2010). KSHV miRNAs miR-K3 and miR-K11 also target the B cell TFs MYB, Ets-1, and C/EBPα, which activate the ORF50/RTA promoter (Gottwein et al. 2011; Haecker et al. 2012; Plaisance-Bonstaff et al. 2014). Similarly, EBV miR-BHRF1-2 targets PRDM1/BLIMP1, a master regulator of the B cell transcriptional program during B cell differentiation (Skalsky et al. 2012; Ma et al. 2016). BLIMP1 can activate the EBV immediate-early BZLF1 and BRLF1 promoters via multiple BLIMP1-responsive elements (Reusch et al. 2015). While a BHRF1-2 miRNA knockout virus does not spontaneously reactivate in vitro (Feederle et al. 2011b), regulation of BLIMP1 by miR-BHRF1-2 at a key stage of infection in vivo may help promote latency.

Herpesvirus miRNAs regulate many other TFs that drive transcriptional reprogramming of host cells and consequently, can impact viral gene expression patterns through more complicated layers of transcriptional control. A clear example of this are the KSHV miRNAs, in particular, miR-K11 and miR-K6, that can induce endothelial cell reprogramming by targeting the bZIP transcription factor MAF which mediates terminal differentiation of several cell types including lymphatic endothelial cells (Hansen 2010b). Additionally, NF-kB, AP-1, and other signaling pathways that result in transcriptional activation are regulated at multiple steps by KSHV, EBV, and other gamma-herpesvirus miRNAs (Gottwein et al. 2007; Skalsky et al. 2014, 2016; Puig-Barbera et al. 2016; Abend et al. 2012). Both NF-kB and AP-1 are activated upon gamma-herpesvirus infection and control multiple cellular processes including proliferation and differentiation as well as cytokine responses and inflammation. These and other TFs are also involved in the latent/lytic switch since the TFs bind response elements within promoter regions of several herpesvirus immediate early genes (Caposio et al. 2007; Kenney and Mertz 2014).

Individual viral miRNAs often exert subtle effects on protein expression levels. These subtle effects of single miRNAs can be amplified through TF targeting (Sun et al. 2014), and may be further amplified through the additive or combinatorial effects of multiple viral miRNAs acting in concert with cellular miRNAs coopted by the virus. Subtle inhibition of specific host TFs may also lead to dramatic changes in TF site occupancy- particularly in cases where an array of multiple TFs are needed to govern gene expression. EBV EBNA2, for example, forms super-enhancers with multiple B cell master TFs that contribute to B cell proliferation; transcripts for several of these super-enhancer TFs are RISC-associated and contain binding sites for multiple EBV miRNAs as well as host miRNAs such as miR-155, miR-146a, and miR-17 that are upregulated during EBV infection (Zhou et al. 2015; Gottwein et al. 2011; Riley et al. 2012; Skalsky et al. 2012).

6 Mimicry of Host MiRNA Function

While most herpesvirus miRNAs exhibit little or no sequence conservation to other viral miRNAs (exceptions being closely related primate gamma-herpesviruses) or their cellular counterparts, there are interesting examples of homology to host miRNA seed sequences (nt 2–7 or 2–8). Mimicking of this effector seed region may allow viral miRNAs to act in part as functional homologs and tie into host miRNA regulatory networks through the binding of cognate sites on cellular transcripts [reviewed in (Kincaid and Sullivan 2012)]. Such mimicry was first demonstrated for KSHV miR-K11 which bears seed homology to the miR-155 oncomiR. miR-155 is a key regulator of lymphocyte differentiation and is dysregulated in many hematopoietic cancers. Notably, 5′ nucleotides 1–8 of KSHV miR-K11 and miR-155 are identical, and initial gene expression studies demonstrated that miR-K11 and miR-155 could regulate an overlapping set of target cellular transcripts (Skalsky et al. 2007; Gottwein et al. 2007). Further in vivo experiments in humanized NOD/LtSz-scid IL2Rγ(null) mice demonstrated that KSHV miR-K11 and miR-155 could both induce expansion of splenic B lymphocytes, in part through inhibition of the transcriptional regulator C/EBPβ (Boss et al. 2011). Intriguingly, the avian lymphotrophic alpha-herpesvirus MDV-1 also encodes a miR-155 homolog, miR-M4. The presence of this miRNA was shown to contribute to T cell lymphoma formation in chickens (Zhao et al. 2011) and recent RISC-capture studies revealed many conserved miR-M4 and miR-155 targets that may play a role in lymphomagenesis (Parnas et al. 2014).

KSHV encodes additional cellular miRNA mimics including a mimic of miR-23a/b (KSHV miR-K3) and a mimic of miR-142-3p (miR-K10a) (Manzano et al. 2013, 2015; Gottwein et al. 2011). miRNA targetome studies demonstrate that at least some targets of miR-142-3p and miR-K10a overlap (Gottwein et al. 2011), and functional assays have related these back to the regulation of transforming growth factor (TGF) β signaling (Lei et al. 2012). miR-23 family members are highly expressed in endothelial cells and tightly linked to hematopoeitic lineage commitment; recent reports demonstrate opposing functions of miR-23a and miR-23b related to vascular angiogenesis and endothelial cell permeability (Li et al. 2016). For B cells, miR-23a is an inhibitor of B cell lymphopoesis and blocks expression of B cell TFs such as EBF1 and PAX5 (Kurkewich et al. 2016). KSHV infects both endothelial and B cells, and at present, it is unclear in which of these cell types that miR-K3 activity has the most impact (Manzano et al. 2013). Manzano and colleagues hypothesized that miR-K3 might substitute for miR-23 function in B cells where the miRNA absent or at very low levels; this may also be the case for miR-K11 which is detected is KSHV-infected B cell lymphoma types that do not express high levels of miR-155 (Manzano et al. 2013; Gottwein et al. 2007; Skalsky et al. 2007). Recent studies demonstrate that closely related non-human primate rhadinoviruses, RRV and JMRV, encode mimics of the miR-17/20/106 seed family (Skalsky et al. 2016). Functional analysis demonstrated that JMRV miR-J8, specifically, can block cytokine-mediated NF-kB activation as well as target 3′ UTRs bearing miR-17 cognate target sites. While this activity can dampen anti-viral inflammatory responses, miR-17 is also associated with B cell cancers and thus, the mimicking of miR-17 family members may contribute to the oncogenicity of these NHP rhadinoviruses.

7 Cataloging Targets Reveals Diversity in Networks Regulated by Viral MiRNAs

While sequence homology to host miRNAs or perfect complementarity to viral protein coding transcripts provides relatively easy insight into a handful of targets for herpesvirus miRNAs, comprehensively identifying targets for viral miRNAs has been a daunting challenge. Target recognition requires only limited sequence complementarity (generally, ~7–8 nt in the miRNA seed) between a miRNA and target site, thus a single miRNA can bind an estimated 150–200 unique sites (Bartel 2009). This is further complicated by the fact that viral miRNA effector seed sequences are not nearly as conserved as the cellular miRNAs; thus, miRNA target prediction programs that rely on evolutionary scoring measures do not necessarily work for predicting viral miRNA targets.

Significant advancements have been made over the last five to seven years in the ability to globally capture and catalog miRNA targets in a high-throughput manner. Cross-linking and immunoprecipitation (CLIP) approaches, such as RIP-Chip, PAR-CLIP, or HITS-CLIP, whereby RISC-associated RNAs are immunopurified directly, analyzed by either microarray or RNA sequencing, and analyzed further by bioinformatics were first applied to human B cells infected with KSHV and/or EBV (Dolken et al. 2010b; Gottwein et al. 2011; Skalsky et al. 2012, 2014; Riley et al. 2012). Large-scale proteomic approaches, such as SILAC (stable isotope labeling by amino acids in cell culture), have yielded additional insight into viral miRNA targets. To identify KSHV miRNA targets by SILAC, miRNA mimics were transfected into human umbilical vein endothelial cells (HUVECs), cells were pulse-labeled and analyzed by mass spectrometry alongside control cells to determine changes in the proteome (Gallaher et al. 2013). Collectively, these CLIP and proteomics studies have revealed that at least 15% of host protein coding transcripts are subject to viral miRNA regulation during latent infection. Not surprisingly, the viral miRNA targets identified in these studies are related to innate anti-viral immunity, cell survival, and multiple signaling pathways such as NF-kB, Wnt, and MAPK signaling that control cell proliferation and differentiation through distinct transcriptional programs (Fig. 3). Recent temporal analysis of the HCMV miRNA targetome during productive infection identified enriched networks such as metabolic and immune processes, trafficking and transport, and DNA replication that are subject to miRNA regulation during early, intermediate, and late times post-infection, respectively (Kim et al. 2015). Similar to the gamma-herpesviruses, apoptosis and intracellular signal pathways were influenced by HCMV miRNAs throughout all time points, reinforcing common networks that are critical to the herpesvirus life cycle in general.

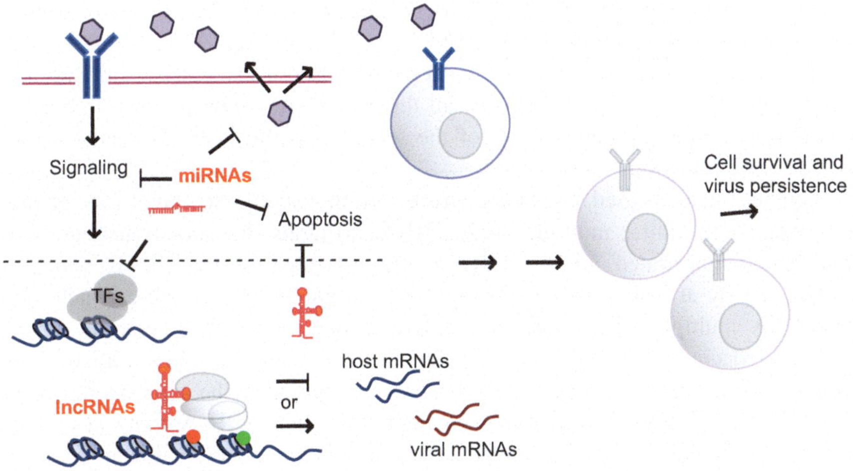

Fig. 3 Intrinsic and extrinsic modulation of host signaling pathways by viral ncRNAs. Pleiotropic herpesvirus post-transcriptional silencers (miRNAs) and nuclear modifiers (lncRNAs) of the internal host cell transcriptional environment can subsequently alter the external micro-environment through the manipulation of cell surface receptors, signaling pathways that activate or repress transcription, and subsequent trafficking and display or secretion of effector factors (such as cytokines) that are capable of influencing neighboring uninfected cells. The combined activities of viral ncRNAs (and viral proteins) provide a fit environment to support prolonged virus persistence

8 Combinatorial Targeting and MiRNA Synergy

miRNAs are often thought to have relatively minor effects on protein expression levels, (Bartel 2004, 2009), and while in many cases this can have a profound effect on cellular processes, redundancies and overlapping functions may be capable of overcoming the effect of a single miRNA on a single target protein. One means to overcome this issue is for a viral miRNA to target several components within a cellular process or signaling pathway. There are several examples of HMCV, in particular, utilizing this approach to block innate immune signaling, cell cycle regulation and secretory pathways.

Grey et al. (2010) were the first to suggest that viral miRNAs could cooperatively target multiple components within a cellular pathway, and identified several HCMV miR-US25-1 targets related to cell cycle control including cyclin E2, BRCC3, EID1, MAPRE2 and CD147. Interestingly, miR-US25-1 target sites were identified within 5′ UTRs, rather than the 3′ UTRs of these targets. In addition, HCMV encodes at least three miRNAs (miRs-US5-1, US5-2 and UL112-1) that target multiple components of the cellular secretory pathway, including RAB5C, RAB11A, SNAP23 and CDC42 (Hook et al. 2014). Hook and colleagues showed that infection of cells with viral mutants lacking these miRNAs results in significantly increased secretion of IL6 and TNFa as well as aberrant VAC formation and increased secretion of non-infectious virus particles (Hook et al. 2014). Importantly, the full effect was only evident when all three miRNAs were mutated, indicating that targeting multiple components of the secretory pathway is essential for productive viral infection.

Another method used by herpesviruses to maximally downregulate protein expression is to utilize multiple viral miRNAs to target the same transcript. This was first demonstrated for the THBS1 (Thrombospondin 1) 3′ UTR which is co-targeted by at least four KSHV miRNAs; inhibition of this cell-cell adhesion protein by multiple KSHV miRNAs may contribute to KS pathogenesis as this results in decreased TGFβ activity (Samols et al. 2007). HCMV utilizes both miR-US5-1 and miR-US5-2 to synergistically downregulate the expression of US7 (Tirabassi et al. 2011). Some evidence exists that viral miRNAs may also work synergistically with cellular miRNAs in order to maximize protein downregulation. For example, cellular miRNA miR-376a acts in concert with HCMV miR-UL112-1 to downregulate the stress-induced ligand MICB (Nachmani et al. 2010). CLIP studies have demonstrated that in addition to host miR-17, the EBV LMP1 3′ UTR is targeted by multiple EBV BART miRNAs (Riley et al. 2012; Skalsky et al. 2012; 2014; Lo et al. 2007). EBV BHRF1 transcripts may also be synergistically regulated since binding sites within the 3′ UTR for EBV miR-BART10 as well as the miR-17 family and miR-142-3p have been captured by PAR-CLIP (Skalsky et al. 2012, 2014).

Co-targeting and cooperativity between HCMV miRNAs and human miRNAs (i.e. particularly, those induced upon HCMV infection) has recently been observed on a more widespread level; following de novo infection of human fibroblasts, these miRNAs collectively regulate genes involved in cell cycle networks and RNA/protein processing which may enhance virus replication and virion production (Kim et al. 2015). HITS-CLIP studies on EBV + Burkitt's lymphoma cell lines revealed that members of the miR-17/92 cluster and EBV miRNAs may also act in concert to co-target several host genes related to apoptosis or involved in B cell lymphomagenesis (Riley et al. 2012). Advancements in network analysis and system-wide approaches to capture miRNA targets at different infection stages will facilitate the realization of additional cooperativity between host and viral miRNAs.

9 Unique Interplay Between Herpesviruses and Cellular MiRNAs

Cellular miRNAs can have anti-viral functions (Seo et al. 2013) and it is clear that herpesviruses have evolved to modulate the overall levels of cellular miRNAs by transcriptional and post-transcriptional methods (Santhakumar et al. 2010). In an interesting twist, Buck et al. showed that herpesvirus transcripts could act as decoys for cellular miRNAs, functionally decreasing the miRNA levels in infected cells in a post-transcriptional manner (Buck et al. 2010). MCMV infection of mouse fibroblasts mediates the rapid downregulation of miR-27, a miRNA capable of limiting MCMV replication when overexpressed (Buck et al. 2010). A canonical miR-27 binding site was discovered that maps to the 3′ UTR of m169, an ORF of unknown function (Buck et al. 2010). Mutation of the miR-27 binding site is capable of retargeting the transcript to other cellular miRNAs (Marcinowski et al. 2012). Importantly, Marcinowski et al. (2012) showed a growth defect in multiple organs in vivo using viral mutants incapable of binding miR-27. Similar results with infection of SCID BALB/c mice indicate the defect is independent of the adaptive immune response. Why MCMV targets miR-27 for degradation is unclear, although the miRNA has roles in cell cycle progression, a process highly manipulated by CMVs. Interestingly, the miR-27 binding site is highly conserved in MCMV species, but not the closely related RCMV (McCaskill et al. 2015).

Lee et al. (2013) discovered an HCMV intergenic RNA sequence between UL144 and UL145 that functions to downregulate members of the miR-17–92 cluster termed the miRNA decay element (miRDE). HCMV IE1 and IE2 induce the transcription of pri-miR-17–92, however of the six miRNAs encoded within this cluster (Mogilyansky and Rigoutsos 2013) only mature miR-17 and miR-20a are selectively downregulated by the miRDE. In fact, levels of pre-miRNA, as well as the passenger miR-17* and miR-20a* strands, increase during infection. Unlike what is observed with MCMV and miR-27, binding of the miRNAs to the miRDE is non-canonical and does not affect miRDE stability. Like the m169 ORF of MCMV, the miRDE sequence can be re-engineered to target other cellular miRNAs. The miR-17 binding site within the intergenic region between UL144 and UL145 is highly conserved in clinical strains of HCMV and other Old World primates (McCaskill et al. 2015), however the consequence of downregulating miR-17 and miR-20a during HCMV infection is unclear. Peak viral titers are slightly decreased when the miRDE is mutated, a defect that can be rescued using antisense oligomers to miR-17 and miR-20a. The authors suggest that since miR-17 and miR-20a target cell cycle regulators, the virus may prevent cell cycle arrest, although this awaits experimental confirmation.

As another example of miRNA sponging by herpesviruses, Cazalla et al. (2010) determined that HSURs 1 and 2 of HVS contain binding sites for three cellular miRNAs: miR-16, miR-27 and miR-142-3p. Co-immunoprecipitation experiments identified a direct association between HSUR snRNPs and the cellular miRNAs in HVS-transformed cells. Using plasmids expressing individual HSURs, a direct interaction between miR-27 and HSUR1 was determined. Analysis of T cell lines transformed with WT or HSUR-deleted viruses showed that miR-27 levels inversely correlate with HSUR1 expression. HSUR1 contains an AU-rich RNA decay element (ARE), which is involved in turnover of HSUR1. Mutation of the ARE sequence results in increased levels of HSUR1 and further decreased levels of miR-27, indicating that miR-27 turnover may occur through the ARE pathway. Increased levels of FOXO1, a confirmed target of miR-27 (Guttilla and White 2009), was observed when HSURs 1 and 2 were expressed. Further investigation into miR-27 targets identified GRB2 and other proteins involved in T cell receptor signaling that are modulated by HSUR1 expression, suggesting that HVS manipulation of miR-27 levels may contribute to the T cell transformation phenotype (Guo et al. 2014). The importance of HSUR binding of miR-16, a miRNA involved in cell cycle control (Liu et al. 2008) and miR-142-3p, a miRNA involved in hematopoietic differentiation (Lu et al. 2013) awaits investigation. miRNA decoys have not yet been demonstrated for human gamma-herpesviruses, although this has been alluded to for EBV (Cao et al. 2015).

10 In Vivo Phenotypes of Herpesvirus NcRNAs

Initial observations of EBV miRNAs provided some hints that herpesvirus miRNAs are not essential for basic aspects of the viral replication cycle and rather function as virulence factors to enhance viral persistence within a host. 17 of the 22 EBV BART miRNAs are absent from the prototypical EBV B95-8 laboratory strain and were in fact identified only after sequencing small RNAs from cells infected with patient-derived EBV strains (Pfeffer et al. 2004; Grundhoff et al. 2006; Chen et al. 2010). EBV B95-8 effectively immortalizes primary B cells and establishes latent infection comparable to full-length, wildtype virus in vitro. Deletion of all the BHRF1 miRNAs from EBV B95-8 has only a moderate effect on the outgrowth of LCLs in vitro (Feederle et al. 2011a). While difficult to test in vivo, in humanized mice, this BHRF1 miRNA mutant virus exhibits a significant delay in systemic infection compared to wild-type, with lower levels of mutant virus appearing in circulation following injection of the spleen (Wahl et al. 2013). Interestingly,

EBV + tumors develop in these mice several weeks after primary infection with no measurable differences between wild-type B95-8 and the BHRF1 miRNA mutant (Wahl et al. 2013). Recent in vitro studies suggest that the BHRF1 miRNAs may play a more pertinent role during acute stages of infection during the transition to latency when robust T cell responses must be counteracted for effective long-term persistence (Albanese et al. 2016; Tagawa et al. 2016).

Overexpressing viral miRNAs in cells and injecting these cells into mice have tested phenotypes for EBV or KSHV miRNAs in the absence of infection. EBV-negative nasopharyngeal carcinoma (NPC) cells were engineered to express EBV miR-BART7-3p and injected into irradiated mice (Cai et al. 2015). Three weeks later, lymph nodes were examined for metastases; ~83% of mice injected with miR-BART7-3p expressing cells developed metastases compared to only ~17% of mice injected with control NPC cells (Cai et al. 2015). EBV miR-BART7-3p was subsequently shown to target PTEN (phosphatase and tensin homolog) and alter PI3 K/Akt signaling events that may contribute to epithelial-to-mesenchymal transition and enhance the metastatic potential of NPC cells (Cai et al. 2015). Ectopic EBV BART miRNA expression in AGS gastric carcinoma cells produced slightly different results when injected into the nasopharyngeal epithelial tissue of immunodeficient NSG mice (Qiu et al. 2015). In this xenograft mouse model, 5/5 mice injected with BART miRNA expressing cells developed large, aggressive tumors that required sacrifice within 74 days compared to 2/5 control mice that developed small tumors and remained healthy. In contrast to Cai et al. (2015), no significant differences in invasion or metastases were observed, which could be due to the overall expression levels of the individual EBV BART miRNAs within each model. As mentioned earlier, injection of human hematopoietic progenitor cells (HPCs) ectopically expressing KSHV miR-K11 into NOD/LtSz-scid IL2Rγ(null) mice induced expansion of human CD19 + B lymphocytes in the spleen (Boss et al. 2011). Confirming these observations, expression of miR-K11 in HPCs in a different mouse model resulted in expansion of peripheral mature B cells as well as increased numbers of immature pre-B cells in bone marrow (Dahlke et al. 2012). Thus, in the absence of virus, these studies demonstrate that individual gamma-herpesvirus miRNAs can potentiate oncogenesis. One outstanding question is what functions these viral miRNAs and other ncRNAs have with respect to pathogenesis during the course of natural infection or at least within the genetic background of a natural host.

10.1 Viral Small NcRNA Phenotypes in Natural Hosts

Loss of function studies using miRNA mutant herpesviruses in small animal models that are natural hosts have demonstrated that many viral miRNAs are dispensable for acute replication in vivo. Similar to HSV-1, swine PRV encodes multiple miRNAs within the LAT region. PRV is neurotropic, establishes latency in the trigeminal ganglia, and expresses an 8.4 kb polyadenylated large latency transcript

(LLT) which is spliced to yield a 4.6 kb stable LLT intron encoding the PRV miRNAs. German Landrace pigs were inoculated intra-nasally with a LLT-deletion virus (spanning 9 of 11 PRV pre-miRNAs) (Mahjoub et al. 2015). Compared to wildtype virus, no differences in viral replication were observed during acute infection and surprisingly, the mutant virus established latency with no evidence of promiscuous lytic reactivation (Mahjoub et al. 2015). Recently, two groups investigated MHV68 small ncRNA (miRNA and TMER) knockout viruses compared to wild-type or revertant viruses (Feldman et al. 2014, 2016; Diebel et al. 2015). In vitro, viruses lacking the MHV68 small ncRNAs replicate with the same efficiency as wild-type virus. During acute infection in immunocompetent mice, there were no discernable differences in viral replication. Analysis of a panel of TMER-mutant viruses in vivo revealed that MHV68 TMER4 can contribute to viral dissemination independent of the viral miRNAs, which is important for establishment of latency in the periphery (Feldman et al. 2016).

The first in vivo phenotype for a beta-herpesvirus miRNA mutant was demonstrated during MCMV infection in mice (Dolken et al. 2010a). When two of the most highly expressed MCMV miRNAs [miR-M23-2 and miR-m21-1 (Dolken et al. 2007)] were mutated, the resultant virus showed a ~100-fold reduction in replication in the salivary gland at 14 days post-infection. This defect was dependent on mouse strain and viral dose and was not observed in other tissues. The salivary gland replication defect could be partially restored through combined depletion of NK and CD4 + T cells; however, targets of miR-M23-2 and miR-m21-1 have not been identified. Given that the salivary gland is an immunologically unique site that is important for horizontal transmission, insights into the targets of these miRNAs would aid in understanding how CMV alters the cellular environment to establish a persistent infection while evading the host immune response.

To date, the only non-human primate model for which a viral miRNA mutant has been tested is RhCMV infection of Indian rhesus macaques. Similar to observations made in small animal models, deletion of a large region (RhCMV US2-US11) spanning the HCMV miR-US5-2 homolog, RhCMV miR-183-1, and other RhCMV miRNAs (Skalsky, Hancock et al., unpublished observations) showed no changes in viral replication or dissemination (Hansen et al. 2010a). For humans, epidemiological studies where viral genomic information is available can provide important links between viral miRNAs and disease phenotypes. As a notable example, natural KSHV strains contain SNPs in several of the pre-miRNA regions which affect miRNA processing (Han et al. 2013; Marshall et al. 2007); several pre-miRNA SNPs track with increased disease risk for AIDS-KS patients or are associated with Multicentric Castleman's disease (Marshall et al. 2010; Ray et al. 2012).

10.2 In Vivo Phenotypes of LncRNA Knockouts

For herpesvirus lncRNAs, in vivo phenotypes are more difficult to pinpoint as being a direct effect of the lncRNA since abrogation of the lncRNA requires larger deletion(s) from the virus that could disrupt protein and other non-coding regions of potentially unknown functions. Attempts have been made to characterize several herpesvirus lncRNAs in vivo. A homolog of HCMV RNA5.0, MCMV RNA7.2, is an extremely long-lived stable intron during MCMV infection (Kulesza and Shenk 2006; Schwarz et al. 2013). Although mutation of the RNA7.2 region had no effect on replication in vitro, RNA7.2 mutant viruses fail to persist in the mouse salivary gland at 14 days post-infection (Schwarz et al. 2013; Kulesza and Shenk 2006) suggesting that RNA7.2 is a virulence factor.

HSV LAT is the most abundantly transcribed RNA in latently infected cells in humans (Croen et al. 1987) and in small animal models (Stevens et al. 1987), and as such likely plays an important role during latency. Surprisingly then, it has been observed that LAT-negative viruses are still capable of establishing and reactivating from latency (Perng et al. 1994; Chen et al. 1997). However, a decrease in the number of latently infected trigeminal ganglia upon infection of mice with LAT- viruses (Thompson and Sawtell 1997; Devi-Rao et al. 1994) suggests at least a contributing role for the transcript in establishment or maintenance of latency. Many of the ascribed functions of LAT, including blocking apoptosis and heterochromatinization of the lytic promoters may contribute to this phenotype. Additionally, LAT has been shown to repress lytic gene expression (Chen et al. 1997), likely since the primary LAT transcript encodes for many of the HSV miRNAs that target ICP0 and ICP4 (Umbach et al. 2010; Tang et al. 2009; Umbach et al. 2008). Finally, LAT seems to be important for promoting spontaneous reactivation in rabbits (Perng et al. 1994; Hill et al. 1990), although this is not observed in mice (Margolis et al. 2007; Carr et al. 1998), indicating species-specific phenotypes (Margolis et al. 2007). The role of LAT in HSV latency in small animal models is complicated by the relatively short lifespan of these non-natural hosts and thus, the essential functions of LAT in humans may as yet be identified.

Phenotypes for EBV lncRNAs have also been difficult to elucidate in vivo. EBER1 and EBER2 knockout viruses were examined recently in a humanized mouse model. Compared to EBER-expressing viruses, there were no significant differences in viral loads in spleen and peripheral blood or reproducible phenotypes for CD8 + T cell responses (Gregorovic et al. 2015). Historical phenotypes associated with deletions or mutations in herpesvirus genomes may yet be linked to viral lncRNA functions as additional molecular and transcriptomic studies are performed. Furthermore, as more non-conventional molecular interactions are uncovered for viral lncRNAs, this information can guide the exploration for specific phenotypes in vivo.

11 Summary and Outlook

Much progress has been made in defining the ncRNAs that are produced by herpesviruses at various stages of their complex life cycles, and viral ncRNA functions are progressively being elucidated. The lncRNAs that are abundantly expressed during gamma-herpesvirus lytic replication play important roles at the epigenetic layer in modulating chromatin structure; presumably, this allows access to the viral genome for lytic replication and/or lytic viral gene expression. For alpha and beta herpesviruses, lncRNAs have important roles in recruiting chromatin remodelers to silence viral gene expression; these chromatin marks can poise the lytic promoters for reactivation stimuli. Likely, this is further aided through the myriad of activities of the viral miRNAs (Fig. 3).

Many studies demonstrate that herpesviruses utilize their own miRNAs as well as co-opt host miRNAs to regulate the kinetics and level of viral transcripts; this contributes primarily towards the silencing of viral gene expression to promote an environment for latency. Viral and host miRNAs are further utilized to manipulate the host proteome which subsequently impacts signaling pathways, cell networks, and the micro-environment (Fig. 3). Important particularly for beta- and gamma-herpesviruses, and directly related to viral pathogenesis, many of these miRNA-regulated networks control cell lineage commitment and differentiation. The combinatorial targeting of signaling components as well as targeting of the transcription factors governing cell processes by viral miRNAs amplifies the subtle effects of individual miRNAs; furthermore, many viral miRNAs are expressed as clusters and thus can act simultaneously within a specific stage of the viral life cycle. Thus, these collective actions induce significant changes in the host transcriptome, amplified initially at the transcriptional level and further amplified at the post-translational level. In short, these tiny molecules have big impacts in providing a fit cellular environment to effectively support herpesvirus persistence.

New technologies to capture detailed molecular interactions between ncRNAs and binding partners should yield insight into novel functions during herpesvirus infection. Genome-wide meta-approaches to integrate the multiple layers of epigenetic, transcriptional, and post-transcriptional control facilitated by herpesvirus ncRNAs will help to further dissect the complexities of herpesvirus-host interactions and identify innovative opportunities for therapeutic intervention.

Questions Moving Forward

1. What viral ncRNAs are present and functional in vivo?
2. What is the viral ncRNA influence on the extracellular environment and how much does this contribute to pathogenesis?
3. How does the cell context contribute to viral ncRNA function(s)?
4. How can viral ncRNAs be used in novel ways to understand functions of host ncRNAs or probe unknown cellular functions?
5. How can viral ncRNAs be used as effective therapeutic targets?

Acknowledgements We thank the past and present members of Jay Nelson's laboratory at VGTI for discussions, and apologize to the many scientists whose work we could not mention due to space limitations. This work was supported by a Pathway to Independence Award CA175181 from the National Cancer Institute to R.L.S. Support for M.H. comes from NIH grant AI21640.

References

Abend JR, Ramalingam D, Kieffer-Kwon P, Uldrick TS, Yarchoan R, Ziegelbauer JM (2012) KSHV microRNAs target two components of the TLR/IL-1R signaling cascade, IRAK1 and MYD88, to reduce inflammatory cytokine expression. J Virol. doi:10.1128/JVI.01147-12

Albanese M, Tagawa T, Bouvet M, Maliqi L, Lutter D, Hoser J, Hastreiter M, Hayes M, Sugden B, Martin L, Moosmann A, Hammerschmidt W (2016) Epstein-Barr virus microRNAs reduce immune surveillance by virus-specific CD8 + T cells. Proc Natl Acad Sci U S A 113 (42):E6467–E6475. doi:10.1073/pnas.1605884113

Ambros V (2004) The functions of animal microRNAs. Nature 431(7006):350–355. doi:10.1038/nature02871

Amelio AL, Giordani NV, Kubat NJ, O'Neil JE, Bloom DC (2006) Deacetylation of the herpes simplex virus type 1 latency-associated transcript (LAT) enhancer and a decrease in LAT abundance precede an increase in ICP0 transcriptional permissiveness at early times postexplant. J Virol 80(4):2063–2068. doi:10.1128/JVI.01528-10

Amoroso R, Fitzsimmons L, Thomas WA, Kelly GL, Rowe M, Bell AI (2011) Quantitative studies of Epstein-Barr virus-encoded microRNAs provide novel insights into their regulation. J Virol 85(2):996–1010. doi:10.1128/JVI.01528-10

Arias C, Weisburd B, Stern-Ginossar N, Mercier A, Madrid AS, Bellare P, Holdorf M, Weissman JS, Ganem D (2014) KSHV 2.0: a comprehensive annotation of the Kaposi's sarcoma-associated herpesvirus genome using next-generation sequencing reveals novel genomic and functional features. PLoS Pathog 10(1):e1003847. doi:10.1371/journal.ppat.1003847

Arvey A, Tempera I, Tsai K, Chen HS, Tikhmyanova N, Klichinsky M, Leslie C, Lieberman PM (2012) An atlas of the Epstein-Barr virus transcriptome and epigenome reveals host-virus regulatory interactions. Cell Host Microbe 12(2):233–245. doi:10.1016/j.chom.2012.06.008

Bartel DP (2004) MicroRNAs: genomics, biogenesis, mechanism, and function. Cell 116 (2):281–297. S0092867404000455

Bartel DP (2009) MicroRNAs: target recognition and regulatory functions. Cell 136 (2):215–233. S0092-8674(09)00008-7

Barth S, Pfuhl T, Mamiani A, Ehses C, Roemer K, Kremmer E, Jaker C, Hock J, Meister G, Grasser FA (2008) Epstein-Barr virus-encoded microRNA miR-BART2 down-regulates the viral DNA polymerase BALF5. Nucleic Acids Res 36(2):666–675. doi:10.1093/nar/gkm1080

Bellare P, Ganem D (2009) Regulation of KSHV lytic switch protein expression by a virus-encoded microRNA: an evolutionary adaptation that fine-tunes lytic reactivation. Cell Host Microbe 6(6):570–575. doi:10.1016/j.chom.2009.11.008

Bogerd HP, Karnowski HW, Cai X, Shin J, Pohlers M, Cullen BR (2010) A Mammalian Herpesvirus Uses Noncanonical Expression and Processing Mechanisms to Generate Viral MicroRNAs. Mol Cell 37:135–142

Boss IW, Nadeau PE, Abbott JR, Yang Y, Mergia A, Renne R (2011) A Kaposi's sarcoma-associated herpesvirus-encoded ortholog of microRNA miR-155 induces human splenic B-cell expansion in NOD/LtSz-scid IL2Rgammanull mice. J Virol 85(19):9877–9886. doi:10.1128/JVI.05558-11

Britt W (2008) Manifestations of human cytomegalovirus infection: proposed mechanisms of acute and chronic disease. Curr Top Microbiol Immunol 325:417–470

Bruce AG, Ryan JT, Thomas MJ, Peng X, Grundhoff A, Tsai CC, Rose TM (2013) Next-generation sequence analysis of the genome of RFHVMn, the macaque homolog of Kaposi's sarcoma (KS)-associated herpesvirus, from a KS-like tumor of a pig-tailed macaque. J Virol 87(24):13676–13693. doi:10.1128/JVI.02331-13

Buck AH, Perot J, Chisholm MA, Kumar DS, Tuddenham L, Cognat V, Marcinowski L, Dolken L, Pfeffer S (2010) Post-transcriptional regulation of miR-27 in murine cytomegalovirus infection. RNA 16(2):307–315. doi:10.1261/rna.1819210

Buck AH, Santoyo-Lopez J, Robertson KA, Kumar DS, Reczko M, Ghazal P (2007) Discrete clusters of virus-encoded micrornas are associated with complementary strands of the genome and the 7.2-kilobase stable intron in murine cytomegalovirus. J Virol 81(24):13761–13770. doi:10.1128/JVI.01290-07

Cai LM, Lyu XM, Luo WR, Cui XF, Ye YF, Yuan CC, Peng QX, Wu DH, Liu TF, Wang E, Marincola FM, Yao KT, Fang WY, Cai HB, Li X (2015) EBV-miR-BART7-3p promotes the EMT and metastasis of nasopharyngeal carcinoma cells by suppressing the tumor suppressor PTEN. Oncogene 34(17):2156–2166. doi:10.1038/onc.2014.341

Cai X, Lu S, Zhang Z, Gonzalez CM, Damania B, Cullen BR (2005) Kaposi's sarcoma-associated herpesvirus expresses an array of viral microRNAs in latently infected cells. Proc Natl Acad Sci U S A 102(15):5570–5575. doi:10.1073/pnas.0408192102

Cai X, Schafer A, Lu S, Bilello JP, Desrosiers RC, Edwards R, Raab-Traub N, Cullen BR (2006) Epstein-Barr virus microRNAs are evolutionarily conserved and differentially expressed. PLoS Pathog 2(3):e23. doi:10.1371/journal.ppat.0020023

Campbell M, Kung HJ, Izumiya Y (2014) Long non-coding RNA and epigenetic gene regulation of KSHV. Viruses 6(11):4165–4177. doi:10.3390/v6114165

Cao S, Strong MJ, Wang X, Moss WN, Concha M, Lin Z, O'Grady T, Baddoo M, Fewell C, Renne R, Flemington EK (2015) High-throughput RNA sequencing-based virome analysis of 50 lymphoma cell lines from the Cancer Cell Line Encyclopedia project. J Virol 89(1):713–729. doi:10.1128/JVI.02570-14

Caposio P, Luganini A, Hahn G, Landolfo S, Gribaudo G (2007) Activation of the virus-induced IKK/NF-kappaB signalling axis is critical for the replication of human cytomegalovirus in quiescent cells. Cell Microbiol 9(8):2040–2054. doi:10.1111/j.1462-5822.2007.00936.x

Carr DJ, Halford WP, Veress LA, Noisakran S, Perng GC, Wechsler SL (1998) The persistent elevated cytokine mRNA levels in trigeminal ganglia of mice latently infected with HSV-1 are not due to the presence of latency associated transcript (LAT) RNAs. Virus Res 54(1):1–8

Cazalla D, Yario T, Steitz JA (2010) Down-regulation of a host microRNA by a Herpesvirus saimiri noncoding RNA. Science 328(5985):1563–1566. doi:10.1126/science.1187197

Cech TR, Steitz JA (2014) The noncoding RNA revolution-trashing old rules to forge new ones. Cell 157(1):77–94. doi:10.1016/j.cell.2014.03.008

Chandriani S, Xu Y, Ganem D (2010) The lytic transcriptome of Kaposi's sarcoma-associated herpesvirus reveals extensive transcription of noncoding regions, including regions antisense to important genes. J Virol 84(16):7934–7942. doi:10.1128/JVI.00645-10

Chen SH, Kramer MF, Schaffer PA, Coen DM (1997) A viral function represses accumulation of transcripts from productive-cycle genes in mouse ganglia latently infected with herpes simplex virus. J Virol 71(8):5878–5884

Chen SJ, Chen GH, Chen YH, Liu CY, Chang KP, Chang YS, Chen HC (2010) Characterization of Epstein-Barr virus miRNAome in nasopharyngeal carcinoma by deep sequencing. PLoS ONE 5(9):0001. doi:10.1371/journal.pone.0012745

Chinen M, Tani T (2012) Diverse functions of nuclear non-coding RNAs in eukaryotic gene expression. Front Biosci (Landmark Ed) 17:1402–1417

Cliffe AR, Coen DM, Knipe DM (2013) Kinetics of facultative heterochromatin and polycomb group protein association with the herpes simplex viral genome during establishment of latent infection. MBio 4(1):0001. doi:10.1128/mBio.00590-12

Cliffe AR, Garber DA, Knipe DM (2009) Transcription of the herpes simplex virus latency-associated transcript promotes the formation of facultative heterochromatin on lytic promoters. J Virol 83(16):8182–8190. doi:10.1128/JVI.00712-09

Concha M, Wang X, Cao S, Baddoo M, Fewell C, Lin Z, Hulme W, Hedges D, McBride J, Flemington EK (2012) Identification of new viral genes and transcript isoforms during Epstein-Barr virus reactivation using RNA-Seq. J Virol 86(3):1458–1467. doi:10.1128/JVI.06537-11

Conrad NK (2016) New insights into the expression and functions of the Kaposi's sarcoma-associated herpesvirus long noncoding PAN RNA. Virus Res 212:53–63. doi:10.1016/j.virusres.2015.06.012

Conrad NK, Steitz JA (2005) A Kaposi's sarcoma virus RNA element that increases the nuclear abundance of intronless transcripts. EMBO J 24(10):1831–1841. doi:10.1038/sj.emboj.7600662

Croen KD, Ostrove JM, Dragovic LJ, Smialek JE, Straus SE (1987) Latent herpes simplex virus in human trigeminal ganglia. Detection of an immediate early gene "anti-sense" transcript by in situ hybridization. N Engl J Med 317(23):1427–1432. doi:10.1056/NEJM198712033172302

Dahlke C, Maul K, Christalla T, Walz N, Schult P, Stocking C, Grundhoff A (2012) A microRNA encoded by Kaposi sarcoma-associated herpesvirus promotes B-cell expansion in vivo. PLoS ONE 7(11):e49435. doi:10.1371/journal.pone.0049435

Derrien T, Johnson R, Bussotti G, Tanzer A, Djebali S, Tilgner H, Guernec G, Martin D, Merkel A, Knowles DG, Lagarde J, Veeravalli L, Ruan X, Ruan Y, Lassmann T, Carninci P, Brown JB, Lipovich L, Gonzalez JM, Thomas M, Davis CA, Shiekhattar R, Gingeras TR, Hubbard TJ, Notredame C, Harrow J, Guigo R (2012) The GENCODE v7 catalog of human long noncoding RNAs: analysis of their gene structure, evolution, and expression. Genome Res 22(9):1775–1789. doi:10.1101/gr.132159.111

Devi-Rao GB, Bloom DC, Stevens JG, Wagner EK (1994) Herpes simplex virus type 1 DNA replication and gene expression during explant-induced reactivation of latently infected murine sensory ganglia. J Virol 68(3):1271–1282

Diebel KW, Oko LM, Medina EM, Niemeyer BF, Warren CJ, Claypool DJ, Tibbetts SA, Cool CD, Clambey ET, van Dyk LF (2015) Gammaherpesvirus small noncoding RNAs are bifunctional elements that regulate infection and contribute to virulence in vivo. MBio 6(1):e01670–e01614. doi:10.1128/mBio.01670-14

Diebel KW, Smith AL, van Dyk LF (2010) Mature and functional viral miRNAs transcribed from novel RNA polymerase III promoters. RNA 16(1):170–185. doi:10.1261/rna.1873910

Dolken L, Krmpotic A, Kothe S, Tuddenham L, Tanguy M, Marcinowski L, Ruzsics Z, Elefant N, Altuvia Y, Margalit H, Koszinowski UH, Jonjic S, Pfeffer S (2010a) Cytomegalovirus microRNAs facilitate persistent virus infection in salivary glands. PLoS Pathog 6(10):e1001150. doi:10.1371/journal.ppat.1001150

Dolken L, Malterer G, Erhard F, Kothe S, Friedel CC, Suffert G, Marcinowski L, Motsch N, Barth S, Beitzinger M, Lieber D, Bailer SM, Hoffmann R, Ruzsics Z, Kremmer E, Pfeffer S, Zimmer R, Koszinowski UH, Grasser F, Meister G, Haas J (2010b) Systematic analysis of viral and cellular microRNA targets in cells latently infected with human gamma-herpesviruses by RISC immunoprecipitation assay. Cell Host Microbe 7(4):324–334. doi:10.1016/j.chom.2010.03.008

Dolken L, Perot J, Cognat V, Alioua A, John M, Soutschek J, Ruzsics Z, Koszinowski U, Voinnet O, Pfeffer S (2007) Mouse cytomegalovirus microRNAs dominate the cellular small RNA profile during lytic infection and show features of posttranscriptional regulation. J Virol 81(24):13771–13782. doi:10.1128/JVI.01313-07

Dunn W, Trang P, Zhong Q, Yang E, van Belle C, Liu F (2005) Human cytomegalovirus expresses novel microRNAs during productive viral infection. Cell Microbiol 7(11):1684–1695. doi:10.1111/j.1462-5822.2005.00598.x

Edwards RH, Marquitz AR, Raab-Traub N (2008) Epstein-Barr virus BART microRNAs are produced from a large intron prior to splicing. J Virol 82(18):9094–9106. doi:10.1128/JVI.00785-08

Estep RD, Hansen SG, Rogers KS, Axthelm MK, Wong SW (2013) Genomic characterization of Japanese macaque rhadinovirus, a novel herpesvirus isolated from a nonhuman primate with a

spontaneous inflammatory demyelinating disease. J Virol 87(1):512–523. doi:10.1128/JVI.02194-12
Feederle R, Haar J, Bernhardt K, Linnstaedt SD, Bannert H, Lips H, Cullen BR, Delecluse HJ (2011a) The members of a viral miRNA cluster co-operate to transform B lymphocytes. J Virol. doi:10.1128/JVI.05100-11
Feederle R, Linnstaedt SD, Bannert H, Lips H, Bencun M, Cullen BR, Delecluse HJ (2011b) A viral microRNA cluster strongly potentiates the transforming properties of a human herpesvirus. PLoS Pathog 7(2):e1001294. doi:10.1371/journal.ppat.1001294
Feldman ER, Kara M, Coleman CB, Grau KR, Oko LM, Krueger BJ, Renne R, van Dyk LF, Tibbetts SA (2014) Virus-encoded microRNAs facilitate gammaherpesvirus latency and pathogenesis in vivo. MBio 5(3):e00981–e00914. doi:10.1128/mBio.00981-14
Feldman ER, Kara M, Oko LM, Grau KR, Krueger BJ, Zhang J, Feng P, van Dyk LF, Renne R, Tibbetts SA (2016) A Gammaherpesvirus Noncoding RNA Is Essential for Hematogenous Dissemination and Establishment of Peripheral Latency. mSphere 1 (2). doi:10.1128/mSphere.00105-15
Flores O, Nakayama S, Whisnant AW, Javanbakht H, Cullen BR, Bloom DC (2013) Mutational inactivation of herpes simplex virus 1 microRNAs identifies viral mRNA targets and reveals phenotypic effects in culture. J Virol 87(12):6589–6603. doi:10.1128/JVI.00504-13
Fok V, Mitton-Fry RM, Grech A, Steitz JA (2006) Multiple domains of EBER 1, an Epstein-Barr virus noncoding RNA, recruit human ribosomal protein L22. RNA 12(5):872–882. doi:10.1261/rna.2339606
Forte E, Luftig MA (2011) The role of microRNAs in Epstein-Barr virus latency and lytic reactivation. Microbes and infection/ Institut Pasteur 13(14–15):1156–1167. doi:10.1016/j.micinf.2011.07.007
Fu M, Gao Y, Zhou Q, Zhang Q, Peng Y, Tian K, Wang J, Zheng X (2014) Human cytomegalovirus latent infection alters the expression of cellular and viral microRNA. Gene 536(2):272–278. doi:10.1016/j.gene.2013.12.012
Gallaher AM, Das S, Xiao Z, Andresson T, Kieffer-Kwon P, Happel C, Ziegelbauer J (2013) Proteomic screening of human targets of viral microRNAs reveals functions associated with immune evasion and angiogenesis. PLoS Pathog 9(9):e1003584. doi:10.1371/journal.ppat.1003584
Gatherer D, Seirafian S, Cunningham C, Holton M, Dargan DJ, Baluchova K, Hector RD, Galbraith J, Herzyk P, Wilkinson GW, Davison AJ (2011) High-resolution human cytomegalovirus transcriptome. Proc Natl Acad Sci U S A 108(49):19755–19760. doi:10.1073/pnas.1115861108
Gawn JM, Greaves RF (2002) Absence of IE1 p72 protein function during low-multiplicity infection by human cytomegalovirus results in a broad block to viral delayed-early gene expression. J Virol 76(9):4441–4455
Gottwein E, Cai X, Cullen BR (2006) A novel assay for viral microRNA function identifies a single nucleotide polymorphism that affects Drosha processing. J Virol 80(11):5321–5326. doi:10.1128/JVI.02734-05
Gottwein E, Corcoran DL, Mukherjee N, Skalsky RL, Hafner M, Nusbaum JD, Shamulailatpam P, Love CL, Dave SS, Tuschl T, Ohler U, Cullen BR (2011) Viral microRNA targetome of KSHV-infected primary effusion lymphoma cell lines. Cell Host Microbe 10(5):515–526. doi:10.1016/j.chom.2011.09.012
Gottwein E, Mukherjee N, Sachse C, Frenzel C, Majoros WH, Chi JT, Braich R, Manoharan M, Soutschek J, Ohler U, Cullen BR (2007) A viral microRNA functions as an orthologue of cellular miR-155. Nature 450(7172):1096–1099. doi:10.1038/nature05992
Gregorovic G, Boulden EA, Bosshard R, Elgueta Karstegl C, Skalsky R, Cullen BR, Gujer C, Ramer P, Munz C, Farrell PJ (2015) Epstein-Barr Viruses (EBVs) Deficient in EBV-Encoded RNAs Have Higher Levels of Latent Membrane Protein 2 RNA Expression in Lymphoblastoid Cell Lines and Efficiently Establish Persistent Infections in Humanized Mice. J Virol 89 (22):11711–11714. doi:10.1128/JVI.01873-15

Grey F (2015) Role of microRNAs in herpesvirus latency and persistence. J Gen Virol 96(Pt 4):739–751. doi:10.1099/vir.0.070862-0

Grey F, Antoniewicz A, Allen E, Saugstad J, McShea A, Carrington JC, Nelson J (2005) Identification and characterization of human cytomegalovirus-encoded microRNAs. J Virol 79 (18):12095–12099. doi:10.1128/JVI.79.18.12095-12099.2005

Grey F, Meyers H, White EA, Spector DH, Nelson J (2007) A human cytomegalovirus-encoded microRNA regulates expression of multiple viral genes involved in replication. PLoS Pathog 3 (11):e163. doi:10.1371/journal.ppat.0030163

Grey F, Tirabassi R, Meyers H, Wu G, McWeeney S, Hook L, Nelson JA (2010) A viral microRNA down-regulates multiple cell cycle genes through mRNA 5′ UTRs. PLoS Pathog 6 (6):e1000967. doi:10.1371/journal.ppat.1000967

Grundhoff A, Sullivan CS (2011) Virus-encoded microRNAs. Virology 411(2):325–343. doi:10.1016/j.virol.2011.01.002

Grundhoff A, Sullivan CS, Ganem D (2006) A combined computational and microarray-based approach identifies novel microRNAs encoded by human gamma-herpesviruses. RNA 12 (5):733–750. doi:10.1261/rna.2326106

Guo YE, Riley KJ, Iwasaki A, Steitz JA (2014) Alternative capture of noncoding RNAs or protein-coding genes by herpesviruses to alter host T cell function. Mol Cell 54(1):67–79. doi:10.1016/j.molcel.2014.03.025

Guttilla IK, White BA (2009) Coordinate regulation of FOXO1 by miR-27a, miR-96, and miR-182 in breast cancer cells. J Biol Chem 284(35):23204–23216. doi:10.1074/jbc.M109.031427

Ha M, Kim VN (2014) Regulation of microRNA biogenesis. Nat Rev Mol Cell Biol 15(8):509–524. doi:10.1038/nrm3838

Haecker I, Gay LA, Yang Y, Hu J, Morse AM, McIntyre LM, Renne R (2012) Ago HITS-CLIP Expands Understanding of Kaposi's Sarcoma-associated Herpesvirus miRNA Function in Primary Effusion Lymphomas. PLoS Pathog 8(8):e1002884. doi:10.1371/journal.ppat.1002884

Han SJ, Marshall V, Barsov E, Quinones O, Ray A, Labo N, Trivett M, Ott D, Renne R, Whitby D (2013) Kaposi's sarcoma-associated herpesvirus microRNA single-nucleotide polymorphisms identified in clinical samples can affect microRNA processing, level of expression, and silencing activity. J Virol 87(22):12237–12248. doi:10.1128/JVI.01202-13

Hancock MH, Tirabassi RS, Nelson JA (2012) Rhesus cytomegalovirus encodes seventeen microRNAs that are differentially expressed in vitro and in vivo. Virology 425(2):133–142. doi:10.1016/j.virol.2012.01.009

Hansen A, Henderson S, Lagos D, Nikitenko L, Coulter E, Roberts S, Gratrix F, Plaisance K, Renne R, Bower M, Kellam P, Boshoff C (2010a) KSHV-encoded miRNAs target MAF to induce endothelial cell reprogramming. Genes Dev 24:195–205

Hansen SG, Powers CJ, Richards R, Ventura AB, Ford JC, Siess D, Axthelm MK, Nelson JA, Jarvis MA, Picker LJ, Fruh K (2010b) Evasion of CD8 + T cells is critical for superinfection by cytomegalovirus. Science 328(5974):102–106. doi:10.1126/science.1185350

Hill JM, Sedarati F, Javier RT, Wagner EK, Stevens JG (1990) Herpes simplex virus latent phase transcription facilitates in vivo reactivation. Virology 174(1):117–125

Hook LM, Grey F, Grabski R, Tirabassi R, Doyle T, Hancock M, Landais I, Jeng S, McWeeney S, Britt W, Nelson JA (2014) Cytomegalovirus miRNAs target secretory pathway genes to facilitate formation of the virion assembly compartment and reduce cytokine secretion. Cell Host Microbe 15(3):363–373. doi:10.1016/j.chom.2014.02.004

Hutchinson NI, Tocci MJ (1986) Characterization of a major early gene from the human cytomegalovirus long inverted repeat; predicted amino acid sequence of a 30-kDa protein encoded by the 1.2-kb mRNA. Virology 155(1):172–182

Hutzinger R, Feederle R, Mrazek J, Schiefermeier N, Balwierz PJ, Zavolan M, Polacek N, Delecluse HJ, Huttenhofer A (2009) Expression and processing of a small nucleolar RNA from the Epstein-Barr virus genome. PLoS Pathog 5(8):e1000547. doi:10.1371/journal.ppat.1000547

Iizasa H, Wulff BE, Alla NR, Maragkakis M, Megraw M, Hatzigeorgiou A, Iwakiri D, Takada K, Wiedmer A, Showe L, Lieberman P, Nishikura K (2010) Editing of EBV-encoded BART6 microRNAs controls their dicer targeting and consequently affects viral latency. J Biol Chem. doi:10.1074/jbc.M110.138362

Iyer MK, Niknafs YS, Malik R, Singhal U, Sahu A, Hosono Y, Barrette TR, Prensner JR, Evans JR, Zhao S, Poliakov A, Cao X, Dhanasekaran SM, Wu YM, Robinson DR, Beer DG, Feng FY, Iyer HK, Chinnaiyan AM (2015) The landscape of long noncoding RNAs in the human transcriptome. Nat Genet 47(3):199–208. doi:10.1038/ng.3192

Javier RT, Stevens JG, Dissette VB, Wagner EK (1988) A herpes simplex virus transcript abundant in latently infected neurons is dispensable for establishment of the latent state. Virology 166(1):254–257

Jiang X, Chentoufi AA, Hsiang C, Carpenter D, Osorio N, BenMohamed L, Fraser NW, Jones C, Wechsler SL (2011) The herpes simplex virus type 1 latency-associated transcript can protect neuron-derived C1300 and Neuro2A cells from granzyme B-induced apoptosis and CD8 T-cell killing. J Virol 85(5):2325–2332. doi:10.1128/JVI.01791-10

Jung YJ, Choi H, Kim H, Lee SK (2014) MicroRNA miR-BART20-5p Stabilizes Epstein-Barr Virus Latency by Directly Targeting BZLF1 and BRLF1. J Virol 88(16):9027–9037. doi:10.1128/JVI.00721-14

Jurak I, Hackenberg M, Kim JY, Pesola JM, Everett RD, Preston CM, Wilson AC, Coen DM (2014) Expression of herpes simplex virus 1 microRNAs in cell culture models of quiescent and latent infection. J Virol 88(4):2337–2339. doi:10.1128/JVI.03486-13

Juranic Lisnic V, Babic Cac M, Lisnic B, Trsan T, Mefferd A, Das Mukhopadhyay C, Cook CH, Jonjic S, Trgovcich J (2013) Dual analysis of the murine cytomegalovirus and host cell transcriptomes reveal new aspects of the virus-host cell interface. PLoS Pathog 9(9):e1003611. doi:10.1371/journal.ppat.1003611

Kang D, Skalsky RL, Cullen BR (2015) EBV BART MicroRNAs Target Multiple Pro-apoptotic Cellular Genes to Promote Epithelial Cell Survival. PLoS Pathog 11(6):e1004979. doi:10.1371/journal.ppat.1004979

Kapranov P, Cheng J, Dike S, Nix DA, Duttagupta R, Willingham AT, Stadler PF, Hertel J, Hackermuller J, Hofacker IL, Bell I, Cheung E, Drenkow J, Dumais E, Patel S, Helt G, Ganesh M, Ghosh S, Piccolboni A, Sementchenko V, Tammana H, Gingeras TR (2007) RNA maps reveal new RNA classes and a possible function for pervasive transcription. Science 316 (5830):1484–1488. doi:10.1126/science.1138341

Kenney SC, Mertz JE (2014) Regulation of the latent-lytic switch in Epstein-Barr virus. Semin Cancer Biol 26:60–68. doi:10.1016/j.semcancer.2014.01.002

Khalil AM, Guttman M, Huarte M, Garber M, Raj A, Rivea Morales D, Thomas K, Presser A, Bernstein BE, van Oudenaarden A, Regev A, Lander ES, Rinn JL (2009) Many human large intergenic noncoding RNAs associate with chromatin-modifying complexes and affect gene expression. Proc Natl Acad Sci U S A 106(28):11667–11672. doi:10.1073/pnas.0904715106

Kim S, Seo D, Kim D, Hong Y, Chang H, Baek D, Kim VN, Lee S, Ahn K (2015) Temporal Landscape of MicroRNA-Mediated Host-Virus Crosstalk during Productive Human Cytomegalovirus Infection. Cell Host Microbe 17(6):838–851. doi:10.1016/j.chom.2015.05.014

Kincaid RP, Sullivan CS (2012) Virus-encoded microRNAs: an overview and a look to the future. PLoS Pathog 8(12):e1003018. doi:10.1371/journal.ppat.1003018

Krug LT (2013) Complexities of gammaherpesvirus transcription revealed by microarrays and RNAseq. Curr Opin Virol 3(3):276–284. doi:10.1016/j.coviro.2013.04.006

Kulesza CA, Shenk T (2004) Human cytomegalovirus 5-kilobase immediate-early RNA is a stable intron. J Virol 78(23):13182–13189. doi:10.1128/JVI.78.23.13182-13189.2004

Kulesza CA, Shenk T (2006) Murine cytomegalovirus encodes a stable intron that facilitates persistent replication in the mouse. Proc Natl Acad Sci U S A 103(48):18302–18307. doi:10.1073/pnas.0608718103

Kurkewich JL, Bikorimana E, Nguyen T, Klopfenstein N, Zhang H, Hallas WM, Stayback G, McDowell MA, Dahl R (2016) The mirn23a microRNA cluster antagonizes B cell development. J Leukoc Biol 100(4):665–677. doi:10.1189/jlb.1HI0915-398RR
Kwiatkowski DL, Thompson HW, Bloom DC (2009) The polycomb group protein Bmi1 binds to the herpes simplex virus 1 latent genome and maintains repressive histone marks during latency. J Virol 83(16):8173–8181. doi:10.1128/JVI.00686-09
Lan K, Kuppers DA, Verma SC, Robertson ES (2004) Kaposi's sarcoma-associated herpesvirus-encoded latency-associated nuclear antigen inhibits lytic replication by targeting Rta: a potential mechanism for virus-mediated control of latency. J Virol 78(12):6585–6594. doi:10.1128/JVI.78.12.6585-6594.2004
Lau B, Poole E, Krishna B, Sellart I, Wills MR, Murphy E, Sinclair J (2016) The Expression of Human Cytomegalovirus MicroRNA MiR-UL148D during Latent Infection in Primary Myeloid Cells Inhibits Activin A-triggered Secretion of IL-6. Sci Rep 6:31205. doi:10.1038/srep31205
Lee N, Moss WN, Yario TA, Steitz JA (2015) EBV noncoding RNA binds nascent RNA to drive host PAX5 to viral DNA. Cell 160(4):607–618. doi:10.1016/j.cell.2015.01.015
Lee N, Pimienta G, Steitz JA (2012) AUF1/hnRNP D is a novel protein partner of the EBER1 noncoding RNA of Epstein-Barr virus. RNA 18(11):2073–2082. doi:10.1261/rna.034900.112
Lee N, Yario TA, Gao JS, Steitz JA (2016) EBV noncoding RNA EBER2 interacts with host RNA-binding proteins to regulate viral gene expression. Proc Natl Acad Sci U S A 113 (12):3221–3226. doi:10.1073/pnas.1601773113
Lee S, Song J, Kim S, Kim J, Hong Y, Kim Y, Kim D, Baek D, Ahn K (2013) Selective degradation of host MicroRNAs by an intergenic HCMV noncoding RNA accelerates virus production. Cell Host Microbe 13(6):678–690. doi:10.1016/j.chom.2013.05.007
Lei X, Bai Z, Ye F, Huang Y, Gao SJ (2010) Regulation of herpesvirus lifecycle by viral microRNAs. Virulence 1(5):433–435. doi:10.4161/viru.1.5.12966
Lei X, Zhu Y, Jones T, Bai Z, Huang Y, Gao SJ (2012) A Kaposi's sarcoma-associated herpesvirus microRNA and its variants target the transforming growth factor beta pathway to promote cell survival. J Virol 86(21):11698–11711. doi:10.1128/JVI.06855-11
Lerner MR, Andrews NC, Miller G, Steitz JA (1981) Two small RNAs encoded by Epstein-Barr virus and complexed with protein are precipitated by antibodies from patients with systemic lupus erythematosus. Proc Natl Acad Sci U S A 78(2):805–809
Li CX, Shan GZ, Fan B, Tao PZ, Zhao LX, Jiang JD, Li YH, Li ZR (2011) Synthesis and antiviral activities of geldanamycin analog TC-GM in vitro. Yao Xue Xue Bao 46(6):683–687
Li J, Zhao Y, Lu Y, Ritchie W, Grau G, Vadas MA, Gamble JR (2016) The Poly-cistronic miR-23-27-24 Complexes Target Endothelial Cell Junctions: Differential Functional and Molecular Effects of miR-23a and miR-23b. Molecular therapy Nucleic acids 5(8):e354. doi:10.1038/mtna.2016.62
Lin Z, Wang X, Strong MJ, Concha M, Baddoo M, Xu G, Baribault C, Fewell C, Hulme W, Hedges D, Taylor CM, Flemington EK (2013) Whole-genome sequencing of the Akata and Mutu Epstein-Barr virus strains. J Virol 87(2):1172–1182. doi:10.1128/JVI.02517-12
Liu Q, Fu H, Sun F, Zhang H, Tie Y, Zhu J, Xing R, Sun Z, Zheng X (2008) miR-16 family induces cell cycle arrest by regulating multiple cell cycle genes. Nucleic Acids Res 36 (16):5391–5404. doi:10.1093/nar/gkn522
Lo AK, To KF, Lo KW, Lung RW, Hui JW, Liao G, Hayward SD (2007) Modulation of LMP1 protein expression by EBV-encoded microRNAs. Proc Natl Acad Sci U S A 104(41):16164–16169. doi:10.1073/pnas.0702896104
Lu CC, Li Z, Chu CY, Feng J, Sun R, Rana TM (2010) MicroRNAs encoded by Kaposi's sarcoma-associated herpesvirus regulate viral life cycle. EMBO Rep 11(10):784–790. doi:10.1038/embor.2010.132
Lu X, Li X, He Q, Gao J, Gao Y, Liu B, Liu F (2013) miR-142-3p regulates the formation and differentiation of hematopoietic stem cells in vertebrates. Cell Res 23(12):1356–1368. doi:10.1038/cr.2013.145

Lung RW, Tong JH, Sung YM, Leung PS, Ng DC, Chau SL, Chan AW, Ng EK, Lo KW, To KF (2009) Modulation of LMP2A expression by a newly identified Epstein-Barr virus-encoded microRNA miR-BART22. Neoplasia 11(11):1174–1184

Lung RW, Tong JH, To KF (2013) Emerging roles of small Epstein-Barr virus derived non-coding RNAs in epithelial malignancy. Int J Mol Sci 14(9):17378–17409. doi:10.3390/ijms140917378

Ma J, Nie K, Redmond D, Liu Y, Elemento O, Knowles DM, Tam W (2016) EBV-miR-BHRF1-2 targets PRDM1/Blimp1: potential role in EBV lymphomagenesis. Leukemia 30(3):594–604. doi:10.1038/leu.2015.285

Mahjoub N, Dhorne-Pollet S, Fuchs W, Endale Ahanda ML, Lange E, Klupp B, Arya A, Loveland JE, Lefevre F, Mettenleiter TC, Giuffra E (2015) A 2.5-kilobase deletion containing a cluster of nine microRNAs in the latency-associated-transcript locus of the pseudorabies virus affects the host response of porcine trigeminal ganglia during established latency. J Virol 89 (1):428–442. doi:10.1128/JVI.02181-14

Manzano M, Forte E, Raja AN, Schipma MJ, Gottwein E (2015) Divergent target recognition by coexpressed 5'-isomiRs of miR-142-3p and selective viral mimicry. RNA 21(9):1606–1620. doi:10.1261/rna.048876.114

Manzano M, Shamulailatpam P, Raja AN, Gottwein E (2013) Kaposi's sarcoma-associated herpesvirus encodes a mimic of cellular miR-23. J Virol 87(21):11821–11830. doi:10.1128/JVI.01692-13

Marcinowski L, Tanguy M, Krmpotic A, Radle B, Lisnic VJ, Tuddenham L, Chane-Woon-Ming B, Ruzsics Z, Erhard F, Benkartek C, Babic M, Zimmer R, Trgovcich J, Koszinowski UH, Jonjic S, Pfeffer S, Dolken L (2012) Degradation of cellular mir-27 by a novel, highly abundant viral transcript is important for efficient virus replication in vivo. PLoS Pathog 8(2):e1002510. doi:10.1371/journal.ppat.1002510

Margolis TP, Elfman FL, Leib D, Pakpour N, Apakupakul K, Imai Y, Voytek C (2007) Spontaneous reactivation of herpes simplex virus type 1 in latently infected murine sensory ganglia. J Virol 81(20):11069–11074. doi:10.1128/JVI.00243-07

Marquitz AR, Mathur A, Chugh PE, Dittmer DP, Raab-Traub N (2014) Expression profile of microRNAs in Epstein-Barr virus-infected AGS gastric carcinoma cells. J Virol 88(2):1389–1393. doi:10.1128/JVI.02662-13

Marshall V, Martro E, Labo N, Ray A, Wang D, Mbisa G, Bagni RK, Volfovsky N, Casabona J, Whitby D (2010) Kaposi sarcoma (KS)-associated herpesvirus microRNA sequence analysis and KS risk in a European AIDS-KS case control study. J Infect Dis 202(7):1126–1135. doi:10.1086/656045

Marshall V, Parks T, Bagni R, Wang CD, Samols MA, Hu J, Wyvil KM, Aleman K, Little RF, Yarchoan R, Renne R, Whitby D (2007) Conservation of virally encoded microRNAs in Kaposi sarcoma–associated herpesvirus in primary effusion lymphoma cell lines and in patients with Kaposi sarcoma or multicentric Castleman disease. J Infect Dis 195(5):645–659. doi:10.1086/511434

Mayr C, Bartel DP (2009) Widespread shortening of 3′ UTRs by alternative cleavage and polyadenylation activates oncogenes in cancer cells. Cell 138(4):673–684. doi:10.1016/j.cell.2009.06.016

McCaskill J, Praihirunkit P, Sharp PM, Buck AH (2015) RNA-mediated degradation of microRNAs: A widespread viral strategy? RNA Biol 12(6):579–585. doi:10.1080/15476286.2015.1034912

McDonough SH, Spector DH (1983) Transcription in human fibroblasts permissively infected by human cytomegalovirus strain AD169. Virology 125(1):31–46

McDonough SH, Staprans SI, Spector DH (1985) Analysis of the major transcripts encoded by the long repeat of human cytomegalovirus strain AD169. J Virol 53(3):711–718

McSharry BP, Tomasec P, Neale ML, Wilkinson GW (2003) The most abundantly transcribed human cytomegalovirus gene (β2. 7) is non-essential for growth in vitro. J Gen Virol 84 (9):2511–2516. doi:10.1099/vir.0.19298-0

Meshesha MK, Bentwich Z, Solomon SA, Avni YS (2016) In vivo expression of human cytomegalovirus (HCMV) microRNAs during latency. Gene 575(1):101–107. doi:10.1016/j.gene.2015.08.040

Meyer C, Grey F, Kreklywich CN, Andoh TF, Tirabassi RS, Orloff SL, Streblow DN (2011) Cytomegalovirus microRNA expression is tissue specific and is associated with persistence. J Virol 85(1):378–389. doi:10.1128/JVI.01900-10

Mogilyansky E, Rigoutsos I (2013) The miR-17/92 cluster: a comprehensive update on its genomics, genetics, functions and increasingly important and numerous roles in health and disease. Cell Death Differ 20(12):1603–1614. doi:10.1038/cdd.2013.125

Moss WN, Steitz JA (2013) Genome-wide analyses of Epstein-Barr virus reveal conserved RNA structures and a novel stable intronic sequence RNA. BMC Genom 14:543. doi:10.1186/1471-2164-14-543

Murphy E, Vanicek J, Robins H, Shenk T, Levine AJ (2008) Suppression of immediate-early viral gene expression by herpesvirus-coded microRNAs: implications for latency. Proc Natl Acad Sci U S A 105(14):5453–5458. doi:10.1073/pnas.0711910105

Nachmani D, Lankry D, Wolf DG, Mandelboim O (2010) The human cytomegalovirus microRNA miR-UL112 acts synergistically with a cellular microRNA to escape immune elimination. Nat Immunol 11(9):806–813. doi:10.1038/ni.1916

O'Connor CM, Vanicek J, Murphy EA (2014) Host microRNA regulation of human cytomegalovirus immediate early protein translation promotes viral latency. J Virol 88(10):5524–5532. doi:10.1128/JVI.00481-14

O'Grady T, Cao S, Strong MJ, Concha M, Wang X, Splinter Bondurant S, Adams M, Baddoo M, Srivastav SK, Lin Z, Fewell C, Yin Q, Flemington EK (2014) Global bidirectional transcription of the Epstein-Barr virus genome during reactivation. J Virol 88(3):1604–1616. doi:10.1128/JVI.02989-13

Pan D, Flores O, Umbach JL, Pesola JM, Bentley P, Rosato PC, Leib DA, Cullen BR, Coen DM (2014) A neuron-specific host microRNA targets herpes simplex virus-1 ICP0 expression and promotes latency. Cell Host Microbe 15(4):446–456. doi:10.1016/j.chom.2014.03.004

Parnas O, Corcoran DL, Cullen BR (2014) Analysis of the mRNA targetome of microRNAs expressed by Marek's disease virus. MBio 5(1):e01060–e01013. doi:10.1128/mBio.01060-13

Perng GC, Ghiasi H, Kaiwar R, Nesburn AB, Wechsler SL (1994) An improved method for cloning portions of the repeat regions of herpes simplex virus type 1. J Virol Methods 46(2):111–116

Pfeffer S, Sewer A, Lagos-Quintana M, Sheridan R, Sander C, Grasser FA, van Dyk LF, Ho CK, Shuman S, Chien M, Russo JJ, Ju J, Randall G, Lindenbach BD, Rice CM, Simon V, Ho DD, Zavolan M, Tuschl T (2005) Identification of microRNAs of the herpesvirus family. Nat Methods 2(4):269–276. doi:10.1038/nmeth746

Pfeffer S, Zavolan M, Grasser FA, Chien M, Russo JJ, Ju J, John B, Enright AJ, Marks D, Sander C, Tuschl T (2004) Identification of virus-encoded microRNAs. Science 304(5671):734–736. doi:10.1126/science.1096781304/5671/734

Plaisance-Bonstaff K, Choi HS, Beals T, Krueger BJ, Boss IW, Gay LA, Haecker I, Hu J, Renne R (2014) KSHV miRNAs decrease expression of lytic genes in latently infected PEL and endothelial cells by targeting host transcription factors. Viruses 6(10):4005–4023. doi:10.3390/v6104005

Puig-Barbera J, Burtseva E, Yu H, Cowling BJ, Badur S, Kyncl J, Sominina A (2016) Influenza epidemiology and influenza vaccine effectiveness during the 2014-2015 season: annual report from the Global Influenza Hospital Surveillance Network. BMC Public Health 16(Suppl 1):757. doi:10.1186/s12889-016-3378-1

Qiu J, Smith P, Leahy L, Thorley-Lawson DA (2015) The Epstein-Barr virus encoded BART miRNAs potentiate tumor growth in vivo. PLoS Pathog 11(1):e1004561. doi:10.1371/journal.ppat.1004561

Ray A, Marshall V, Uldrick T, Leighty R, Labo N, Wyvill K, Aleman K, Polizzotto MN, Little RF, Yarchoan R, Whitby D (2012) Sequence analysis of Kaposi sarcoma-associated herpesvirus (KSHV) microRNAs in patients with multicentric Castleman disease and

KSHV-associated inflammatory cytokine syndrome. J Infect Dis 205(11):1665–1676. doi:10.1093/infdis/jis249

Reeves MB, Davies AA, McSharry BP, Wilkinson GW, Sinclair JH (2007) Complex I binding by a virally encoded RNA regulates mitochondria-induced cell death. Science 316(5829):1345–1348. doi:10.1126/science.1142984

Rennekamp AJ, Lieberman PM (2011) Initiation of Epstein-Barr virus lytic replication requires transcription and the formation of a stable RNA-DNA hybrid molecule at OriLyt. J Virol 85 (6):2837–2850. doi:10.1128/JVI.02175-10

Reusch JA, Nawandar DM, Wright KL, Kenney SC, Mertz JE (2015) Cellular differentiation regulator BLIMP1 induces Epstein-Barr virus lytic reactivation in epithelial and B cells by activating transcription from both the R and Z promoters. J Virol 89(3):1731–1743. doi:10.1128/JVI.02781-14

Riley KJ, Rabinowitz GS, Steitz JA (2010) Comprehensive analysis of Rhesus lymphocryptovirus microRNA expression. J Virol 84(10):5148–5157. doi:10.1128/JVI.00110-10

Riley KJ, Rabinowitz GS, Yario TA, Luna JM, Darnell RB, Steitz JA (2012) EBV and human microRNAs co-target oncogenic and apoptotic viral and human genes during latency. EMBO J 31(9):2207–2221. doi:10.1038/emboj.2012.63

Rossetto CC, Pari G (2012) KSHV PAN RNA associates with demethylases UTX and JMJD3 to activate lytic replication through a physical interaction with the virus genome. PLoS Pathog 8 (5):e1002680. doi:10.1371/journal.ppat.1002680

Rossetto CC, Tarrant-Elorza M, Pari GS (2013) Cis and trans acting factors involved in human cytomegalovirus experimental and natural latent infection of CD14 (+) monocytes and CD34 (+) cells. PLoS Pathog 9(5):e1003366. doi:10.1371/journal.ppat.1003366

Samols MA, Hu J, Skalsky RL, Renne R (2005) Cloning and identification of a microRNA cluster within the latency-associated region of Kaposi's sarcoma-associated herpesvirus. J Virol 79 (14):9301–9305. doi:10.1128/JVI.79.14.9301-9305.2005

Samols MA, Skalsky RL, Maldonado AM, Riva A, Lopez MC, Baker HV, Renne R (2007) Identification of cellular genes targeted by KSHV-encoded microRNAs. PLoS Pathog 3(5): e65. doi:10.1371/journal.ppat.0030065

Sandberg R, Neilson JR, Sarma A, Sharp PA, Burge CB (2008) Proliferating cells express mRNAs with shortened 3' untranslated regions and fewer microRNA target sites. Science 320 (5883):1643–1647. doi:10.1126/science.1155390

Santhakumar D, Forster T, Laqtom NN, Fragkoudis R, Dickinson P, Abreu-Goodger C, Manakov SA, Choudhury NR, Griffiths SJ, Vermeulen A, Enright AJ, Dutia B, Kohl A, Ghazal P, Buck AH (2010) Combined agonist-antagonist genome-wide functional screening identifies broadly active antiviral microRNAs. Proc Natl Acad Sci U S A 107(31):13830–13835. doi:10.1073/pnas.1008861107

Saxena A, Carninci P (2011) Long non-coding RNA modifies chromatin: epigenetic silencing by long non-coding RNAs. BioEssays 33(11):830–839. doi:10.1002/bies.201100084

Scarpini CG, May J, Lachmann RH, Preston CM, Dunnett SB, Torres EM, Efstathiou S (2001) Latency associated promoter transgene expression in the central nervous system after stereotaxic delivery of replication-defective HSV-1-based vectors. Gene Ther 8(14):1057–1071. doi:10.1038/sj.gt.3301497

Schafer A, Cai X, Bilello JP, Desrosiers RC, Cullen BR (2007) Cloning and analysis of microRNAs encoded by the primate gamma-herpesvirus rhesus monkey rhadinovirus. Virology 364(1):21–27. doi:10.1016/j.virol.2007.03.029

Schwarz TM, Volpe LA, Abraham CG, Kulesza CA (2013) Molecular investigation of the 7.2 kb RNA of murine cytomegalovirus. Virol J 10:348. doi:10.1186/1743-422X-10-348

Seo GJ, Kincaid RP, Phanaksri T, Burke JM, Pare JM, Cox JE, Hsiang TY, Krug RM, Sullivan CS (2013) Reciprocal inhibition between intracellular antiviral signaling and the RNAi machinery in mammalian cells. Cell Host Microbe 14(4):435–445. doi:10.1016/j.chom.2013.09.002

Shen ZZ, Pan X, Miao LF, Ye HQ, Chavanas S, Davrinche C, McVoy M, Luo MH (2014) Comprehensive analysis of human cytomegalovirus microRNA expression during lytic and quiescent infection. PLoS ONE 9(2):e88531. doi:10.1371/journal.pone.0088531

Skalsky RL, Barr SA, Jeffery AJ, Blair T, Estep R, Wong SW (2016) Japanese Macaque Rhadinovirus Encodes a Viral MicroRNA Mimic of the miR-17 Family. J Virol 90(20):9350–9363. doi:10.1128/JVI.01123-16

Skalsky RL, Corcoran DL, Gottwein E, Frank CL, Kang D, Hafner M, Nusbaum JD, Feederle R, Delecluse HJ, Luftig MA, Tuschl T, Ohler U, Cullen BR (2012) The viral and cellular microRNA targetome in lymphoblastoid cell lines. PLoS Pathog 8(1):e1002484. doi:10.1371/journal.ppat.1002484

Skalsky RL, Cullen BR (2015) EBV Noncoding RNAs. Curr Top Microbiol Immunol 391:181–217. doi:10.1007/978-3-319-22834-1_6

Skalsky RL, Kang D, Linnstaedt SD, Cullen BR (2014) Evolutionary conservation of primate lymphocryptovirus microRNA targets. J Virol 88(3):1617–1635. doi:10.1128/JVI.02071-13

Skalsky RL, Samols MA, Plaisance KB, Boss IW, Riva A, Lopez MC, Baker HV, Renne R (2007) Kaposi's sarcoma-associated herpesvirus encodes an ortholog of miR-155. J Virol 81 (23):12836–12845. doi:10.1128/JVI.01804-07

Stark TJ, Arnold JD, Spector DH, Yeo GW (2012) High-resolution profiling and analysis of viral and host small RNAs during human cytomegalovirus infection. J Virol 86(1):226–235. doi:10.1128/JVI.05903-11

Stern-Ginossar N, Saleh N, Goldberg MD, Prichard M, Wolf DG, Mandelboim O (2009) Analysis of human cytomegalovirus-encoded microRNA activity during infection. J Virol 83 (20):10684–10693. doi:10.1128/JVI.01292-09

Stern-Ginossar N, Weisburd B, Michalski A, Le VT, Hein MY, Huang SX, Ma M, Shen B, Qian SB, Hengel H, Mann M, Ingolia NT, Weissman JS (2012) Decoding human cytomegalovirus. Science 338(6110):1088–1093. doi:10.1126/science.1227919

Stevens JG, Wagner EK, Devi-Rao GB, Cook ML, Feldman LT (1987) RNA complementary to a herpesvirus alpha gene mRNA is prominent in latently infected neurons. Science 235 (4792):1056–1059

Sun Z, Wang L, Eckloff BW, Deng B, Wang Y, Wampfler JA, Jang J, Wieben ED, Jen J, You M, Yang P (2014) Conserved recurrent gene mutations correlate with pathway deregulation and clinical outcomes of lung adenocarcinoma in never-smokers. BMC Med Genomics 7:32. doi:10.1186/1755-8794-7-32

Tagawa T, Albanese M, Bouvet M, Moosmann A, Mautner J, Heissmeyer V, Zielinski C, Lutter D, Hoser J, Hastreiter M, Hayes M, Sugden B, Hammerschmidt W (2016) Epstein-Barr viral miRNAs inhibit antiviral CD4 + T cell responses targeting IL-12 and peptide processing. J Exp Med 213(10):2065–2080. doi:10.1084/jem.20160248

Tang S, Bertke AS, Patel A, Margolis TP, Krause PR (2011) Herpes simplex virus 2 microRNA miR-H6 is a novel latency-associated transcript-associated microRNA, but reduction of its expression does not influence the establishment of viral latency or the recurrence phenotype. J Virol 85(9):4501–4509. doi:10.1128/JVI.01997-10

Tang S, Bertke AS, Patel A, Wang K, Cohen JI, Krause PR (2008) An acutely and latently expressed herpes simplex virus 2 viral microRNA inhibits expression of ICP34.5, a viral neurovirulence factor. Proc Natl Acad Sci U S A 105(31):10931–10936. doi:10.1073/pnas.0801845105

Tang S, Patel A, Krause PR (2009) Novel less-abundant viral microRNAs encoded by herpes simplex virus 2 latency-associated transcript and their roles in regulating ICP34.5 and ICP0 mRNAs. J Virol 83(3):1433–1442. doi:10.1128/JVI.01723-08

Thompson RL, Sawtell NM (1997) The herpes simplex virus type 1 latency-associated transcript gene regulates the establishment of latency. J Virol 71(7):5432–5440

Tirabassi R, Hook L, Landais I, Grey F, Meyers H, Hewitt H, Nelson J (2011) Human cytomegalovirus US7 is regulated synergistically by two virally encoded microRNAs and by two distinct mechanisms. J Virol 85(22):11938–11944. doi:10.1128/JVI.05443-11

Toth Z, Maglinte DT, Lee SH, Lee HR, Wong LY, Brulois KF, Lee S, Buckley JD, Laird PW, Marquez VE, Jung JU (2010) Epigenetic analysis of KSHV latent and lytic genomes. PLoS Pathog 6(7):e1001013. doi:10.1371/journal.ppat.1001013

Tuddenham L, Jung JS, Chane-Woon-Ming B, Dolken L, Pfeffer S (2012) Small RNA deep sequencing identifies microRNAs and other small noncoding RNAs from human herpesvirus 6B. J Virol 86(3):1638–1649. doi:10.1128/JVI.05911-11

Tycowski KT, Guo YE, Lee N, Moss WN, Vallery TK, Xie M, Steitz JA (2015) Viral noncoding RNAs: more surprises. Genes Dev 29(6):567–584. doi:10.1101/gad.259077.115

Tycowski KT, Shu MD, Borah S, Shi M, Steitz JA (2012) Conservation of a triple-helix-forming RNA stability element in noncoding and genomic RNAs of diverse viruses. Cell Rep 2(1):26–32. doi:10.1016/j.celrep.2012.05.020

Umbach JL, Kramer MF, Jurak I, Karnowski HW, Coen DM, Cullen BR (2008) MicroRNAs expressed by herpes simplex virus 1 during latent infection regulate viral mRNAs. Nature 454 (7205):780–783. doi:10.1038/nature07103

Umbach JL, Nagel MA, Cohrs RJ, Gilden DH, Cullen BR (2009) Analysis of human alphaherpesvirus microRNA expression in latently infected human trigeminal ganglia. J Virol 83(20):10677–10683. doi:10.1128/JVI.01185-09

Umbach JL, Wang K, Tang S, Krause PR, Mont EK, Cohen JI, Cullen BR (2010) Identification of viral microRNAs expressed in human sacral ganglia latently infected with herpes simplex virus 2. J Virol 84(2):1189–1192. doi:10.1128/JVI.01712-09

Wahl A, Linnstaedt SD, Esoda C, Krisko JF, Martinez-Torres F, Delecluse HJ, Cullen BR, Garcia JV (2013) A cluster of virus-encoded microRNAs accelerates acute systemic Epstein-Barr virus infection but does not significantly enhance virus-induced oncogenesis in vivo. J Virol 87(10):5437–5446

Walz N, Christalla T, Tessmer U, Grundhoff A (2010) A global analysis of evolutionary conservation among known and predicted gammaherpesvirus microRNAs. J Virol 84(2):716–728. doi:10.1128/JVI.01302-09

Wang QY, Zhou C, Johnson KE, Colgrove RC, Coen DM, Knipe DM (2005) Herpesviral latency-associated transcript gene promotes assembly of heterochromatin on viral lytic-gene promoters in latent infection. Proc Natl Acad Sci U S A 102(44):16055–16059. doi:10.1073/pnas.0505850102

Wang Y, Li H, Tang Q, Maul GG, Yuan Y (2008) Kaposi's sarcoma-associated herpesvirus ori-Lyt-dependent DNA replication: involvement of host cellular factors. J Virol 82(6):2867–2882. doi:10.1128/JVI.01319-07

Xie M, Zhang W, Shu MD, Xu A, Lenis DA, DiMaio D, Steitz JA (2015) The host Integrator complex acts in transcription-independent maturation of herpesvirus microRNA 3' ends. Genes Dev 29(14):1552–1564. doi:10.1101/gad.266973.115

Xing L, Kieff E (2011) cis-Acting effects on RNA processing and Drosha cleavage prevent Epstein-Barr virus latency III BHRF1 expression. J Virol 85(17):8929–8939. doi:10.1128/JVI.00336-11

Zhao Y, Xu H, Yao Y, Smith LP, Kgosana L, Green J, Petherbridge L, Baigent SJ, Nair V (2011) Critical role of the virus-encoded microRNA-155 ortholog in the induction of Marek's disease lymphomas. PLoS Pathog 7(2):e1001305. doi:10.1371/journal.ppat.1001305

Zhou C, Xie Z, Gao L, Liu C, Ai J, Zhang L, Shen K (2015) Profiling of EBV-Encoded microRNAs in EBV-Associated Hemophagocytic Lymphohistiocytosis. Tohoku J Exp Med 237(2):117–126. doi:10.1620/tjem.237.117

The manufacturer's authorised representative in the EU is Springer Nature Customer Service Centre GmbH, Europaplatz 3, 69115 Heidelberg, Germany. If you have any concerns regarding our products, please contact ProductSafety@springernature.com

Printed and bound by CPI Group (UK) Ltd, Croydon, CR0 4YY
30/09/2026
02225443-0013